Big Data Application in Power Systems

Big Data Application in Power Systems

Edited by

Reza Arghandeh

Assistant Prof. in Electrical Engineering,
Department of Electrical & Computer Engineering,
Florida State University

Yuxun Zhou

PhD candidate, Department of Electrical Engineering
and Computer Sciences, UC Berkeley

Elsevier
Radarweg 29, PO Box 211, 1000 AE Amsterdam, Netherlands
The Boulevard, Langford Lane, Kidlington, Oxford OX5 1GB, United Kingdom
50 Hampshire Street, 5th Floor, Cambridge, MA 02139, United States

Library of Congress Cataloging-in-Publication Data
A catalog record for this book is available from the Library of Congress

British Library Cataloguing-in-Publication Data
A catalogue record for this book is available from the British Library

ISBN: 978-0-12-811968-6

For information on all Elsevier publications
visit our website at https://www.elsevier.com/books-and-journals

Working together
to grow libraries in
developing countries

www.elsevier.com • www.bookaid.org

Publisher: Joe Hayton
Acquisition Editor: Lisa Reading
Editorial Project Manager: Ana Claudia A. Garcia
Production Project Manager: Vijayaraj Purushothaman
Cover Designer: Mark Rogers

Typeset by SPi Global, India

Contents

Contributors

Reza Arghandeh   UC Berkeley and Florida State University, Tallahassee, FL, United States

Mohammad Babakmehr   Colorado School of Mines, Golden, CO, United States

Ricardo J. Bessa   INESC Technology and Science—INESC TEC, Porto, Portugal

Saverio Bolognani   Automatic Control Laboratory ETH Zürich, Zürich, Switzerland

Angelo Cenedese   University of Padova, Padova, Italy

Michael Chertkov   Los Alamos National Laboratory, Los Alamos, NM, United States

Deepjyoti Deka   Los Alamos National Laboratory, Los Alamos, NM, United States

Roy Dong   University of California, Berkeley, Berkeley, CA, United States

Feng Gao   Tsinghua University Energy Internet Research Institute, Beijing, China

Madeleine Gibescu   Eindhoven University of Technology, Eindhoven, The Netherlands

Bri-Mathias Hodge   National Renewable Energy Laboratory, Golden, CO, United States

Gabriela Hug   ETH Zurich, Power Systems Laboratory, Zurich, Switzerland

Jeffrey S. Katz   IBM, Hartford, CT, United States

Stephan Koch   ETH Zurich; Adaptricity AG, c/o ETH Zurich, Power Systems Laboratory, Zurich, Switzerland

Hanif Livani   University of Nevada Reno, Reno, NV, United States

Mehrdad Majidi   University of Nevada, Reno, NV, United States

John D. McDonald   GE Energy Connections-Grid Solutions, Atlanta, GA, United States

Sadaf Moaveninejad   Polytechnic University of Milan, Milan, Italy

Elena Mocanu   Eindhoven University of Technology, Eindhoven, The Netherlands

Ingo Nader   Unbelievable Machine, Vienna, Austria

Behzad Najafi   Polytechnic University of Milan, Milan, Italy

Phuong H. Nguyen   Eindhoven University of Technology, Eindhoven, The Netherlands

Lillian J. Ratliff   University of Washington, Seattle, WA, United States

Fabio Rinaldi   Polytechnic University of Milan, Milan, Italy

Marcelo G. Simoes   Colorado School of Mines, Golden, CO, United States

Matthias Stifter   AIT Austrian Institute of Technology, Center of Energy, Vienna, Austria

Carol L. Stimmel   Manifest Mind, LLC, Canaan, NY, United States

Gian Antonio. Susto   University of Padova, Padova, Italy

Akin Tascikaraoglu   Mugla Sitki Kocman University, Mugla, Turkey

Matteo Terzi   University of Padova, Padova, Italy

Andreas Ulbig   ETH Zurich; Adaptricity AG, c/o ETH Zurich, Power Systems Laboratory, Zurich, Switzerland

Yang Weng Arizona State University, Tempe, AZ, United States
Rui Yang National Renewable Energy Laboratory, Golden, CO, United States
Yingchen Zhang National Renewable Energy Laboratory, Golden, CO, United States
Jie Zhang University of Texas at Dallas, Richardson, TX, United States
Yuxun Zhou UC Berkeley and Florida State University, Tallahassee, FL, United States
Thierry Zufferey ETH Zurich, Power Systems Laboratory, Zurich, Switzerland

About the Editors

Reza Arghandeh is an Assistant Professor in the ECE Department in Florida State University. He is director of the Collaborative Intelligent Infrastructure Lab. He has been a postdoctoral scholar at the University of California, Berkeley's California Institute for Energy and Environment 2013–15. He has 5 years industrial experience in power and energy systems. He completed his PhD in Electrical Engineering with a specialization in power systems at Virginia Tech. He holds Master's degrees in Industrial and System Engineering from Virginia Tech 2013 and in Energy Systems from the University of Manchester 2008. From 2011 to 2013, he was a power system software designer at Electrical Distribution Design Inc. in Virginia. Dr. Arghandeh's research interests include, but are not limited to, data analysis and decision support for smart grids and smart cities using statistical inference, machine learning, information theory, and operations research. He is a recipient of the Association of Energy Engineers (AEE) Scholarship 2012, the UC Davis Green Tech Fellowship 2011, and the best paper award from the ASME 2012 Power Conference and IEEE PESGM 2015. He is the chair of the IEEE Task Force on Big Data Application for Power Distribution Network.

Yuxun Zhou is currently a PhD candidate at Department of EECS, UC Berkeley. Prior to that, he obtained the Diplome d'Ingenieur in applied mathematics from Ecole Centrale Paris and a BS degree from Xi'an Jiaotong University. Yuxun has published more than 30 refereed articles, and has received several student awards. His research interest is on machine learning theories and algorithms for modern sensor rich, ubiquitously connected cyber-physical systems, including smart grid, power distribution networks, smart buildings, etc.

Preface: Objective and Overview of the Book

The term "big data" is fairly new in power systems. Yet, its application and methodologies applied to massive data sets were developed a long time ago for electricity load consumption forecasting. The recent developments in monitoring, sensor networks, and advanced metering infrastructure (AMI) dramatically increase the variety, volume, and velocity of measurement data in electricity transmission and distribution networks. Moreover, the progress in advanced statistics, machine learning (ML), database structure, and data mining methodologies marked by increasing the availability of open source platforms for data analytics is transforming the power system area and turning utilities into data-driven enterprises.

In order to discuss the big data analytics applications for power systems, this book brings together experts from all organizations and institutions impacted including academia and industry. We focus on rapidly modernizing monitoring and analytical approaches to process the high dimensional, heterogeneous, and spatiotemporal data. This book discusses challenges, opportunities, success stories, and pathways for utilizing big data value in smart grids. The dramatic change in the field of scientific computing, microprocessors, and data communications is a burden for electric utilities to understand, follow, and adopt the advanced statistics, computer science, and mathematics concepts. Today's utility engineers need to be more informed of the basic concepts and applications for massive field data analysis. This book's goal is to facilitate the transition to data-driven utilities by providing a comprehensive view on big data issues, methodologies, and their various applications in the power systems area.

Much like the authorship of the chapters in this volume, the intended audience for this book extends from researchers, graduate students, and faculty working in electricity networks and smart grid area to industrial scientists, engineers, data analysis experts, and software developers who are working on electricity networks and advanced technologies for smart grids. This book is also useful for people with less technical expertise in scientific computing. We expect that the reader will have some proficiency in power systems fundamentals and that he/she has had at least one elementary course in statistics. This book can also be useful for senior undergraduate students who have passed courses on power systems.

This book has three sections as follows: I. *Harness the Big Data From Power Systems*, II. *Harness the Power of Big Data*, and III. *Put the Power of Big Data Into Power Systems*. The opening section is an overview of the opportunities and challenges for data-driven utilities in the era of distributed technologies and resources such as Internet of Things (IoT), flexible demand, distributed generation, and energy storage. The second section reviews research trends on ML and artificial intelligence for the power system industry. The final section provides examples of the advanced data analytic applications for the grid operation. Taken together, these three book sections provide an overview of the entire cycle of data analysis in power systems. The book begins with the utility enterprise structure, business model, and privacy issues, then delves into research trends in advanced data analysis, and ends full circle with real-world examples of actual applications of data analytics used daily by utilities.

SECTION ONE: HARNESS THE BIG DATA FROM POWER SYSTEMS

To provide a big picture for electric utilities, this section describes the current and future trends for data mining and data processing in electric utilities. The move toward data-driven utility is possible by a fundamental shift in organizational culture and business processes, as well as data-related technology and practices. Moreover, enriching electric utilities with data requires interoperability across all operational and enterprise units and recognition that maintaining the data privacy, security, and the seamless data flow is highly challenging. The interoperability in holistic data-driven utilities expands to customers through their engagement and continues demand-side management for higher reliability, service quality, and efficiency. Aligning customers' needs and expectations with utilities' business drivers will shape the roadmap to generate, process, and access the data in utilities.

The information and communication technology (ICT) platform is at the heart of the roadmap to data-driven utilities which supports the data flow from customers all the way to the transmission and generation operators. Utilities made aggressive steps toward smartness by adopting the distribution automation (DA) solutions followed by AMI platforms. DA and AMI made a revolution in grid operation. However, the data flood from DA and AMI has created a nightmare for the utilities' ICT infrastructure. A holistic approach for data-driven utilities is needed to openly discuss and clarify the foundational ICT requirements to serve all functions of the electric grid.

Becoming a data-driven utility is inevitable in the age of internet, cloud computing, smart phones, and distributed resources. The advanced data analytics make continuous innovation possible by unlocking insights never seen before. The ML, deep learning, and statistical inference are tools that help utilities to keep up with the torrent of data from different resources. Advanced big data analytics provide estimation, predication, diagnostics, and prognostics conclusions from historical and real-time data flows. As more data becomes available to utilities over time, the ML algorithms provide more refined insights on grid operation planning. However, the synergies between ICT networks, grid components, operators, and customers run the power system into a complex giant for ad hoc data-driven approaches and policies. This section of the book endeavors to deliver the message that a holistic approach based on a foundation of open architecture and standards will ensure the open flow of data and interoperability between devices, systems, databases, and people in order to make data-driven utilities. The all-inclusive approaches to generate, transfer, and handle data also bring tremendous opportunities to break traditional barriers in utility organizations for delivering safe, reliable, and affordable power to their customers.

Chapter 1 by John McDonald introduces these concepts through three case studies. He explores the value of a data-driven utility in terms of asset management and safety, the fundamentals of standards and interoperability, and the enterprises of increased visibility into the transmission and distribution network. The chapter illustrates the holistic data-driven utility and its fundamental business drivers to establish information and communications technology foundation, human resource, customer relation, and data-oriented organizational cultural data in the grid operation and planning.

The data-driven utilities now face greater and more frequent risk of intrusion and/or interruption due to the fact that these networks are merging with cyber networks, resulting in sociotechnical and cyber-physical systems that are creating an infrastructural IoT where all grid components can interact and collaborate. Integrating cyber components into the electric grid also means an incredible increase in security vulnerability and interdependencies among infrastructure components that create the risk of cascading effects after attacks. Moreover, the enhanced observability of the grid thanks to the smart meters' high granular data is making more customers concerned and uncomfortable about data privacy. Carol L. Stimmel discusses state-of-the-art data privacy and security in Chapter 2. She lists a number of actual cases for cyber security attacks on the grid and explains the impact of data-driven approaches in enhancing the data security and privacy. Data-driven utilities function as much more than the operators of the physical grid; utilities are also responsible for massive enterprise systems with financial information, customer data, and a growing network of digital operations under human control. Thus, security strategies must become more nuanced and complex, and should include privacy and other internal information technology controls.

The big data era is changing the utility workforce paradigm. Several major utilities are adding more software developers and data scientists to their R&D and operating groups, as well as power system experts. The power of data in innovation is seen in more smart grid projects and AMI implementations. Some of these projects applied Big Data and Analytics even without adding any new sensors, demonstrating the power of knowing more about what information was already available to the utility through the SCADA systems. The utility innovation movement came from the foresight that discarded data may prove useful. This includes data discarded during the process of developing an analytics strategy, including predictive maintenance programs, thought to be valuable as the design phase began, even though there was no known need for all of the data at the time. Analytics has moved from replicating alarm limits already available, to deep learning for customer behavioral studies and cognitive computing for renewable adoption optimization, as well as numerical methodologies for dynamic electricity market forecasting. Jeffrey Katz from IBM contributes Chapter 3, "The Rule of Big Data and Analytics in Utilities Innovation" that explains how data analytics pave the ground for innovation in utilities by pointing to a number of successful projects in different utilities.

To harvest the advantages of big data, utilities need to employ platforms that can handle high volume, velocity, and volatility of the data. There are commercial and ready-to-use platforms that serve the big data community. It is time for utilities to take the lead in shaping power systems-specific data platforms. The in-memory calculation engine and parallel computing framework, Hadoop/MapReduce and Spark, are ready for handling an extremely large scale of dataset; on the other hand, the stream processing engine, Storm, Streams, and Spark Streaming are built to analyze data in motion and act on information as it is happening. The architecture of big data platforms includes data integration, warehousing, analytics, and combining the demand of smart grids to put forward a set of frameworks such as the Apache Hadoop ecosystem which has excellent computing ability and can adapt to various business requirements. Chapter 4 "Frameworks for Big Data Integration, Warehousing, and Analytics" by Feng Gao discusses different tools and techniques to support the growth of smart grid and big data with high performance computing, with a focus on the platform, data integration, warehousing, and analytics that are particularly adaptive to handle a variety of characteristics of energy industry data within the data lifetime cycle.

SECTION TWO: HARNESS THE POWER OF BIG DATA

This theory-oriented section focuses on big data analytics. In particular, it discusses ML and data mining algorithms, methods, and implementation that are adaptable for data visualization, representation, exploratory analysis, regression, and pattern recognition in power systems. The objectives of this section are twofold. On one hand, both classical and status quo ML paradigms are reviewed and discussed, motivating the proper usage of traditional supervised/unsupervised learning tools and the recent developments of semi-supervised learning, multitask, multiview learning, sparse representation, deep learning, etc., for various tasks in power systems. The hope is that the dramatic progress in ML can be fully harnessed to reform the solution of power system state estimation, load forecasting, event detection, and structure identification. On the other hand, the reversed direction, i.e., the challenges and new problems brought by power system data to ML, is discussed. Similarly to the impact of computer vision, natural language processing, speech recognition, or robot control on the advancement of ML, it is expected that the complexity of the interconnected system, the behavior-related data generating process, as well as the unique sensing and measurement techniques in power systems, would inspire novel theoretical and methodological results for ML.

It is worth pointing out that in this section, the term ML is used in a broader sense, generally referring to a task to improve some performance metric, by executing a series of computation (algorithm) with some training experience (in the form of collected sensor measurement, expert knowledge, survey entries, etc.). Lying at the crossroads of statistics, computer sciences, artificial intelligence, and applied mathematics, the ML methods discussed in this section deserve a comprehensive description from diverse perspectives, including, but not limited to, their underlying probabilistic assumption, theoretical/empirical generalization performance, model selection (hyper-parameter selection), computational complexity, numerical implementation, etc. Although a mathematically rigorous treatment of the above topics is not the focus of this book, useful references are provided to interested readers. More often than not, the proper usage of the state-of-the-art ML algorithm, or a desire to advance ML driven by power system applications, would surprisingly progress both research fields.

More specifically, Chapter 5 starts with a brief discussion of classical supervised and unsupervised learning paradigms. The focus is not to give an extensive review of the field, which is impossible due to its many ramifications, but rather to equip the readers with popular approaches for regression, classification, dimension reduction, among other fundamentals. The chapter then focuses on two important issues, feature engineering and model selection, in some depth to demonstrate the proper usage and systematic tuning of those off-the-shelf ML tools. The rest of this chapter is devoted to the introduction of some recent schemes of ML that seem promising for power system data analysis applications. The topics discussed include semi-supervised learning, multitask learning, transfer learning, multiview learning, information representation, etc.

Following the discussion, Chapter 6 provides a case study on the use of the clustering algorithms for enhanced visibility of the electrical distribution system. Based on smart meter data of more than 30,000 loads in the city of Basel, Switzerland, the authors demonstrate the power of exploratory data analysis using unsupervised learning methods, which successfully reveals hidden structure, property, and geographical consistency from the measurement data. The rich information mined from this analysis can be leveraged by DSOs to support the grid operation.

The rest of the chapters in this section discuss in detail several advanced ML methods for power system applications. Motivated by the unprecedented high volumes of data made available by the growth of home energy management systems and AMI, Dr. Mocanu et al. in Chapter 7 present the deep learning framework to automatically extract knowledge and use it to improve grid operation. The chapter starts with a moderate introduction to the most well-known deep learning concepts, such as deep belief networks and high-order restricted Boltzmann machine, followed by a discussion on their theoretical advantages and limitations, such as computational requirements, convergence, and stability. As a concrete application, two case studies involving building energy prediction using supervised and unsupervised deep learning methods are presented. The chapter concludes with a glimpse into future trends highlighting some open questions as well as new possible applications.

Chapter 8 "Compressive Sensing for Power System Data Analysis," focuses on the applications of another state-of-the-art ML framework, namely compressive sensing-sparse recovery (CS-SR), which has enjoyed great success in other fields like bio-engineering, signal processing, and computer vision, among others. The adaptation of CS-SR in smart power networks monitoring, data analysis, security, and reliability should expect similar successes. The sparse nature of the electrical power grids, as well as electrical signals, can be exploited to introduce alternative mathematical formulations to address some of the most challenging system modeling, that of sparse identification problems in power engineering. The chapter begins with a concise presentation on the theoretical and technical background of CS-SR. Next, the discussion moves to innovative CS-SR applications in smart grid technology. Finally, the CS-SR techniques are explored in depth to propose novel methods for distribution system state estimation (DSSE), single and simultaneous fault location in smart distribution, and transmission networks, and partial discharge (PD) pattern recognition.

The rapid advancement of sensing and measurement technology in power systems has given researchers access to real-time records of system dynamic states. In particular, development of phasor measurement unit (PMU) technology has allowed the continuous monitoring of the transmission line and the connected power systems, and can be complemented with utility monitoring devices, smart meters, and insulation monitoring units to build a thorough picture of the whole grid structure, health, and dynamic behavior. The data collected from these real-time measuring procedures is usually in the form of time series (TS). Hence, in Chapter 9 of this section, Dr. Gian Antonio Susto et al. present an overview about the most recent ML techniques used for TS pattern recognition. The chapter first summarizes existing methods of TS classification and highlights the issue of computational complexity, and then provides discourse on the various dimension reduction and numerosity reduction techniques for a more parsimonious and informative representation of TS data. The chapter concludes with a comprehensive comparison of diverse classification methods in terms of their underlying assumption, performance, computational complexity, flexibility for decentralized execution, and other categories.

SECTION THREE: PUT THE POWER OF BIG DATA INTO POWER SYSTEMS

This final section of the book presents the data-driven approaches unique to the design, operation, and planning of utilities. Moreover, data-driven utilities need new business models for knowledge extraction from data. Some examples are analysis of the demand response (DR) potential of grid users, big data preprocessing from grid sensors, large-scale simulation of electricity markets, and predictive maintenance of electrical equipment. Forecasting of real-time and day ahead market price, load, and renewable generation TS present huge business value for utilities' stakeholders and customers. The big data applications in the distribution and transmission networks are mainly driven by two objectives: firstly, to increase the monitoring and situational awareness capability and develop fast decision-making methods for operators, and secondly, to implement predictive active management strategies that take advantage of flexibility from various technologies in the electricity supply and demand such distributed energy resources, energy storage, and DR.

However, exploiting the full potential of big data in utilities is challenged by lack of statistics and data analytics knowledge in utilities workforce. Moreover, the "ready-to-use" and industry-level ML tools and solutions are not wildly available to utilities which may increase the learning curve and utilities' modernization time. This section provides a collection of modern data-driven solutions such as distributed learning and optimization, spatial-temporal modeling of TS, data reduction, assimilation, and visualization methods for classic power system problems including state estimation, topology detection, fault detection, and load disaggregation. The author hopes this book brings more interests in ML and deep learning applications in power system operation and planning.

Chapter 10, "An Overview of Big Data Application in Power Transmission and Distribution Networks" provides a comprehensive overview of data-driven trends such as feature extraction/reduction and distributed learning to extract knowledge from the power system and market data. Furthermore, it describes the data-driven techniques for dynamic and steady-state analysis and control of distribution and transmission systems.

In Chapter 11, "On Data-Driven Approaches for Demand Response," Akin Tascikaraoglu presents a detailed investigation of the applications and benefits of big data analytics in demand-side management or DR and their roles in providing higher saving potential for both system operators and end users. He also shows some examples of real-world implementations of DR.

Chapters 12 and 13 are devoted to topology detection. Knowledge of the exact topology, the open or closed status of switches and circuit breakers throughout the network, is essential for all aspects of the power system operation. Chapter 12, "Topology Learning in Radial Distribution Grids" presents an acquisitive algorithm to learn the grid topology using voltage measurements collected at a subset of the buses in power distribution networks. Chapter 13, "Grid Topology Identification via Distributed Statistical Hypothesis Testing," proposes an algorithm based on the identification of Markov random fields (graphical models) and conditional correlation properties that characterize voltage measurements in power distribution networks. It shows the correlation of voltage magnitude measurements in a radial distribution feeder with the topology of the grid.

In Chapter 14 entitled "Supervised Learning-Based Fault Location in Power Grid," Dr. Livani, Hanif suggests an SVM network for the classification, identification, and localization of faults in a complex power transmission grid. Based on the high-resolution/high-volume data made available by the proliferation of intelligent electronic devices (IEDs) in smart grids, this method is able to achieve efficient and accurate fault diagnosis for system operators. The lesson learned from this chapter, in particular, is to combine the effort to modify existing ML algorithms with signal processing, and to increase our knowledge about the system itself for handling new problems arising from the complex power system and grid.

To introduce cutting edge tools, packages, and information technology for readers who are interested in developing real-world power system data analysis platforms, the authors of Chapter 15 investigate the usage of recent big data tools and methods in the context of power distribution networks. This chapter illustrates the use of MapReduce functions within R or Java, which is combined with commercial distributed analytics database, the application of affinity graphs for representing collaborative filters, a performance comparison to conventional database concepts, and many other features.

Being able to forecast energy resources, load patterns, and system state are key features of next-generation smart grid technology. An accurate predictive platform would greatly benefit the planning, scheduling, and unit commitment in terms of both efficiency and security. Chapter 16 entitled "Predictive Analytics for Comprehensive Energy System State Estimation" provides an overview and a thorough discussion on predictive ML methods for wind, solar energy forecasting, load prediction, power system state estimation, etc. The ML tools included in the chapter range from classical regression, TS analysis, to kernel method such as support vector regression and Gaussian process.

Finally, Chapters 17 and 18 are devoted to a particular yet important application of big data analytics method to smart grid, namely energy disaggregation or nonintrusive load monitoring (NILM). In essence, the goal is to estimate the power usage of individual appliances from an aggregate electricity consumption measurement. Provided with more precise information including itemized energy consumption profiles, both end users and grid managers can improve their utility in terms of energy consumption prediction, demand side management, and user segmentation. Chapter 17 surveys the existing literature for background, ML methods, and possible applications of energy disaggregation, while Chapter 18 discussed the issue of privacy in the energy disaggregation framework. Both chapters are witness to the combination of cutting edge ML methods and a deep understanding of the system characteristics for the advancement of smart grid technologies.

Acknowledgments

The idea for this book goes back to a few years ago when we were analyzing smart meters and SCADA data from some Californian electric utilities using different machine learning and statistical inferences. Later on, we started to work on phasor measurement units (PMU) and micro-PMU data streams which have much more resolution than the smart meters. The PMU and power quality recording data (120 Hz to 30 kHz and beyond) plus highly spatial distributed data from smart meters marked the advent of big data in power systems. Utilities are already dealing with big data challenges considering the lack of knowledge in workforce and the lack of suitable infrastructure to handle and process the massive data. We are sure that some of our readers have a similar experience. On top of that, in the near future every house may have rooftop solar panels, controllable loads, smart appliances, electric vehicles, and various software-enabled hardware that will be more connected in the era of Internet of Things.

This book is a step toward data-driven utilities by presenting a combination of the high-level view on utility enterprise architecture, data analysis methodology, and various applications of data analytics in power transmission and distribution networks.

We have been lucky enough to have great maestros in our lives. Our parents Ali & Soodabeh Arghandeh and Yanping & Suxue Zhou, our advisers Prof. Robert Broadwater and Prof. Saifur Rahman at Virginia Tech and Prof. Costas Spanos and Prof. Alexandra von Meier at UC Berkeley.

In this book, we have a collection of highly recognized experts in academia and industry in the field of power systems and data analysis from all around the world. We would like to thank them all for their outstanding contributions. We would like to thank Dr. Heather Paudler for her valuable input on the book. We extend special thanks to Renata R. Rodrigues and Ana C. A. Garcia from the Elsevier editorial team for their countless help and advice during the different stages of preparation for this book. We also appreciate Honoka

Hamano's efforts in designing the book cover, icons for each section, and various other creative graphics inside the book.

Finally, we would like to thank several reviewers for valuable comments on preliminary drafts of this book: Jeffrey S. Katz, Ricardo Bessa, John D. McDonald, Carol L. Stimmel, Mohammad Babakmehr, Elena Mocanu, Madeleine Gibescu, Mehrdad Majidi, Gian Antonio Susto, Deepjyoti Deka, Fabio Rinaldi, Feng Gao, Han Zou, Ming Jin, Ruoxi Jia, Yingchen Zhang, Behzad Najafi, Amin Hassanzadeh, Mihye Ahn, Hanif Livani, Matthias Stifter, Saverio Bolognani, Michael Chertkov, Amirhessam Tahmassebi, Madhavi Konila Sriram, Roy Dong, and Jose Cordova.

We look forward to hearing from our readership; please contact us with any comments, suggestions, and questions.

Reza Arghandeh

Florida State University, Tallahassee, FL, United States

Yuxun Zhou

University of California, Berkeley, CA, United States

Harness the Big Data From Power Systems

CHAPTER 1

A Holistic Approach to Becoming a Data-Driven Utility

John D. McDonald

GE Energy Connections-Grid Solutions, Atlanta, GA, United States

CHAPTER OVERVIEW

The ultimate goal of harnessing big data is to improve customer service and achieve enterprise business goals while increasing the reliability, resiliency, and efficiency of operations. Thus, business drivers should dictate data needs and the technology roadmap to achieve ongoing improvements in these areas. A data-driven utility should first identify its fundamental business drivers to understand precisely what intelligence is needed for operations and the enterprise and what specific technology supports the creation of intelligence and value, both for current business challenges and for future business needs and technology functionalities. Intelligence, and automation, relies on a two-way, integrated communication system based on standards; thus a utility must first develop a "strong" grid by establishing an information and communications technology foundation based on an open architecture and standards. This first step requires that information technology and communications groups work together to understand and support the functional requirements such as network response requirements, bandwidth, and latency, of each disparate data path—from sensor to end user—for current and future systems and applications. Then a data-driven utility should develop a "smart" grid, which requires the convergence of information technology and operations technology and their respective staffs—the beginning of an operations- and enterprise-wide cultural shift to holistic utility management that focuses on value creation and eliminates organizational silos. On the technology side, integration of data-producing devices and systems precedes automation. Determining substation automation applications relies on observing the behavior of data over time (daily, seasonally) and diverse conditions (weather patterns). On the organization side, all operations and enterprise groups should cooperate to identify their data needs to create a data requirements matrix. Information and operations technology personnel can then determine the least number of platforms and the most efficient paths to route data from device to end user, taking security into account. Access and authentication rules ensure that only the right person gets the right data at the right time. A key concept in a data-driven utility is that every internal stakeholder who can create value from data should have secure access to that data. Operational data is routed to the control center in real time, while nonoperational data is extracted from intelligent electronic devices, concentrated and sent across the operations firewall to be stored and processed in a data mart for on-demand access by enterprise groups and their applications. Three case studies illustrate the value of a data-driven utility in terms of asset management and safety, the fundamentals of standards and interoperability, and the enterprise value, in dollars, of increased visibility into the transmission and distribution network.

Big Data Application in Power Systems. https://doi.org/10.1016/B978-0-12-811968-6.00001-2

1 INTRODUCTION

In this digital age, power utilities *must* harness data to achieve the operational and enterprise efficiencies, insights, and flexibility to thrive amid emerging technologies and disruptive market forces. The question is not whether to become a data-driven utility, but how to do so. The opportunities and challenges are many. In the simplest terms, harnessing data in a comprehensive manner will require a transformational journey that will remake every power utility that undertakes the challenge. The process of becoming a data-driven utility requires a fundamental shift in organizational culture and business processes as well as data-related technology and practices. The desired result is not limited to the creation of a more reliable, resilient, and efficient grid. This transformation should also enable enterprise flexibility that supports new utility business models. Becoming a data-driven utility is an endeavor in which philosophy and technology go hand in hand.

The philosophy piece is simple and three-fold. First, data should drive improvements in a power utility's raison d'être. The ultimate, traditional goal of a power utility is to serve customers by delivering power safely, efficiently, and affordably. We are likely to see this fundamental mandate broaden to include customer service options, enabled by data. Harnessing data can support improvements in customer service, enhance customer and stakeholder value and increase the reliability, resiliency, and efficiency of operations. This is true whether a utility is cooperatively owned, municipally owned, or investor owned. Second, the organizational and technological transformations required to become a data-driven utility are so far-reaching that only a holistic approach will serve. Third, and most broadly, current and near-term societal and market trends pose a challenge to utilities' historic, regulated monopoly business and regulatory model. If a utility wants to determine its own fate, it must be proactive. Data is the new enabler of value and its opportunities and challenges must be actively embraced with a sense of urgency.

2 ALIGNING INTERNAL AND EXTERNAL STAKEHOLDERS

One fundamental concept in becoming a data-driven utility is that every internal stakeholder who can create value from data should have secure and timely access to that data. The very process of identifying useful data, collecting, processing, and presenting it or making it accessible on-demand will drive cultural and business process change throughout a utility. Creating a data-driven utility requires cooperation and coordination across all operational and enterprise units and the recognition that silos are obsolete legacies of past practices.

One should not underestimate the fundamental transformation unleashed by pursuing the goal of becoming a data-driven utility.

This observation holds true for external stakeholders as well. On the customer side, data has also become a valuable commodity. Customers are no longer passive ratepayers. Their energy use data belongs to them and, increasingly, they expect value for it. Public utility commissions recognize that customers own their energy use data, that utilities must secure it, and that the individual customer has the prerogative to say how that customer-specific data is used or shared. Whether utilities use data to create service options with value to both utility and customer may well determine their future success as an enterprise. Today, emerging technologies, third parties, and disruptive market forces abound, seeking to provide utility customers with value and service options based on their energy use data. For utilities, data has become not only the means to thrive but also the means to survive.

3 TAKING A HOLISTIC APPROACH

A holistic, methodical approach to becoming a data-driven utility has several common, recognizable steps, though the outcome for any individual utility will likely be unique, due to its existing customer base, business model, and legacy infrastructure. In this introductory chapter and overview of the topic, we will examine the implications of a holistic approach, the technology-related phases it requires, and connect the dots between data-producing sensor and data-reliant end user. A brief synopsis of three case studies will illustrate many of these points.

A holistic approach to becoming a data-driven utility literally takes everything into account. It views transmission and distribution as a single integrated entity. It encompasses the operations and business of delivering power to customers in a manner that achieves customer engagement and satisfaction based on increased system reliability, resiliency, and efficiency. Built on a foundation of open architecture and standards, a holistic approach ensures interoperability between devices, systems, and databases. It enables value creation at operational and enterprise levels. It enables forward and backward compatibility to derive full value from current and future investments in technology while maintaining the value of legacy equipment. In terms of an end-to-end system, a holistic approach provides a means by which all data-producing devices—increasingly, nearly every device in a T&D system—can be mapped to communication channels and networks with the appropriate response requirements, routed to both operations and enterprise sides of the organization, and presented and/or made accessible on-demand to the right people in the right time and place for value creation.

A holistic approach aligns customer needs and expectations with utility business drivers and depends on a technology roadmap for grid modernization that supports this alignment. In terms of utility culture and organization, a holistic approach eliminates silos and demands utility-wide cooperation and coordination to avoid redundant systems and costs. Thus it provides the basis for prudent, well-vetted investments that will create customer and stakeholder value and benefits that increase over time, meet future needs, and are likely to win regulatory approval.

In an era in which the utility business model requires review and transformation and digital technology produces an increasing granularity, quality, and quantity of data, a holistic approach to becoming a data-driven utility offers the richest opportunity for success.

4   "STRONG" FIRST, THEN "SMART"

Aligning customer needs and expectations with utility operational and business drivers should dictate how data is generated, collected, stored, processed, presented, or accessed, and how actionable intelligence is applied. A data-driven utility should review its current and mid-term operational and business models and identify its customer needs and fundamental business drivers. This will help in understanding precisely what actionable intelligence—and, thus, data—is needed for both operations and the enterprise to meet its self-determined goals of improving customer service and pursuing value creation.

To optimize current practices and enable future flexibility in reaching operations and enterprise goals, a utility must first develop a "strong" grid before pursuing a "smart" grid. This can only be achieved by establishing an information and communications technology (ICT) foundation based on open architecture and industry standards. The development of operational intelligence (and automation) and enterprise value relies on a two-way, standards-based, integrated communication system [1].

This first step requires that information technology (IT) and communications groups work together to understand and support the functional requirements (response requirements, bandwidth, latency) of each disparate data path—from sensor to end user—for current and future systems and applications. This approach requires organization-wide cooperation, which is no small feat. Enabling this fundamental cultural shift requires executive leadership, potentially third-party facilitation, and incentives that reward personnel for organization-wide and customer value creation rather than for individual staff and bailiwick-level achievements.

A foundational ICT platform that links all operational and enterprise aspects of a utility is a prerequisite for enterprise-wide data management. This ICT platform should support full information flow, data management and analytics, and grid monitoring and control. It also comprises the basis for future functionalities that potentially include new consumer services, the integration of distributed energy resources (DERs) and other, yet-to-be-determined needs. The efficacy of this phased approach—seeking a "strong" grid before a "smart" grid—has been affirmed by lessons learned from the stimulus-funded work accomplished under the American Recovery and Reinvestment Act (ARRA) between 2009 and the present. One simple example illustrates this point.

ARRA funding opportunities allowed many utilities to adopt advanced metering infrastructure (AMI). Some of these utilities took a traditional approach by assigning AMI implementation to their metering group alone. As these same utilities later contemplated the implementation of distribution automation (DA), they compounded their original mistake by assigning DA to a distribution engineering group in operations [2].

The direction is positive, but the execution is flawed. DA is the next logical step in grid modernization after AMI and it has the most attractive, stand-alone (i.e., nonsubsidized) business case. But these utilities are finding that their earlier decisions on data networks and IT infrastructure to support AMI do not support DA integration or that implementing DA requires a costly, disruptive workaround. In a holistic approach to data management, all operational and enterprise units would openly discuss their future direction and related projects and set foundational ICT requirements to serve them all. This fundamental step would eliminate redundant efforts and costs—and the creation of two separate data streams—because two or more systems in this example share a need for a service territory-wide communication network. Extrapolate this single example across a utility's many networks, systems, and applications and extend it into the future along a well-plotted technology roadmap. Although it requires daunting cultural change and significant up-front time and effort, a holistic approach ultimately saves time, effort, and money and provides ever-increasing benefits to a future-facing, data-driven utility. In contrast, as this example illustrates, a fragmented, piecemeal approach is likely to result in stranded assets or, at best, time-consuming, costly workarounds at each step in a technology roadmap.

Once a strong ICT foundation has been established, a data-driven utility can proceed to develop a "smart" grid and to map data from sensor to end user. This next step requires the convergence of IT and operations technology (OT) and their respective staffs—the beginning of an operations- and enterprise-wide cultural shift to holistic utility management that focuses on customer-stakeholder-centric value creation and eliminates organizational silos and siloed thinking.

Guidelines for a holistic approach to becoming a data-driven utility:

- Align internal and external stakeholders.
- Think in terms of holistic solutions across the organization.
- Build a strong grid first, with robust ICT performance, then build a smart one.

---

5   INCREASING VISIBILITY WITH IEDs

As readers know, sensors, processing, and the visibility they produce have been applied to the transmission system for some time. The real growth in the need for visibility is downstream in the distribution system, where data-producing sensors and devices in the form of intelligent electronic devices (IEDs) are proliferating. The proliferation of IEDs in the distribution system is enabling utilities to treat T&D as a single entity and is a major enabler for the transformation to a data-driven utility. Yet a lack of visibility in the distribution system remains widespread; for example, only two-thirds of the distribution substations in the United States currently have automation.

IEDs can take the form of standalone sensors or they can be data-producing substation protection and control equipment such as protective relays, load tap changers, and voltage regulators. They produce two streams of data: operational and nonoperational. Operational data is routed in real time to operators in control centers for monitoring and control purposes (see Fig. 1, Types of data: "operational" data). Nonoperational data can provide significant insights

Types of data: "operational" data

- Data that represents the *real-time status, performance, and loading* of power system equipment
- This is the *fundamental information used by system operators* to monitor and control the power system

Examples:
  - Circuit breaker open/closed status
  - Line current (amperes)
  - Bus voltages
  - Transformer loading (real and reactive power)
  - Substation alarms (high temperature, low pressure, intrusion)

FIG. 1

Types of data: "operational" data. From J.D. McDonald, Powerpoint presentation, Enterprise Data Management, slide # 5.

Types of data: "Nonoperational" data

- Data items for which the *primary user is someone other than the system operators* (engineering, maintenance, etc.)
- Note that operators are usually interested in some data that is classified as nonoperational

Examples of "Nonoperational" data:
  - Digital fault recorder records (waveforms) (protection engineer)
  - Circuit breaker contact wear indicator (maintenance)
  - Dissolved gas/moisture content in oil (maintenance)

FIG. 2

Types of data: "nonoperational" data. From J.D. McDonald, Powerpoint presentation, Enterprise Data Management, slide # 6.

for value creation, if properly routed, stored, processed, and made accessible on-demand to both operations personnel and to a utility's enterprise units for use with their applications (see Fig. 2, Types of data: "nonoperational" data). Specifically, nonoperational data can inform enterprise goals for energy efficiency, load shaping, and capital deferral. Metering data, another form of nonoperational data, can support programs aimed at energy efficiency and reliability such as demand response and dynamic pricing [3]. To better understand the defining characteristics of both types of data, see Fig. 3, Characteristics of operational & nonoperational data.

The value of nonoperational data and an example of how it can be overlooked is represented by a protective relay. (See the second bullet point under Examples of nonoperational data, in Fig. 2, Types of data: "nonoperational" data.) A utility's protection group typically buys and installs protective relays for their operational role in detecting faults and tripping circuit breakers. But every IED-enabled relay generates two types of data valuable to the utility's maintenance group. As a fault is detected by a relay, the latter opens a breaker to isolate the fault and quench the arc associated with the opening breaker. The level of energy associated with that arc is captured by the formula i^2t, which is the square of the current (i) flowing through the breaker when the contacts open, multiplied by the time (t) it takes to extinguish the arc. The i^2t data, combined with the breaker's operations counter (how many times it has operated), can tell maintenance when that breaker is due for service. The actual values that indicate a need for maintenance action will vary by the breaker's make and model. But if the protection group or the utility enterprise is not aware of

Characteristics of operational & Nonoperational data		
Characteristic	**Operational data**	**Nonoperational data**
Data format	Usually limited to *individual time sequenced data items*	*Usually a data file* that consists of a collection of related data elements
Real time vs historical	Usually consists of *real-time or near real-time* quantities	Mostly *historical* data: trends over time
Data integration	Easily transportable by conventional SCADA RTUs using *standard (non proprietary) protocols*	Typically use *vendor specific (proprietary) formats* that are not easily transported_ by SCADA communication protocols

FIG. 3

Characteristics of operational & nonoperational data. From J.D. McDonald, Powerpoint presentation, Enterprise Data Management, slide # 7.

the value of that nonoperational data, the maintenance group typically will service the breaker based on time schedules unrelated to its actual condition or when the device fails. Fix-on-fail, of course, is the antithesis of proactive, data-driven asset management [4], so this example underscores the value of nonoperational data and that enterprise users must have ready access to it. Speaking of value, the business case for IEDs themselves only makes sense when both data streams are fully exploited. IEDs range in cost from $5000 to $10,000 apiece and typically they are implemented by the hundreds. But perhaps as much as 75% of their potential value is unrealized if their nonoperational data is not fully utilized.

Every IED has a number of "points" that produce either operational or nonoperational data. The collection of points in a particular IED may be thought of as a "data map." Each IED and its data map must be matched with one or more communication network(s) that provide the response requirements appropriate to the data being transmitted. Operational and nonoperational data each have their own set of communication network response requirements, worthy of a brief review here [5].

6 NETWORK RESPONSE REQUIREMENTS

To make functional sense and support a positive business case, communication networks should be designed to meet the priorities and quality of the data they carry. As different data streams rely on a variety of response requirements, a utility may mix and match various communication networks to achieve diverse functionalities and cost-effectiveness.

IEDs' real-time operational data typically demands the most stringent response requirements, which include reliability, redundancy, speed, latency, band-width, throughput, and cybersecurity. That holds true whether the medium is redundant fiber optic cable laid in rings around a service territory, wireless microwave, or UHF. Still, operational data is heterogeneous and lends itself to a mix-and-match approach to corresponding communication networks. For example, "smart" interval meters record data at 15 min intervals. Integrated Volt/VAr Control (IVVC) requires only 30–60 s to switching on distribution feeder-based capacitor banks. In contrast, Fault Detection, Isolation and service Restoration (FDIR) requires a 2 s response.

Operational data typically has been and is likely to continue to travel over fiber optic networks. The use of licensed wireless spectrum for substations upstream of the "last mile" may also provide a cost-effective network solution. And unli-censed spread-spectrum technology can provide a cost-effective solution for rural "last mile" networks where interference is unlikely.

As mentioned, nonoperational data offers the enterprise a wealth of valuable information that supports the IED business case and is indispensable to a data-driven utility. As noted, nonoperational data can support a shift from time-based to condition-based asset management and it can aid value creation by planning, power quality, asset management, maintenance, engineering, and other enterprise units.

The salient consideration in determining the response requirements of nono-perational data in a communication network is bandwidth, because a digitized waveform, for example, may require a "fat pipe" to reach the enterprise uncor-rupted. Speed, latency, and other metrics are of less importance because non-operational data is often used for after-the-fact analysis and event forensics. Nonoperational data is also heterogeneous, with lower response requirements and security demands than operational data, so it can also benefit from a mix-and-match approach to communication networks.

7 INTEGRATION BEFORE AUTOMATION

Integrating IEDs across the substation and on distribution feeders means prop-erly assigning each data stream to the appropriate communication network and routing those data streams to the control center and/or the enterprise. This is a critical step because IEDs are now synonymous with nearly every piece of power system equipment, including protective relays, meters, transformers, circuit breakers, reclosers, load tap changer controls, voltage regulators, etc.

The accompanying Fig. 4 provides a graphic representation of the integration challenge. The foundation is power system equipment such as transformers

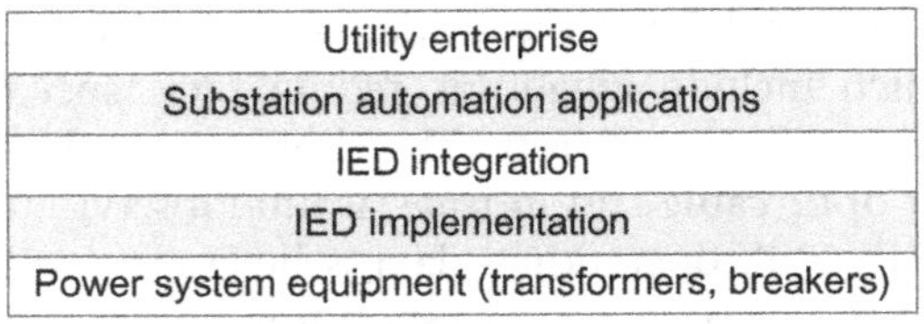

Utility enterprise
Substation automation applications
IED integration
IED implementation
Power system equipment (transformers, breakers)

FIG. 4

Five levels of substation integration and automation. From J.D. McDonald, Substation automation, IEEE Power Energy Mag. (March/April 2003).

and circuit breakers. The next levels include IED implementation, IED integration, and substation automation (SA) applications. The enterprise comprises the fifth and highest level.

In the past, IED integration has too often focused exclusively on operational data such as instantaneous values of voltage, current, and related data, while overlooking nonoperational data and its value. The latter can include on-demand or event-triggered data of logs of events and oscillography that aids diagnostics and forensics on conditions that lead to major events such as outages or equipment failures [6].

Distribution system integration requires that the utility tie together protection, control, and data acquisition functions using the minimum possible number of platforms, thus reducing capital and O&M costs, physical footprint, and eliminating redundant equipment and databases. The integration of data-producing devices and systems precedes SA. Determining SA applications relies on first observing the behavior of data over time (daily, seasonally) and diverse conditions (weather patterns) to arrive at rational, data-based establishment of threshold values that trigger automated responses. SA simply refers to implementing SCADA, alarm processing, and other elements to optimize asset management and operational efficiencies that operate without human intervention.

8   FUNCTIONAL DATA PATHS: KEEP IT SIMPLE

The functional data path for operational data typically sends real-time data on voltage, amps, etc., to the utility's SCADA system every 2–4 s for dispatchers who monitor and control the power system. Ideally, SCADA would also pull nonoperational data from IEDs and route it to data concentrators at the substation level. Then that nonoperational data can be routed over its own communication network(s), across the corporate firewall to data repositories within the enterprise. Business units and personnel, on an authorized basis, can then retrieve needed data on-demand through queries and data mining on the corporate network.

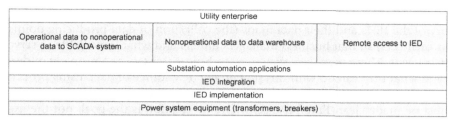

FIG. 5

Three functional data paths from substation to utility enterprise. From J.D. McDonald, Substation automation, IEEE Power Energy Mag. (March/April 2003).

Note that the physical media involved within a substation requires consideration. IED integration must address legacy equipment. Adding Ethernet connections to serial communications within a substation produces a hybrid configuration that may not lend itself to full integration. This potential hurdle is moot if the fundamental ICT platform is in place.

Fig. 5 illustrates three functional data paths from substation to the utility enterprise: the two paths most relevant here are operational data going to the SCADA system, and nonoperational data heading across the operations firewall to a data warehouse within the enterprise. In the latter case, we must now address how the establishment of a data warehouse and, more importantly, a "data mart," is at once a technical, organizational, and cultural challenge [5].

9 FROM SENSOR TO END USER: THE PROCESS

Because the routing and use of operational data is likely to be long familiar to readers, our focus here will remain on the routing and availability of nonoperational data. Though this challenge certainly has a technical component, we must begin with the heavy lifting: people working with people, across legacy silos, for the greater good. This is a step that a data-driven utility cannot avoid. Indeed, if properly understood and implemented, the result will unlock significant value in resolving business challenges, allow a shift to a more effective and less costly condition-based maintenance approach, and support future functionalities. This step also leads to cultural and organizational and business process changes, so it is a fundamentally transformative process that, once unleashed, is irreversible.

In designing an informational architecture to deliver nonoperational data to the authorized enterprise individual or unit for value creation, one must create an enterprise-wide "data requirements matrix." This initial step involves querying business unit managers on the question of who in their bailiwick needs nonoperational data; specifically, what type of data, in what form, and at what specific time intervals. This step should be supported by presenting an inventory of the IEDs and their data maps (the collection of data-producing points on each IED) so that business managers understand what is available to serve their needs. These managers may need technical assistance to properly understand what they can do with this potentially new source of data, using processing, applications, and presentation technology. The purpose of the exercise must be made clear. Enterprise-centric value creation is the goal, not the use of data to support individual or business unit achievements. Silo walls should crumble, not be reinforced.

The inventory of distribution IEDs will include their data maps and attributes and the next step is to determine which points in each data map can serve value creation by stakeholders in the enterprise. This is a complex step. Vendors now differentiate their IEDs by taking sometimes unique approaches to the production of nonoperational data. The attributes associated with each IED data point might differ among devices by different vendors; thus they must be carefully documented. The data sampling rate might vary, IED to IED. An end user might seek a peak value or an average value for each hour of data retrieved. An event might drive an attribute. Data might only be of value when it exceeds a preset threshold. We typically refer to these attributes as the "aspect of value." The key is determining what data the IED produces and the "aspect of value" to the end user.

Once we have the IED template—the sensors and their data maps—and the data requirements matrix—who needs which data and its attributes—mapping the source to the end user informs the network architecture that delivers nonoperational data across the corporate firewall into a data repository/warehouse, where the right person can access the right data on demand. Rigor and accuracy in this phase is critical to a successful outcome, and to future IED additions. When additional IEDs are installed on the distribution network, they are simply added to the existing template, matrix, and map.

Utilities often rely on a number of physical data repositories, which remain useful in the data mart scenario described here. A federated data server can sit atop and access these potentially disparate, legacy repositories, creating a "virtual data mart," which includes both operational and nonoperational data. Fig. 6 illustrates a typical, siloed approach to data management (left) and how that can be transformed (right) so that authorized users across the utility organization can access both operational and nonoperational data.

We have described how nonoperational data is gathered and routed to the enterprise. If operational data is needed, the enterprise end user can access it on-demand from the virtual data mart, which has received the data from the operations (SCADA) historian. The historian has been recording a time series of data at a predetermined sampling rate—a subset of all operational data—for

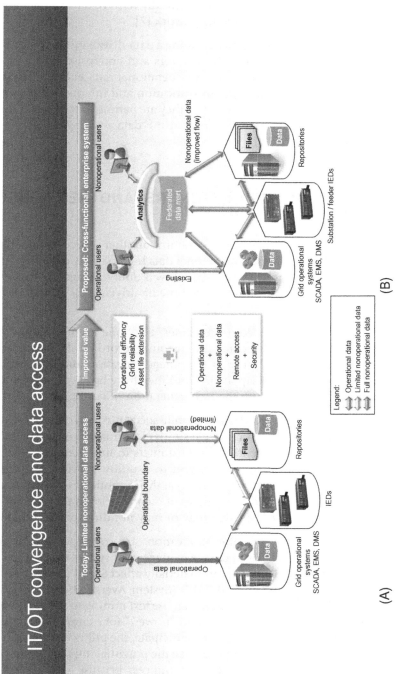

FIG. 6

The siloed arrangement on the left is typical of many utilities' suboptimal approach to data management, which limits access to nonoperational data. On the right, all devices, systems, and data repositories feed into a federated data mart (FDM), which enables organization-wide access to both operational and nonoperational data for improved decision-making. IT/OT convergence and data access. *From J.D. McDonald Powerpoint presentation, Enterprise Data Management, slide # 13.*

export to the enterprise. End users typically retrieve data on-demand by using an application on the corporate network [7].

Thus the holistic approach to becoming a data-driven utility touted throughout this chapter ensures that both operators and enterprise users get authorized access to both operational and nonoperational data either in real time or on demand (Fig. 7). Multifactor authentication and internal controls ensure that only the right people access the data they are permitted to access. Other security controls in the data mart ensure that the data residing there is reliable and accurate.

10   CONSUMERS/CUSTOMERS: ANOTHER SOURCE OF DATA

Let us make a brief digression to examine another source of valuable data that might be classified as nonoperational data but is simply data generated by utility customers themselves. As a data-driven utility engages its customers—as it must—harnessing customers' social media-driven data to improve reliability indices will become commonplace.

In the past, a utility relied on customers to phone in information on outages. Either verbally or by searching the customer record for the customer's address, the utility could get a sense of the outage's location and dimensions. Today, landline phones are disappearing and they have been completely abandoned by millennials, who form the next generation of customers.

Now, with the widespread use of mobile social media, a customer tweet or cluster of customer tweets can provide similar information more swiftly and accurately. A utility can incentivize its customers to link their Twitter tags to their account information so that a tweet to a utility provides the old landline-generated data. Customers who turn on their mobile device's geo-tagging function can deliver the GPS coordinates of their location if they are not home and see the cause of an outage that may or may not directly affect them.

In both cases, newly available applications can connect tweets or other social media-generated data to an outage management system that triggers DA applications such as FDIR, which can materially affect SAIDI (System Average Interruption Duration Index) and SAIFI (System Average Interruption Frequency Index) indices. These applications can use text mining to assess whether a flurry of tweets that mention "outage" and "power," for example, actually refers to a power outage. As more customers participate, the utility benefits accordingly. It is impossible to quantify or generalize the potential impact of using customers' social media on SAIDI and SAIFI indices, because outage factors are so

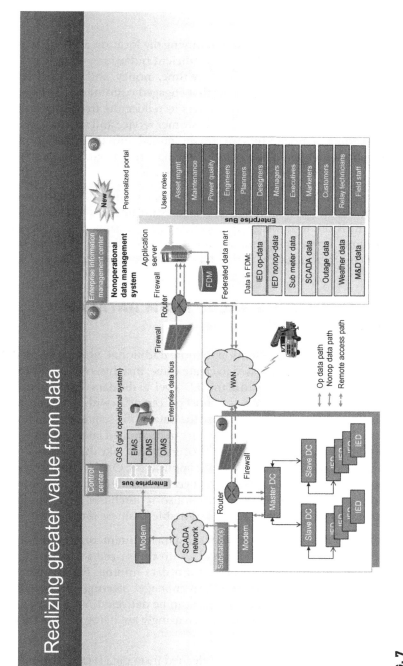

FIG. 7

Realizing greater value from data. This illustration depicts how myriad data sources ideally feed both a control center and an enterprise information management center. On the enterprise side, data in an FDM (examples in horizontal boxes at near-right) can be readily accessed for value creation by diverse end users (user roles in horizontal boxes at far right). *From J.D. McDonald, Powerpoint presentation, Enterprise Data Management, slide # 14.*

dependent on a specific utility, where it is located and its own set of problematic circuits.

Crowdsourcing as a means of identifying the location and extent of outages has several advantages. Swifter, more efficient outage responses can improve reliability indices; fewer truck rolls save time, money, and environmental impacts. Last, but not least, studies show that engaged customers are more satisfied customers and engaging your customers to reduce the frequency and duration of outages has obvious benefits for customer service and satisfaction. Just the shift in customer demographics toward millennials and their abandonment of landline phones and embrace of social media virtually guarantees that crowdsourcing outages will become a common practice.

There is a lot more to understanding and implementing this particular data stream, but it should be mentioned as a significant future trend. Perhaps even more nascent is the technology and resulting data streams potentially gleaned from consumer home energy management systems. These appear inevitable, though their commercialization to date has experienced decidedly mixed results [8].

11   EXTRACTING VALUE FROM DATA, AND PRESENTING IT

Now let us return our attention to more traditional, nonoperational data. Data, of course, must be processed to create value, insights, or actionable intelligence. Processing can occur in situ at the IED level, at the data concentrator level and/or at a desktop level—known as a "host processor"—within the substation. In situ processing helps manage and ease the data traffic sent upstream for central processing. Or, as just noted, processing can be applied later via applications on the corporate network. At that stage, processing can take the form of a calculation, a software application, and/or a logic application. On the enterprise side, business unit managers and their staff need to understand and carefully select applications—either developed in-house or purchased from a third party—that turn data and information into actionable business intelligence.

By linking the IED template with the data requirements matrix, we have routed the right data to the data mart and set-up authorized access for the right people to make data queries and/or perform data mining. Finally, with IT or third-party assistance, the business unit manager steering the process will likely participate in crafting how outputs can be delivered. Visualization aids understanding and, thus, dashboards increasingly are being used to make processed data more easily apprehensible.

The visualization of intelligence gleaned from data is critical and it is an area in progress. This is true of both operational and nonoperational data presentation.

No one should be staring at tables of numbers. Instead, for example, color-coded lines or bars can instantly tell the user that a threshold value has been exceeded and instantly and intuitively convey critical intelligence. This is an area of intense interest and value that will mature over time [7].

12 THE TRANSFORMATION

It is difficult to generalize how de-siloing and a horizontal approach to data needs and applications will affect a given utility. But a few obvious points are worth making.

No individual operational unit should add technology without review by all operational and enterprise units, because the resulting system may serve multiple needs and purposes and redundancy of systems and efforts—unless aimed at security—is no longer justifiable. Any new device, technology, or system is likely to generate operational and nonoperational data and that data must be incorporated into the utility-wide data maps, data requirements matrix, and data mart created under a holistic approach. In a digital, data-driven utility, IT and OT, traditionally leery of one another, must collaborate. In similar fashion, enterprise units and their managers and staff must share data-driven insights with one another to ensure the widest impact of value creation.

In fact, it is likely that a utility embarked upon becoming a data-driven utility will need neutral, third-party assistance in guiding both cultural and organizational transformations and business process change. New incentives to drive new behaviors may be needed. Managers and staff may be rewarded based on their devotion to and achievements in pan-organizational cooperation and value creation.

It has been said that fostering cultural change is far more difficult than implementing new technology. So keep in mind that becoming a data-driven utility is hardly all about technology. Human interoperability is just as important as technology interoperability [5].

13 THREE CASE STUDIES

It is important to understand that the foregoing guidance, though stated in conceptual terms, is not pie-in-the-sky theory. The process described here for becoming a data-driven utility has been successfully implemented in different ways based on different drivers at a variety of utilities. Of course, in the real world, adding new or upgraded technology often involves preserving legacy investments, which sometimes requires complex solutions. Thus, we will conclude this opening chapter by looking at three utilities and how they have successfully managed and applied data and related insights to solve real-world challenges. These three case studies illustrate the value of a data-driven utility in terms of asset management and safety, the fundamentals of standards and interoperability and the enterprise value, in hard dollars, of increased visibility into the transmission and distribution network [9]. To achieve that value, each utility had to embrace one or more of the guidelines for data-driven success: align internal and external stakeholders, think in terms of holistic solutions across the organization, and first build a strong grid (i.e., one with robust ICT capabilities) before building a smart one. The holistic approach is certainly prominent in achieving all three solutions.

13.1   Frankfort, Kentucky, and Greenfield SCADA, SA

A SCADA-related project in Frankfort, Kentucky, illustrates the importance and value of integrating legacy and new equipment and the efficacy of a single user interface.

In the late 1990s, the Frankfort Electric and Water Plant Board (FEWPB) had SCADA plans on hold due to budget constraints when a distribution substation transformer exploded, leading to a significant outage and related costs. Fortunately, no one was injured. The cause: an internal failure in a single-phase regulator had led to erratic voltage fluctuations and, eventually, an explosion, putting thousands of customers out of service. One was a neighbor whose lights had been acting funny. He had thought of calling FEWPB, but did not [10].

Suddenly, a SCADA system and integrated SA was top of mind. Automation to this point had been limited to a set of electro-mechanical and microprocessor-based protective relays. So FEWPB embarked on a dual implementation of SCADA system and integrated SA aimed at improving system reliability, accelerating outage restoration and reducing O&M costs.

Although at that time a lack of SCADA and SA was not uncommon, implementing these technologies in an essentially greenfield situation presented FEWPB with somewhat unique opportunities and challenges. One positive: the utility did not have to integrate new SCADA and SA technologies with a legacy system. Another: FEWPB could develop an architecture that thoroughly networked and integrated the SCADA master and all the utility's 19 substations. A consultant provided an enterprise-wide seminar on SCADA/SA capabilities to introduce the technologies and their capabilities to the utility's relevant personnel.

The utility decided to design a single user interface for both systems, accessible at several locations, including its primary dispatch center, its network operations center (NOC), and several of its largest substations. The unified design meant that several departments could run the system (remotely, if needed) and this also simplified training. Although a work-around could have accommodated the utility's existing array of relays, the utility decided to replace them with new IEDs that produced greater amounts of data and offered better protection at less cost. Sourcing the IEDs from one vendor enabled FEWPB to adopt a standardized data approach that could be applied to all substations.

Having determined its approach to data-generating IEDs, SCADA, and SA, the utility addressed its communication network. Because FEWPB provided a suite of services to its municipal customers—including cable, phone, and Internet—it already had a fiber-optic ring around the city based on Synchronous Optical Network (SONET) technology. This would be ideal for SCADA communications, if not suitable for protection purposes. The fiber ring's backbone provided 100 Mbps transmission capacity at the substations, which could easily handle operational and nonoperational data streams and files.

The SONET network already connected the utility's NOC and its primary dispatch center, and geographically convenient network hubs ran lower capacity network lines to connect neighborhoods to the main ring for cable, phone, and Internet, which substations could piggyback on. Thus the utility's existing fiber ring could be leveraged and designed into the overall SCADA/SA system and reduce costs for the SA part of the project.

To contain costs, FEWPB and its consultant decided to designate six large or new substations as primary substations, which would receive a full suite of IEDs, PCs, or workstations and all SCADA/SA interface components. Another 10 secondary substations would be equipped with only IEDs, integrated with the SA system data concentrator, and linked by fiber to their nearest primary substation. The secondary to primary substation link would transmit data collected from the IEDs and data concentrator to the primary substation for analysis at the SCADA/SA interface. Both primary and secondary substation data would then be sent upstream to the SCADA master over the SONET network.

In seeking a vendor to meet its needs, FEWPB required an open system that relied on commercial off-the-shelf (COTS) hardware and software. The utility initially employed a request for information (RFI) process to generate the greatest interest and favorable, competitive pricing, then moved to the request for proposal (RFP) phase. The project was completed over several years as budgets allowed the inclusion of additional substations.

13.2 Ketchikan, Alaska, Deals With Unsupported, Legacy RTUs

A SCADA-related project in Ketchikan, Alaska, clearly illustrates why proprietary technology and protocols can lead to costly dead ends.

Ketchikan Public Utilities (KPU) in the 1990s found itself in an unenviable position: its SCADA master station and RTU vendor from the 1980s had gone out of business and left KPU without support for the vendor's SCADA system, partly based on a proprietary communication protocol [11].

Compounding the issue of support, KPU sought in 1999 to expand its substation data collection by replacing RTUs with PLCs. In this case, at that time, the PLC choice made sense as they could handle analog and status inputs and control outputs and serve as a platform for SA. But the PLCs would have to talk to the SCADA system over the defunct vendor's proprietary protocol, just as the RTUs had.

KPU managed to find an experienced integrator of SA components, which deciphered the proprietary protocol and programmed its own communication processor box, designed for IED integration and SA. The communication processor box became a slave unit responding to commands from the SCADA master station and a master station to the PLCs—sending commands, retrieving data, and transmitting data back to the SCADA system. The PLC solution worked, but it was costly due to a large number of hard-wired input/output points and the absence of IEDs, which could send and receive the same data digitally.

Over the course of a few years, KPU implemented this strategy in half its distribution substations. But in the middle of the project the utility realized it needed to replace its SCADA master station, which was difficult to support and had reached its data point limit. And the utility knew it would explore an expansion by tying into another power grid to its north, ultimately to provide power to visiting cruise ships, which otherwise would continue to run their diesel engines while in port, unnecessarily polluting the air for visitors and residents.

Here is where geography and service territory factors came into play. Ketchikan is a small town on an island located in far southeastern Alaska. The mountainous terrain means that substations only 20 miles away require hours of driving by truck and, in one case, is only accessible by floatplane. If KPU connected with the grid to the north, the resulting topology and grid could be most efficiently and remotely operated from a central control location in Ketchikan. Just to meet its existing data needs, let alone position itself as a regional control center, KPU needed to upgrade or replace its SCADA system.

KPU researched SCADA and SA technology, and it hired a consultant experienced in utility automation projects. Together they determined that implementing IEDs, which had become more common since KPU's original RTU-related project in the 1980s, made sense. Fewer IEDs could replace many more PLCs at markedly lower cost, while providing more effective data producing and processing capabilities.

In considering a new SCADA master station, KPU wisely determined that it would only invest in one that ran the DNP3 protocol, then a de facto industry standard. To avoid losing its investment in seven substations that used the reverse-engineered proprietary protocol of its defunct vendor, the utility discovered that those communication processor boxes could easily be converted back to their original DNP3 protocol. For communicating with RTUs that remained in place, the utility's new vendor built a converter into its communication processor box that enabled it to receive DNP3 commands from the new SCADA master station and convert those commands to the proprietary protocol required by the RTUs. The process worked in reverse as well for data heading upstream. When KPU had the budget to replace its remaining RTUs, the communication processors running the proprietary protocol could be converted to DNP3.

KPU's research and determination to avoid proprietary solutions and protocols paid off. The result was a system running on an industry standard that cost about a third of its proprietary system, preserved legacy investments in a phased-in approach governed by its budget, and produced an enhanced ability to send and receive data in support of its SCADA system and SA. The utility could look forward, as well, to lower O&M costs through remote monitoring and control and position itself for future expansion.

13.3   North Carolina Agency Pursues New SCADA, Boosts Revenue

A SCADA-related case in North Carolina demonstrates that upgrading or replacing a SCADA system should be considered an investment rather than a cost.

Investments in becoming a data-driven utility should be based on a positive business case. As the North Carolina Municipal Power Agency No. 1 (NCMPA 1) discovered over a decade ago, advancements in SCADA technology have enabled some utilities to replace their existing systems with a rapid ROI, even boosting profits along the way [12].

At the turn of the 21st century, NCMPA 1 had been distributing power on behalf of municipal utilities across North Carolina for two decades. For three quarters of the year, its system experienced peak load of about 600–650 MW, about 250 MW below generation capacity. It sold the excess power on the wholesale market through a power marketing and trading company.

NCMPA 1 installed a SCADA system in 1996 to monitor its distribution system. The utility installed more than four dozen meters to measure instantaneous power and energy usage at 47 substations across the Piedmont region of North Carolina. Each metering site featured RTUs that recorded, processed, and formatted the meter data based on DNP3 and transmitted it to the SCADA master station at NCMPA 1's headquarters in Raleigh, N.C. A frame relay system provided a reliable 56 kbps (kilobits per second) link between substations and the control center. The SCADA master station polled the RTUs for load data once every 5 min. The SCADA system transmitted both peak load and generation data via FTP—in 1 h increments via "block scheduling"—to the power marketing and trading company for near real-time insights into the availability of excess capacity.

Unfortunately, the SCADA system over time exhibited reliability issues tied to local thunderstorms, which led to forecasting errors on excess capacity. NCMPA 1's use of "block scheduling" in 1 h increments added to the inaccuracy of forecasts. To compensate, NCMPA 1 personnel kept unsold power in reserve, to ensure meeting load, which translated to about 25 MW/h of lost revenue.

NCMPA 1 clearly needed to upgrade or replace its SCADA system. And it decided to move to "dynamic scheduling" that would eliminate related errors in forecasting produced by block scheduling. The latter required control area quality SCADA with 4 s scan rate telemetry, more in line with industry average performance. The utility decided the most cost-effective and forward-looking approach would be to replace its SCADA system and, with a consultant, it set about identifying operating parameters and hardware components that could be replaced with newer, more advanced technologies for improved performance and under a relatively tight budget.

The utility and its consultant determined that the SCADA master station would need replacing with one capable of a four-second scan rate and—mindful of budget constraints—that the frame relay system in use could support that rate. The substation meters would be replaced by more accurate and efficient IEDs, linked through frame relay access devices (FRADs) to the communications system, and eliminating a role for the error-prone RTUs. The IEDs used DNP3, a protocol that ensured a more seamless data flow across the system.

The utility and its consultant also decided to make the SCADA system redundant for reliability and resiliency by splitting the master station between two locations linked by high-speed T1 lines. Dial-up communications to each meter site were established as a communications backup.

In its system requirements, codified in its RFP, NCMPA 1 insisted on the use of standards to ensure interoperability. When the utility selected its vendor, it participated in factory tests of the hardware and software, to ensure it met NCMPA 1's specifications and to allow its personnel to become familiar with the new system's O&M needs. This step served as practical training for the utility's SCADA-related personnel.

Once in successful operation, the new SCADA system and related dynamic scheduling allowed the utility to operate without that costly 25 MW buffer. The system's increased accuracy allowed the utility to negotiate a more favorable contract with a new power marketing company. The utility also experienced greater operational efficiencies. All these factors contributed to ROI within an astonishing 6 months.

14 CONCLUSION

Becoming a data-driven utility is an imperative of the digital age. Data-based insights are critical to real-time grid operations and just as crucial to running a utility enterprise in an era rife with disruptive technological and market forces. The interdependence and synergies between sensors, communication networks, software-based systems, and hardware to monitor and control the grid and run the enterprise is too complex for an ad hoc approach. Therefore, a holistic approach based on a foundation of open architecture and standards will ensure interoperability between devices, systems, databases and, not incidentally, people. The processes required in a holistic approach represent an opportunity to eliminate traditional organizational silos and unify utility personnel around the utility's fundamental mission of delivering safe, reliable, affordable power to its customers. The guiding mantra that governs this approach is that every person in a utility who can produce value from data should have access to that data, with proper security safeguards in place.

While operational data currently demands the most attention, nonoperational data—now plentiful with the addition of IEDs to the distribution system—must be exploited as well. The full use of both data streams—and customer-generated social media-related data—offers many benefits to both operations and enterprise. The systematic use of data can make operations safer and more reliable, resilient, and efficient. On the enterprise side, full exploitation of all available data supports a shift from time-based to condition-based asset management and aids value creation by planning, power quality, maintenance, engineering, and other enterprise units. Regulators have affirmed that customers own their energy use data and have the prerogative to share it with third parties that offer value in return. Thus the future viability of the utility enterprise, amid disruptive market forces, appears likely to depend on the creation of customer service options and that eventuality will also rely on the availability and exploitation of data.

To maximize the value of existing assets and guarantee the value of future investments, a foundational ICT platform—a "strong" grid—must be in place before adding intelligence for a "smart" grid. This foundation enables functional data paths with response requirements matched to the data they transmit. Operational data is routed to the control room for real-time monitoring and control, while a SCADA historian sends a subset of operational data across the corporate firewall for enterprise use. Nonoperational data is also sent across the corporate firewall for storage in a virtual data mart, enabling processing and access on-demand by enterprise business units. Easy-to-grasp presentation of results represents a critical final step in this process. Dashboards and other means of presentation remain an area for further innovation.

Ultimately, operations and enterprise personnel should share insights and actionable intelligence gleaned through this holistic approach to data management across the entire organization. The three case studies provided in Chapter 1.1 are just a small sampling of real-world value creation through the use of SCADA-related data based on a positive business case.

14.1   Looking Ahead

This brief overview of the holistic approach to utility data management sets the stage for the contents of this book.

The remaining chapters in Section One will go into more detail on the importance of, and methods to achieve, the initiatives described here. That includes the development of an open, standards-based information architecture for data-driven utilities, frameworks for big data integration, warehousing and analytics, overall management of the data being produced, stored and processed, as well as the data security and privacy challenges facing utilities.

The foregoing overview merely places data analytics and presentation/ visualization in context. Section Two of this book will delve more deeply into the algorithms and mathematics that characterize current data analytics practices, including statistical learning, machine learning, deep learning, and other approaches. Theoretical discussions of these topics will be matched with simple examples that illustrate complex ideas.

A brief overview can only mention the transformational outcomes that a data-driven utility is likely to achieve. Improved operational safety, reliability, and resiliency are among the likely outcomes cited, and the specifics of how those goals are achieved are treated in Section Three. The discussion will cover appropriate, data-driven methods for meeting daily utility challenges such as diagnostics, volt/var optimization, risk management, oscillation mitigation, and market operations. Section Three will also bring readers up to date on new analytical applications that enable DER forecasting, load disaggregation, predictive maintenance, customer behavioral analysis, cyber-attack detection, and other insights.

In sum, this landmark book provides the philosophy, the concepts, and the methods to empower a utility to adopt a proactive, holistic approach to becoming a data-driven utility. The ensuing journey will transform a utility's operational and organizational practices and structure and should provide the flexibility to develop and implement new business models as the market requires. The urgency of this transformational approach cannot be overstated. Carpe diem!

References

[1] For more on "strong" before "smart" grid, see J.D. McDonald, et al., Refining a holistic view of grid modernization, the final chapter in: Smart Grids: Infrastructure, Technology, and Solutions, CRC Press, Boca Raton, in press, 2017. For more on open information architectures and related standards, see J.D. McDonald, Managing Big Data: Challenges and Winning Strategies, T&D Magazine, 2014, pp. 29–30.

[2] J.D. McDonald, Integrating DA With AMI May Be Rude Awakening for Some Utilities, Renew Grid, Oxford, CT, 2013. passim.

[3] For the role of IEDs and non-operational data, see: J.D. McDonald, Extracting Value from Data, Electricity Today (May 2013) passim. For the role of IEDs and non-operational data, see On IED integration, see J.D. McDonald, Substation automation: IED integration and the availability of information, IEEE Power Energy Mag. 99 (2003) 23–24.

[4] J.D. McDonald, Extracting Value from Data, Electricity Today, 2013, 9.

[5] J.D. McDonald, Transformer Monitoring, Communications Networks and Data Marts: Extracting Full Value From Monitoring and Automation Schemes to Aid Enterprise Challenges," Keynote Paper, TechCon Asustralia, 2015.

[6] J.D. McDonald, Substation automation: IED integration and the availability of information, IEEE Power Energy Mag. 99 (2003) 23.

[7] J.D. McDonald, et al., Realizing the power of data marts, IEEE Power Energy Mag. 5 (2007) 64–65. passim.

[8] See both J.D. McDonald, Integrated System, Social Media, Improve Grid Reliability, Customer Satisfaction, Electric Light & Power, Tulsa (Dec. 1, 2012) passim, and J.D. McDonald, Consumers and Home Energy Management: As Standards Emerge, It's No Longer 'if,' but 'when' and 'how', PowerGrid International, Tulsa, (April 15, 2014) passim.

[9] M.S. Thomas, J.D. McDonald, These case studies are summarizedPower System SCADA and Smart Grids, CRC Press, Boca Raton, 2015, pp. 70–73.

[10] D. Carpenter, V. Foster, J.D. McDonald, Kentucky Utility Fires Up Its First SCADA System, T&D World, Overland Park, KS, 2005.

[11] H. Hansen, J.D. McDonald, Ketchikan Public UTILITIES Finds Solutions to Outdated, Proprietary RTUs, vol. 2, Electricity Today, 2004.

[12] J.D. McDonald, North Carolina Municipal Power Agency Boosts Revenues by Replacing SCADA, vol. 7, Electricity Today, 2003.

CHAPTER 2

Emerging Security and Data Privacy Challenges for Utilities: Case Studies and Solutions

Carol L. Stimmel
Manifest Mind, LLC, Canaan, NY, United States

CHAPTER OVERVIEW

Cybersecurity applications are rapidly becoming an integral part of the utility operations, managing and processing millions of events per second with microsecond latency without impacting the underlying grid, operations, or enterprise infrastructures. While there are major weaknesses in the distribution system, which are vulnerable to exploitation for which these applications serve, there are myriad new attack vectors being added every day. Yet, while utilities spend most of their cybersecurity resources building a virtual wall around grid assets, they overlook the source of the most common attack vector on the grid—the utility employee. Utilities are much more than the physical operation of the grid; utilities are also responsible for massive enterprise systems with financial information, customer data, and a growing network of digital operations under human control. Thus, security strategies must become more nuanced and complex, and should include privacy and other internal information technology controls.

1 INTRODUCTION

Cybersecurity and data privacy are major challenges in protecting the utility's critical infrastructure amid the growing population of critical digital assets and consumers within the electric system. Despite the scope of known vulnerabilities, threats, and emerging data analytic approaches that exist to responding to cyberattack against the utility, simulations show that not only are there major weaknesses in the distribution system, but also that a massive cyberattack could leave some parts of even the most advanced systems with outages lasting up to several weeks. In fact, it is grid modernization itself, especially the rapid deployment of distributed energy resources (DER) that have created such a broad swathe of attack vectors.

There is a tendency within the industry to focus on traditional cybersecurity measures, particularly through information technology at the expense of overall resilience. Yet, standard measures of cyberdefense are a poor fit for the electrical system. Consider the fear of an attack that results in a prolonged blackout; from nearly every societal measure, including those of economy, health, and

29

Big Data Application in Power Systems. https://doi.org/10.1016/B978-0-12-811968-6.00002-4

public safety, could cause massive disruption. The power system must always be available, and security countermeasures that impede power availability are just not appropriate. In most other industries, particularly financial services, the confidentiality and integrity of data in the system have a higher precedence over availability. With the grid, the availability of electricity is the preeminent security objective, and its quality and privacy of data transferred are secondary concerns.

There is little conclusive evidence that the traditional cybersecurity tactics currently engaged by many utilities are strategically coherent toward meeting the increased threat levels against the critical electricity infrastructure. What are needed are strategies that are cognizant of the complexities of the digital grid that is comprised in large part by digital technology. As David Kennedy, the CEO of TrustedSec and former Marine Intelligence Officer puts it plainly, "Our grid is definitely vulnerable... The energy industry is pretty far behind most other industries when it comes to security best practices and maintaining systems" [1]. However, what is most needed is not the stacking of more capabilities, but a fundamental rethinking about the problem of cybersecurity far more comprehensively than a traditional security operation, a common approach which is not only shortsighted but also wholly insufficient.

2   CASE STUDIES: THE STATE AND SCOPE OF THE THREAT

While the world faces an array of cyberthreats, the complete dependence that western society has on electricity makes the grid a very ripe target for attack. And with the turn of the 21st century—as billions of connected devices come on line—the face of cyberthreats as a weapon of cyberwar by advanced nation-states brings to bear considerable abilities to probe defenses and coordinate attacks across public and private targets through myriad vectors. There is an accelerating understanding of not just the threats to the utility, but the extensive complex of vulnerabilities that can impact digital networks, machines, and systems, and the information within the system—all which bear consequence to the mission of reliable and secure energy. As the electric grid grows more interconnected with digital and DER, this risk grows exponentially because of the inclusion of physical assets on the grid including rooftop and ground mount solar arrays that may not be controllable by the utility, sensors, and actuators (collectively referred to as the Internet of Things or IoT), smart meters that can link the utility to devices inside home, and the data that is integral to operations, market functions, and customer service.

Given the expanding scope of vulnerability and the demand for a resilient and secure grid, there is surprising little clarity in the industry for understanding cybersecurity and cyberterrorism. Factors that contribute to this murkiness include a lack of regulatory clarity, governance, uniqueness of utilities across markets and geographic regions, low overall investment, and growing public confusion and doubt about not just the gravity of the issue of cybersecurity, but it is very existence. Yet despite an increase in capability for automated attacks, improved policy and governance, education, and application whitelisting, such attacks have declined precipitously in favor of the softest and easiest target—the human being.

The 2016 Human Factor cybersecurity assessment found, "Attackers shifted away from automated exploits and instead engaged people to do the dirty work—infecting systems, stealing credentials, and transferring funds. Across all vectors and in attacks of all sizes, threat actors used social engineering to trick people into doing things that once depended on malicious code" [2]. Perhaps the infamous hacker Kevin Mitnick was correct when he said, "The biggest threat to the security of a company is not a computer virus, an unpatched hole in a key program or a badly installed firewall. In fact, the biggest threat could be you. What I found personally to be true was that it's easier to manipulate people rather than technology. Most of the time organizations overlook that human element." [3]. To a large extent, cybersecurity in the electric grid is a matter of transparent relationships and engagement between the utility and its employees, partners, and customers, including how information and data are transferred, secured, used, and analyzed.

There are several modern-day cases that stand out as remarkable demonstrations of this finding and why we might expect this trend to persist: Burlington Electric, Aramco, and Ukrainian Kyivoblenergo.

2.1 Coordinated Cyberattack Causes Outage in the Ukraine

In December of 2015, two separate power distribution companies in the Ukraine, Prykarpattyaoblenergo electric utility and Kyivoblenergo, announced that they had been hacked. The hack had caused a blackout from lost power to distribution regions that served more than 80,000 people. Further, the hackers also sabotaged several operator computers which made it difficult to restore electricity service. Because of the damage to the operational system, utility company workers had to travel to substations to manually reset the breakers which the hack had remotely opened. It was a fairly short-lived event, but it was remarkable since it is the first known electricity blackout that has ever been caused by a known cyberattack [4].

The attack had several levels of coordination apart from remotely opening the circuit breakers. First, the hackers blinded the operational staff by freezing data on the screens, where unbeknownst to the operators, situational intelligence reported that the power system was functioning properly. Secondly, the hackers launched a denial-of-service attack against the call center to prevent customers from reporting a service outage. The center was left answering bogus calls which prevented real customers from getting through to the operators. Apparently not satisfied, once the operators began trying to repair the outage, the enterprise functions of the company were shut down when a program called KillDisk caused company computers to crash, wiping out the master boot records and preventing a reboot.

Nothing is entirely clear about the execution of the breach except that it began with a spearfishing campaign in March of 2015, which likely was the point of entry for malware known as BackEnergy2. This malware can open a backdoor to the system enabling further injection of malware or other applications and data. Robert M. Lee, a former Cyber Warfare Operations Officer for the US Air Force, underscored how even a simple attack can be devastating when it is coordinated with others, "The capabilities used weren't particularly sophisticated but the logistics, planning, use of three methods of attack, coordinated strike against key sites, etc. was extremely well sophisticated" [5].

2.2 Severe Financial Impacts at Saudi Aramco

In mid-2012, a computer IT specialist working at Saudi Aramco opened a phishing email and clicked on the link. Chris Kubecka, a former security advisor to Saudi Aramco described the attack, "It started sometime in mid-2012… One of the computer technicians on Saudi Aramco's information technology team opened a scam email and clicked on a bad link. The hackers were in" [6]. Half a year later, "weird things" started happening—screens flickered, files began to be wiped from the drives, and some systems fully shut down. Unable to control the onslaught, the IT team had to physically rip cables out of the backs of servers at their international data centers and take every office around the world physically offline.

Still, while the company managed to keep up production, the supply lines faltered, contracts could not be fulfilled, financial transactions halted, office lines, email, and ability to execute new deals that required signatures stopped cold until more utilitarian processes (such as fax machines) were brought online.

Three weeks later, Saudi Aramco was forced to begin giving oil away for free to keep production flowing. At the final tally, in coping with one employee clicking an infected link, the company had to purchase 50,000 new hard drives at a premium price (which ultimately constrained the world supply of hard drives) and it took five months before the company was fully back online. The email spearfishing attack had threatened 10% of the world's oil and would surely have bankrupt other less moneyed entities.

2.3 The Misunderstood Near Miss: Burlington Electric and Grizzly Steppe

There is a reason humans are the best attack vectors, as they tend to ignore safety warnings and procedural controls. The most compelling thing about the case of Burlington Electrics is not that a mail lure was used (and which was found and disabled due to revelations from recently uncovered political campaign hacks), but the public response to the attack which discounted it as fake. This is particularly chilling when malware campaign trends indicate that optimized campaigns from advanced threat actors are sure to increase.

To fully understand this result, this study requires a brief review of the 2016 US presidential campaign. The hacking at Burlington Electric was discovered due to a warning that emerged in the immediate days after the contentious 2016 US presidential election in which state-sponsored hacking by the Russians was determined to have occurred. Of the Russian meddling, the Office of the US Director of National Intelligence concluded, "Russia's intelligence services conducted cyber operations against targets associated with the 2016 US presidential election, including targets associated with both major US political parties" to assist in carrying out an influence campaign to displace candidates who were perceived as hostile to the Kremlin [7]. The importance of this connection is not immediately obvious, but the Burlington Electric hacking concerns occurred at a time of heightened sensitivity to Russian state actors.

Given the sharp divide among the US electorate during the election, citizen response to the revelations cut across party lines, with those who voted for the president elect remaining sanguine about the legitimacy of the election [8]. Thus, the response to this report was met with hostility by many in the US public, either calling it hysteria or accepting the reported details of the event uncritically. The impact was surprising, but Russia was indeed high in the popular imagination. Studies done in the weeks after the election also showed a vivid shift in sentiment from 2014 to 2016 toward the Russian president, Vladimir Putin, which swung dramatically from a net negative of -66 points to a positive 10 [9]. Thus, when the account of Russians breaking into Burlington Electric was released, an event that would normally have garnered slight interest among the broad public became political fodder.

On December 30th 2016, the Washington Post reported that a Russian hacking group—*the very same that was suspected of trying to influence the US presidential election*—had struck the electric with a malware injection, that put the grid at serious risk. At first, a story in the Washington Post claimed that the Russians had penetrated the grid, in fact Grizzly Steppe, the same malware that was reportedly used to influence the 2016 US presidential elections. The malware was discovered after US intelligence agencies released software for a Russian hacker group named Grizzly Steppe to allow utilities and other to search for the digital signatures of the malware to isolate and remove it from their systems. The headline that many in the utilities industry had dreaded for years read, "Russian hackers penetrated US electricity grid through a utility in Vermont," a story (and headline). However, as the investigation progressed, it was found that the malware had been contained to a single employee computer. While the story was reworked the story for greater accuracy, including statements that the grid was never compromised, the headline ultimately came to read, "Russian operation hacked a Vermont utility, showing risk to US electrical grid security, officials say" [10].

Because of the politicization of Russian hacking, the Washington Post took a serious drubbing for an inaccurate story, which would normally have been forgiven as a technical misunderstanding of grid vulnerabilities and overly enthusiastic reporting in covering a potentially catastrophic event. And while there were issues related to the sourcing of the story, as Brian Harrell, the former director of critical infrastructure protection programs at NERC, raised concerns over the handling of attacking of the grid in the future, "Unfortunately, [the leak] may be seen as a reason to hesitate or pause when wanting to communicate with those on the outside… Every utility is monitoring and taking notes as to how this has played out in the media. The result will be hesitancy when reporting in the future" [11].

Short of an employee plugging their computer into the wrong network, Burlington Electric was not brought down. While Forbes dismissed the story as a ridiculous fairytale of "Russian hackers burrowed deep within the US electrical grid, ready to plunge the nation into darkness at the flip of a switch," the Department of Homeland Security had the final world, stating that a Russian hacking cell—the very one that reportedly worked to influence the 2016 presidential election had indeed breached the computer system and dumped malicious software on a computer [12,13]. Clearly, that is only a factor of immanence, as the corrected story says, "It is unclear if the penetration was an attempt to disrupt the utility or simply a test" [13]. Surely, there is little comfort in that conclusion if one understands the scope of danger that human failure brings to cyberdefense.

2.4   Impact on Practices in the Utility Industry

The impact of this shift in attitude should be made clear to those who work to keep systems safe: every year, millions of dollars are spent in information technologies to prevent malicious attack, but it is of diminishing value if employees cannot be convinced that there is a reason to follow security practices. These case studies demonstrate that utilities, like all companies, are going to continue to face relentless targeting by social-engineered emails with malicious URLs to infiltrate automated exploits into a user's system with a malware payload that can steal their information, open a backdoor, or replicate itself and move through the system. The lures will only improve and overwhelmingly, URL-based campaigns, spearfishing for credentials, infected attachments, and invitations to use shared files and images will continue to be offered at an alarming rate.

3 THE DIGITIZED NETWORK INCREASES VULNERABILITY

Utilities will never be immune to cyber, physical, or blended attacks; however the problems related to sabotage is unique. What began sometime around 2009 with "Stuxnet," the digital worm assumed by many to be a joint Israeli-US project, designed to sabotage Iran's uranium enrichment program by damaging centrifuges, was unleashed, ultimately demonstrating the destructive power of the world's first widely known cyberweapon. Certainly, while the notoriety and widespread damage that could occur from a successful attack on the grid are of interest to political hackers, lone wolves, and state-sponsored hacker gangs alike, the specter of state-sponsored disruption of the electrical system is a capability taken directly from the cold war. Government Security News of Russian efforts to install malware into the US grid, write "…the BlackEnergy hacking campaign has been ongoing since 2011, but no attempt has been made to activate the malware to damage, modify, or otherwise disrupt affected systems." ICS-CERT officials believe that Russian intelligence agencies helped place the malware in key US systems as a threat or a deterrent to a US cyberattack on Russian systems—mutual assured destruction from a cold war-era playbook [14].

This stance is sincere: Because electric energy is generated and consumed almost instantaneously, system operators must continuously balance the generation and consumption of power. The smart grid, DER, including grid-scale batteries, which allow this to happen with reliability requires a digital two-way communicating infrastructure which increases the number of vulnerable points in the system. As our two case studies and one near-miss show, disruption of the infrastructure at a single or small number of points in the grid can have substantial negative impacts that can quickly result in a cascading effect across transmission and distribution networks.

As described in Fig. 1, the smart grid layers technology to create the electricity infrastructure comprised of a network of sensors, meters, controls, data

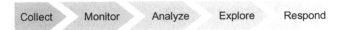

FIG. 1

Situational intelligence from collection to response. Reproduced with permission from C.L. Stimmel, Big Data Analytics Strategies for the Smart Grid, CRC Press, Boca Raton, 2014.

communications infrastructure, and the operational applications and enterprise intelligence systems which make a variety of exploits possible. As the integration of smart devices enabled by computers, software, networks, and the enterprise continues, the danger of attack increases—in both intentional and unintentional ways—and continues to scale with the increased digitization of the system (Table 1).

3.1 Attack Scenarios

The consequences of a cyberattack on the grid infrastructure include potentially massive and large-scale outages that could ravage the power grid with mechanical overstress and breakdown as with Stuxnet, or even a series of coordinated simple attacks which can create chaos, bankrupt a massive company as in the

Table 1 Description of Common Exploits That Could Occur Within a Utility System

Utility System	Example Functions	Possible Exploits
Communications	Data transport, such as over broadband over power line (BPL), cellular, wireless, or satellite networks	Passive wiretapping Man-in-the-middle attacks Data modification Internet Protocol (IP) spoofing
Advanced components	Smart switches, storage devices, smart appliances, transformers	Routing attacks Denial-of-service attacks Node subversion Message corruption Botnets
Automated control systems	Monitoring and control systems such as voltage regulators and substation and distribution equipment	Zero-day exploits Modifications on controllers Spearfishing
Sensing and measurement	Smart meters and phasor measurement units (PMUs)	Wardriving Node capture Routing attacks Node subversion
Decision support	Operational applications to manage the electricity system	Structured Query Language (SQL) injection Buffer overflow Cross-site scripting Cross-site request forgery
Customer-facing systems	Web-based systems that provide account access to customers	SQL injection Cross-site scripting Denial-of-service attack Impersonation attacks

Reproduced with permission from C.L. Stimmel, Big Data Analytics Strategies for the Smart Grid, CRC Press, Boca Raton, 2014.

near-disaster of Aramco, or dangerously fail to raise the concern or perception of threat level in a complacent, distracted, and skeptical society.

The following are the high-level grid-specific scenarios for cyberattack that were identified in the author's 2014 book, *Big Data Analytics Strategies for the Smart Grid*:

1. Reprogramming of critical electricity infrastructure components, resulting in major power delivery disruption
2. Theft of sensitive digital information used to mount later, more coordinated attacks
3. Blended threats using a combination of hacking with a physical attack such as a fire or bombing (p. 128).

As mentioned earlier in the chapter, most societal and institutional fears tend to the Pearl Harbor level of attack, although hacktivism, privacy violations, and other forms of sabotage are still concerns under traditional models. Yet still, many in the popular media find it difficult to shake stereotypical characterizations of hackers as junk-food eating, Red Bull-drinking man, children with digital bolt cutters; potential attackers may include script kiddies, but also revenge seekers, organized criminals, and state-sponsored cyberwarriors. As we have discussed, the common perception of hacking is tired, dated, and dangerous. But still, no matter the threat agent, as much as 80% of information technology breaches are caused or assisted by people "inside" the enterprise, and as was demonstrated in the discussion, spearfishing. Either willingly or unwillingly and with or without mal intent, security breaches are created by people within the organization [15].

This does not mean that the utility should cease efforts to protect against external cyberattacks. It does mean, however, that assessing and managing risk with tools must include accounting for every attack vectors, including those from all nodes within the outside plant, the corporate network, and operations.

4 THE ROLE OF DATA ANALYTICS

Fast-expanding concerns for cybersecurity threats in the digital network have brought attention from regulators and governments that are charged with producing laws and standards for the utility. Industry standards development for cybersecurity have been most prolific in North America, particularly in the United States and Canada; though globally, the lack of fast progress on security and privacy issues has slowed smart grid deployments in some regions. Standard initiatives have progressed quickly, but comprehensive development of coherent cybersecurity plans among utilities has been disparate; utilities need to do more to understand, engage, and comply with the NERC CIP (North

American Electric Reliability Corporation critical infrastructure protection standards).

NERC is the electricity sector's coordinator for CIP, and the firm provides standards development, compliance enforcement, and extensive technical material and subject-matter expertise. The NERC CIP standards are the only fully fledged cybersecurity standards in place to address the security and reliability of the electricity grid. The standards include mandates for incident reporting, authorization protocols, minimum security management controls, and disaster recovery. NERC CIP does much to reduce risk and improve the security posture of North American bulk electricity systems. However, it is impossible to address every security risk, and this is precisely why the opportunity for data-driven cybersecurity analytics holds such profound value for more advanced security controls that utilize the massive spatiotemporal data produced by smart meters, SCADA, and synchrophasors.

As our case studies clearly show, the idea that we can build a virtual wall around our grid assets is crude, rudimentary, and ultimately creates a brittle and vulnerable network. Further, utilities are much more than the physical operation of the grid; utilities are also responsible for massive enterprise systems with financial information, customer data, and a growing network of digital operations. Thus, security strategies are quickly becoming more nuanced. A program of security data analytics may be the best option for proactively and cost-effectively containing threats from the field, the enterprise, and the physical plant as the volume of actuating sensors explodes. Analytics may be the key to usher analytics from chronic vulnerability to a proactive posture.

Analytics for cybersecurity allows for the use of pattern-detection algorithms with both structured and unstructured data sources, including forensic capabilities, to identify both internal and external threats. This allows the utility to ask new questions that have not been possible before, regarding their defensive posture. As Fig. 1 describes, an integrated analytics approach provides closed-loop, continuous learning that furnishes situational intelligence previously unavailable to security programs within the utility.

The following are useful analytical models that can contribute to cybersecurity and resiliency of the digital grid:

1. Descriptive: Situational intelligence
2. Diagnostic: Quantification of threat levels and their characteristics
3. Predictive: Identifying and preventing threat levels and characteristics
4. Prescriptive: Designing action response to future incidents

Traditional security models are largely passive defense systems and primarily focus on detection. Unfortunately, they often fall under the hand of a persistent hacker with a grab bag of cheap exploits. They need only find a single point of entry, and it is virtually free to try all day. Big data analytics on the other hand provides more predictive and prescriptive tools that provide the ability to stop attackers during an attack. For example, some big data models can leverage massive volumes of data, and therefore quite effective at recognizing attack patterns and other anomalous patterns (anomalies which hackers are becoming effective at covering up, as in the frozen screen in the Ukraine case study).

4.1 The Role of Privacy

While a well-conceived cybersecurity program in the utility provides comprehensive situational awareness across the grid, and the enterprise to ultimately respond to and contain emerging threats, it is also very important to recognize the role of data privacy. Securing data, in fact, is the most fundamental step in a full-fledged utility security profile. Further, the ability to properly contextualize collected information to facilitate a cybersecurity program can include detailed models at a granular and personal level. Political, culture, organizational management, and strong data governance are crucial to protecting consumer data, especially generating consumers who are part of the operational milieu.

Ann Cavoukian, the former information and privacy commissioner of Ontario, Canada, and champion of the Privacy by Design (PbD) framework, identified seven foundational principles that are more salient than ever:

- Privacy must be proactive not reactive by design, as in preventive not corrective
- Privacy is always the default and expected setting
- Privacy should be embedded into design plans, not bolted on as an afterthought
- Privacy rights are positive-sum, not zero-sum, meaning rights are not earned at an expense to the user
- End-to-end security—full lifecycle protection exists from collection to sharing, storing, and ultimately destruction
- Visibility and transparency—Personal and private data uses must be evident to users
- Always provide respect for user privacy—it must be user-centric and controllable by the user [16]

PbD should be applied as a standard practice in many areas of design and should considered in the scope of all utility operations, including surveillance, biometrics, the smart grid, near field communications (NFC), sensing, remote services, big-data analytics, and location services. It is important to note that PbD has been widely accepted by international bodies, and is one the most specific and actionable approaches to treating the exploding sphere of privacy challenges. Dr. Cavoukian's work is precise in its position that privacy is always about control—who has it and who does not. Thus, the essence of true privacy is the ability to specify and limit the uses of one's personal information. There is nothing about these principles that would or should undermine a utility operator to maintain their grid, provide services, or run their enterprise, even in competitive markets.

5 CONCLUSION

While the utility implements classic cybersecurity tactics to protect grid operations, big data analytics platforms that combine security intelligence with powerful processing capabilities allow the utility to maintain a proactive posture. Big data analytics platforms combine security intelligence with powerful processing capabilities. The goal of these programs is to provide advanced pattern detection and machine learning that can analyze massive streams of network traffic, millions of unique device data nodes, transport and communication characteristics, and user behavior to understand the machine and human linkages within the entirety of the utility systems. These techniques are especially useful in identifying anomalous activity on a network that is high-volume, high-velocity data traffic as found within the grid's command-and-control systems.

Cybersecurity applications will become an integral part of the utility operations, managing and processing millions of events per second with microsecond latency without impacting the underlying grid, operations, or enterprise infrastructures. These applications will have multiple outputs, including traditional reports, situational intelligence dashboards, predictive models, content analytics, and other virtual reality applications.

DER will cause a variety of communication protocols and standards with varying levels of cybersecurity fitness to be introduced into the grid at an accelerating rate, especially in the secondary distribution grid. Also, with DER, there are many interactions where there may be no central controlling body either for either functional or nonfunctional purposes. Because of the myriad stakeholders, the challenges are complex. A NIST report written in 2014 on the topic stated presciently, "It is not just the utilities who must take responsibility for achieving this resilience goal. Many stakeholders are involved in the design, implementation, and operation of DER systems, including manufacturers, integrator/installers, users, information and communication technology (ICT) providers, security managers, testing and maintenance personnel, and ultimately utility regulators. However, given this new cyber-physical environment, often these stakeholders do not fully understand or appreciate the types of cybersecurity and engineering strategies that could or should be used" [17].

This may understate the scope of the cybersecurity problem; while technology in the DER domain is evolving at an accelerated rate with systems are being interconnected every day, standards, policies, and governance are still being hotly debated. And while the grid advances technologically, and our societal dependency on it grows, it lags far behind in coping with the complexity of the human and physical linkages of the system. Ongoing discussion of cybersecurity must focus not only on the rapidly expanding grid under distributed ownership but also on political, cultural, and behavioral implications, studies, and research.

References

[1] J. Pagliery, Hackers Attacked the U.S. Energy Grid 79 Times This Year, Retrieved on 15 January 2017 from, http://money.cnn.com/2014/11/18/technology/security/energy-grid-hack/, 2014.

[2] The Human Factor, 2016. Retrieved 08 May 2017 from https://www.proofpoint.com/us/human-factor-2016.

[3] SANS Institute InfoSec Reading Room, The Threat of Social Engineering and Your Defense Against It, Retrieved on 15 January 2016 from, https://www.sans.org/reading-room/whitepapers/engineering/threat-social-engineering-defense-1232, 2003.

[4] E-ISAC, Attack on the Ukrainian Power Grid: Defense Use Case, Retrieved on 14 January 2017 from, http://www.nerc.com/pa/CI/ESISAC/Documents/E-ISAC_SANS_Ukraine_DUC_18Mar2016.pdf, 2016.

[5] K. Zetter, Everything We Know About Ukraine's Power Plant Hack, Wired Magazine, 2016. https://www.wired.com/2016/01/everything-we-know-about-ukraines-power-plant-hack/.

[6] J. Pagliery, The Inside Story of the Biggest Hack in History, Retrieved 15 January 2017 from, http://money.cnn.com/2015/08/05/technology/aramco-hack/, 2015.

[7] Department of National Intelligence, Assessing Russian Activities and Intentions in Recent US Elections, Retrieved on 15 January 2017 from, https://www.dni.gov/files/documents/ICA_2017_01.pdf, 2016.

[8] E. Bradner, Poll: 55% of Americans Bothered by Russian Election Hacking, Retrieved on 15 January 2016 from, http://www.cnn.com/2016/12/18/politics/poll-russian-hacking/, 2016.

[9] M. Nussbaum, More Republicans View Putin Favorably, Retrieved 14 January 2017 from, http://www.politico.com/story/2016/12/gop-russia-putin-support-232714, 2016.

[10] J. Eilperin, A. Entous, Russian Operation Hacked a Vermont Utility, Showing Risk to US Electrical Grid Security, Officials say, Retrieved on 15 January 2017 from, http://wpo.st/EqlR2, 2016.

[11] R. Walton, What Electric Utilities Can Learn from the Vermont Hacking Scare, Retrieved 15 January 2017 from, http://www.utilitydive.com/news/what-electric-utilities-can-learn-from-the-vermont-hacking-scare/433426/, 2016.

[12] K. Leetaru, 'Fake News' and How the Washington Post Rewrote Its Story on Russian Hacking of the Power Grid, Retrieved on 15 January 2017 from, http://www.forbes.com/sites/kalevleetaru/2017/01/01/fake-news-and-how-the-washington-post-rewrote-its-story-on-russian-hacking-of-the-power-grid/#270bf2b4291e, 2017.

[13] McCullum, Russian Hackers Strike Burlington Electric with Malware, Retrieved on 15 January 2017 from, http://www.burlingtonfreepress.com/story/news/local/vermont/2016/12/30/russia-hacked-us-grid-through-burlington-electric/96024326/, April 2016.

[14] Government Security News, Black Energy Threatens U.S. Infrastructure, Retrieved on 14 January 2017 from, http://gsnmagazine.com/node/42887, 2014.

[15] M.B.R. Greene, in: CIO, G-6 Headquarters, New York Guard, A Statement Presented at the GovSec 2013 Conference in Washington, DC, USA During the Session "Critical Infrastructure Protection: The Enemy Within, 2013.

[16] A. Cavoukian, Operationalizing Privacy by Design: A Guide to Implementing Strong Privacy Practices, Retrieved 15 January 2017 from, http://gpsbydesign.org/resources-item/operationalizing-privacy-by-design-a-guide-to-implementing-strong-privacy-practices/, 2012.

[17] NIST, CIP for Grids with Interconnected DER Systems: Executive Summary, Retrieved on 14 January 2017 from, https://www.nist.gov/sites/default/files/documents/2016/09/16/xanthus_rfi_response.pdf, 2016.

CHAPTER 3

The Role of Big Data and Analytics in Utility Innovation

Jeffrey S. Katz
IBM, Hartford, CT, United States

CHAPTER OVERVIEW

The computational technology known as big data and its subsequent processing, analytics, are driving innovation in electric power system integration of renewable energy, outage prediction, processing of increasing volumes of smart grid data, and velocity of such data. In the age of cybersecurity, the veracity of these data is also a factor. The almost concurrent rise of cognitive computing gives new importance to unstructured data such as images and text, and the intelligent connection of real-time numerical data with written and visual data gives rise to even more innovation. The benefits of high-precision weather modeling on power demand, grid damage, and solar- and wind-based generation are also considered.

1 INTRODUCTION OF BIG DATA AND ANALYTICS AS AN ACCELERATOR OF INNOVATION

Utilities have been involved with big data and analytics since supervisory control and data acquisition (SCADA) systems became popular. The myriad of devices spread across a vast geographic area provides a huge amount of monitoring data, potentially accumulated over decades. The smart grid [1] era brought even more data, not only from the addition of sensors but also from the increasing amount of embedded computing in traditional power equipment. This was not limited to more operational data, since as device intelligence increased, monitoring of the status of the local real-time computing systems became more important in order to have a coordinating distributed computing environment. While this primarily applies to transmission and distribution, there is expansion of the scope in both directions. The distribution side sensing now creeps in to monitoring of customer-owned equipment, such as solar panel and inverter health. Aspects of demand response systems may be allowed to look at data from home energy control systems or individual appliances, in order to effect a smoother demand response action. On the other side, growing past the transmission side, modern power plant distributed control systems have extended gateways to other plant computers, plus today's effective in-plant wireless sensors reduce the installation cost of sensors, which has

Big Data Application in Power Systems. https://doi.org/10.1016/B978-0-12-811968-6.00003-6

always exceeded the price of the sensor itself. The generating plants themselves now are more distributed, with utility-scale wind turbine farms and solar installations. Unlike traditional large fossil or nuclear power stations, these renewable energy farms may have no local staff. That isolation, combined with the tighter monitoring requirements due to their power output variability, provide a new data deluge. Therefore, the premise of innovation in power systems being driven by big data and analytics is even more promising.

While there is plenty to say about the amount of big data, there are equally significant aspects in analytics that contribute to innovation. The power of personal computing, often driven by gaming needs, is impressive even to those who see the end of Moore's law. The computing power available to the individual power engineer though is far from limited, given the community-computing phenomenon called cloud computing. The days of living with limitations of a single computer running generic mathematical tools are in the past. Easily accessible cloud computing, sometimes needing only a credit card, empowers the solution of complex problems, at flexible scale, such as those transmission analyses solved by huge sparse matrix techniques. Cloud computing has become significant enough to power system applications that the International Electrical and Electronics Engineers (IEEE)'s Power and Energy Society (PES) General Meetings in 2014 and 2015 had cloud panels [2]. Moreover, cloud-based analytics platforms aimed at electric utilities [3,4], remove much of the "IT" development prior to obtaining business results. The irritations of system integration of software modules are abstracted in these integrated platforms. Easier access to tools for statistical analysis, business intelligence, and signal processing provide an easier method to extract knowledge from today's big data. Furthermore, some platforms, by means of uniform organization and modeling of data, along with a standard-based electric component data model, eliminate the bane of power engineers, data curation. Going further, the need for decades of experience to understand the data is made a bit easier through advances in visualization. In fact, some agile development teams begin with visualization and automated modeling of the data, in discovery mode, to see what patterns might be learned. While this does not displace the need for standard IEEE analytics, it does assist in innovating new algorithms to detect unusual operational or maintenance patterns.

The earlier reference to generic mathematical tools does not mean their usefulness is supplanted. On the contrary, many new ideas come from individuals trying out an idea with these types of tools. However, newer, more complex software can assist in uncovering interesting correlations, and most platforms provide interfaces such as through Open Database Connectivity (ODBC) to keep these personal tools relevant and easily connected to these types of utility analytics platforms. Moreover, there are a new generation of data science exploration tools, and easier to implement programmatic interfaces for more complex situations. An essential point though is that these tools are focused on numerical data, which are a subclass of structured data. All of the data referred to so far, in fact, are structured. With the era of cognitive computing blooming, the unstructured data, from maintenance reports in enterprise asset management applications [5,6], to high-resolution still images from helicopters or drones, to acoustic and thermal data, all expand the field of data resources available for analytics processing in utilities. The need for many people to be the arbiter of relevance for unstructured data matched to structured data is now reduced. Software that can detect features in unstructured data and correlate them with measured data gives the opportunity for many new insights. Eventually, insights lead to new understanding for engineers and then sometimes to improved operational guidance or even smarter closed loop control. This is the true promise of big data and analytics-driven innovation.

2 APPROACHES TO DATA DRIVEN INNOVATION

The electric grid is one of the most tightly coupled, high-speed critical infrastructure networks. Today's innovations with big data are often made possible from faster communication of the data, PMUs being an excellent example. Larger computer memory and processing power to analyze the data and visualization techniques to support discoveries within the data have opened the path wider. Several major utilities are adding more software developers and data scientists to their R&D groups, as well as power system experts. The power of data in innovation is seen in some earlier approaches to smart grid. These "smart grid zero" projects applied big data and analytics without adding any new sensors, demonstrating the power of knowing more about what the utility already knew. Some early adopters of big data and analytics in utilities made their first priority to expand the storage associated with SCADA systems. This came from the foresight that data discarded while in the process of developing an analytics strategy, including predictive maintenance programs, would be valuable as the design phase began, even if there was no known need for all of the data at the time. Data storage is inexpensive compared to the loss of information from which to learn.

Analytics has moved from replicating alarm limits already available to online FFTs, cognitive computing, and numerical and algorithmic methodologies. Utilities are being encouraged to adopt some agile software methods, such as a period of data exploration, to see the art of the possible, rather than limiting analytics to already conceived ideas of what needed to be examined in the data. Fig. 1 is a representative architecture for big data and analytics in an electric power distribution utility.

In the current world of computer science, more data accumulated by the equipment owner also become more concentrated data that may attract the cybersecurity offender activity. This requires consideration of cybersecurity during the

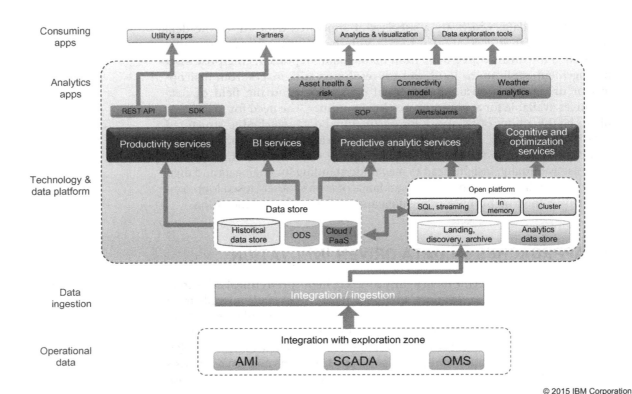

© 2015 IBM Corporation

FIG. 1

Example energy and utilities analytics platform to power innovation.

thrust for innovation and thinking about security measures in newer protocols, in the choice of communication medium, and in appropriate use of ubiquitous cloud computing. Physical security to protect from interception of data or from the planting of false data generators in remote locations and image processing on surveillance cameras are also generating big data for operational support. In fact, modern cybersecurity is looking at the whole asset, and the utility's security posture becomes another measurement feeding a different analytic. Security can now be thought of as another dimension of real-time asset monitoring, just as voltage and current are. The first detection of a cyberattack may well be inferred from insights gleaned from the traditional SCADA data, and not in an alert from a firewall. Cybersecurity in electric power systems has become a rising concern, as cited in many federal government documents and brought to the forefront by the US National Institute of Standards and Technology's (NIST) cybersecurity framework for critical infrastructure. The need for innovation in security becomes a new upper echelon analytics need, along with integration of renewable energy. The reader has likely seen news of substations that may have to process local acoustic or vibration data to see if they are being shot at.

3 INTEGRATION OF RENEWABLE ENERGY

On the top of many big data and analytics development lists is the need for innovation in the era of renewable energy. This is especially true when examining fast variability in wind turbines and solar panel outputs. These fluctuating generators impose changes on the grid that needs data-driven optimization to maintain the system safe, in balance, and meet environmental friendly intentions. This newer equipment, which may not all be utility owned, controlled, or even monitored, contains, due its more recent design, more embedded computing and sensors than traditional generation. Much of the generation source data have to be processed to assist in learning new maintenance patterns, feed weather simulations of solar flux and cloud coverage, manage power quality from inverters, and ensure safe operation. Some believe that human-in-the-loop control, as in traditional control rooms, cannot achieve the utilization goals expected of large-scale renewable energy deployment. The temporal variations in the optimization goals of balancing demand, conventional generation, renewable power production, and storage systems may be possible only with the considered application of big data and analytics.

The need for energy balancing analytics has led to several innovative projects. The Danish Energy Association's flexible clearinghouse (FleCH) is an example. The Internet of things makes it increasingly possible for energy resources to interact with the market. Throughout Europe, both R&D and business initiatives around flexibility are growing at a fast pace. There has been recent work to create a market model study in Belgium for the enablement of distribution level flexibility. There is a consortium called the Universal Smart Energy Framework (USEF) [7], which is a framework for the market design for smart energy products and services that enables the trade of flexible energy. Seven Dutch and Belgium energy market participants established the USEF. The goal is to help all participants in the energy system to benefit from flexible production, storage, and use of energy. While there is an operational need for renewable energy optimization, an equally sized problem is the market participation of this generation, and often, the market puts the demands on the operation. As mentioned in Section 1, security is always a shadow over concentrated collections of big data, and where there is money involved as in market operations, the security risk is even higher.

Another interesting innovation in power system applications of big data and analytics is the work done in the US Department of Energy's ARRA Pacific

Northwest regional demonstration project [8]. This brought 12 utilities across five states into a full-scale transactive energy project, in which big data and advanced analytics are used to create a common "currency" of transactive energy signals, from both suppliers and consumers, to manage power and access. Other work at the Pacific Northwest National Lab that is relevant is the Grid Operations and Planning Technology Integrated Capabilities Suite (Grid OPTICS) [9]. This work, done by the PNNL Future Power Grid Initiative, has been discussed in the four workshops held to date on "next-generation analytics for the future power grid." There are additional industrial research projects applying big data and advanced analytics, done in cooperation with electric power utilities, as described in [10,11].

4   GRID OPERATIONS

Innovations in grid operations, sometimes vaguely referred to today as smart grid 2.0, include taking advantage of near real-time communications, extensive sensor networks, and smarter equipment to get an improved view of the grid. This is step one; the operational improvement from improved fault detection and localization, failure predictions, faster rerouting on the distribution system, incorporation of smart meter data, and applying cognitive and algorithmic techniques is the next major step to a more resilient grid.

One aspect of improved grid operations comes from advances in phasor measurement unit technology. Not only are PMUs more pervasive, but also there is a utility lead cooperation in wide-area PMU networks, in order to understand what might be the ISO or national level events. The North American Synchrophasor Initiative is a utility organized group [12]; obviously, when sampling three phases at twice line frequency, a PMU is a major generator of big data. Just as relevant though is the fact that it is now easier to obtain high-speed data communication networks, to allow analysis of phasor angle differences over hundreds of miles of transmission system, making the analytics more actionable. PMU data are used not only for transmission disturbances but also for security indications. The existence of Internet 2 and the existence of Eastern Interconnect Data Sharing Network are examples of such new, efficient, and effective data networks for the power system. Of course, GPS-synchronized time metadata enhance the value of the measurements. When coordinated with existing electric power transmission system SCADA and energy management systems (EMS), there is plenty of room for innovative analysis. Recently, a large Canadian utility was developing novel real-time, predictive analytics, and visualization tools that may help generate early warnings for geomagnetic and other large-scale disturbances. If there is a next step in this innovation, it might be to look at the use of machine-learning techniques, so that system improves its recognition of events over time. Examining geomagnetic storm influences is a problem that is significant to high-latitude power utilities. The motivation is a need for an integrated system that supports decision-making from raw PMU data, whereas present systems are focused more on "monitoring" rather than decision-making.

To expand on examples of big data and analytics proof-of-concept projects from [11]: these illustrate innovations possible when vast amounts of data are organized and correlated across traditional utility department boundaries.

Asset risk management and optimized repair-rehab-replace (ARMOR3): This research project applied predictive and prescriptive analytics on big data to identify, quantify, and ultimately optimize infrastructure maintenance and planning for electric assets including transformers, cables, poles, and circuits. ARMOR3 converted data into information, insight, and foresight with the aim of providing decision support across the complete electric infrastructure. The solution aimed for the ability to run a broad set of scenarios on the same detailed data, prioritizing across multiple teams/groups. It offered predictive maintenance to identify and fix the next failure before it happens and generated asset risk and investment profiles to enable 100% utilization (useful life) of the asset while taking into account resource constraints.

Connectivity models: Using advanced analytics on advanced metering infrastructure (AMI), or smart meter, measurements, the connectivity model's pilot application inferred customer phase and customer-to-transformer connectivity, which is generally inaccurate or unknown. In fact, many good ideas for using smart meter data in fault location and failure analysis were slowed down when the accuracy levels of meter connectivity information were discovered to be insufficient for the algorithms. An accurate and sustainable connectivity model is a key enabler of capabilities needed to improve the reliability and efficiency of the distribution grid. Utility efforts to build and verify their connectivity database are labor- and resource-intensive. This analytics approach could help radically lower the cost of such processes.

Customer intelligence: Through data-driven analytics, the customer intelligence research project provided advanced customer segmentation capabilities, for utilities to better understand their customers and the impact on utility operations. Such customer insights could help a utility transform the relationship with customers, improving the effectiveness of marketing campaigns from electric power retail and pilot programs by smarter targeting. This intelligence could also help grid stability, by understanding changes in customer dynamics, such as demand response behavior, adoption of renewable energy, and usage of plug-in electric vehicles. In addition, the utility might gain additional revenue protection by more accurately detecting energy theft.

Outage prediction and response optimization: This proof of concept used advanced weather prediction, predictive damage estimates, optimized crew positioning, and response planning to try to improve a utility's preparation for, and response to, weather-related power outages. With more than $14B in total annual lost value of service due to storms in the United States alone, improvements in outage restoration and reduction in operational costs could lead to significant value for the utility, in terms of both economic value and improved customer satisfaction.

Transactive energy: Expanding on the example of the PNNL DoE project [13], transactive energy management is the use of economic and control mechanisms that could allow the dynamic balance of supply and demand across the entire electric infrastructure, using "value" as a key economic operational parameter. All business and operational objectives and constraints can be assigned positive or negative "values" and be incorporated into these transactive signals.

Vermont renewable energy integration: The initial application of the Vermont Weather Analytics Center (VTWAC) [14] research project focused on integration of renewable power (wind and solar). The system was composed of several components, which used an advanced weather prediction capability as a foundation. This physics-based weather model was coupled to data-driven models of electricity demand, wind power, and solar power. The outputs of these probabilistic models were used to assess the uncertainty in the predictions and to drive a stochastic engine to assist the utility in avoiding congestion and improving the stability of the transmission network.

Risk analytics for critical energy: This project had a goal to enable reliable operation of gas pipeline network critical infrastructure by providing holistic analytics capabilities, including leak detection and condition-based maintenance. As more gas turbines become peak demand-time supply units for renewable energy intermittency and some coal and oil plants are converted to gas due to current pricing, the relationship between gas supply and power grid has intensified. For leak detection, the system employed physical and data analytics on sensor data from SCADA, and the system provided early warnings of impending rupture events, capabilities to detect small leak events (which cannot be detected from SCADA systems), and provided localization capability of the leaks and ruptures. For condition-based maintenance, the system uses an advanced predictive optimization engine, and the condition-based maintenance planning provided prediction of a condition deterioration curves for assets, leveraging data from a programmable logic controller (PLC) and past maintenance hours. There was also prediction of future utilization of an asset and computation of an effective condition assessment metric, such as effective run hours. Multiobjective optimization of maintenance and operation plans used the condition deterioration curve, predicted future utilization, and condition assessment metrics. A what-if scenario allowed comparison of alternative maintenance plans.

Wide-area situational awareness: This research project used descriptive and prescriptive analytics to interpret and summarize electric events in the transmission system and provided insights during postevent analysis. It also used predictive analytics to provide early warning indicators of complex events that could affect grid stability and operations. The system sought to identify grid anomalies and alert operators to act before disturbances, such as geomagnetic induced current (GIC) events, lead to grid collapse. The application also provided low-latency and high-throughput monitoring, archiving, reporting, advanced querying, and visualizing of the grid state.

In general, PMU analysis would aspire to the following design goals. In order to achieve high performance, the system would provide for the collection of large amount of measurement data from the PMU/PDC with a high sample rate. As utility size networks become ISO size and sometimes national in scope, there should be unlimited scale-up. For example, 1000 PMUs, at 60 Hz, yield 37.5 K messages per second. Such goals typically require specialized streaming software and databases optimized for time-series. Storing the data is but one aspect. For validation, analysis, and routing in real time, the design should ensure that the content is valid and the analytic results are correct, be able to check for reliable data that have not been manipulated, and filter and route messages according to the needs of the application layer. Important big data also requires storage and synchronization. There should be immediate storage to protect against data loss, storage in a real-time database and an historical database, and synchronizing in case of failover with recovered data. Typical applications are real-time analytics for island state detection and power swing recognition. With the rising emphasis on cognitive analytics, algorithms and correlation of different data sources are based on platform capabilities. In the critical situations that draw upon PMU data to make optimal decisions, there is very little time for the operating engineer to process multiple UIs or navigate incompatibilities between tools. Therefore, a consistent system configuration and data model, driven by a single source solution, may be best in this mission critical application.

5 COGNITIVE COMPUTING ON BIG DATA

Cognitive computing is a term from which people tend to infer their own meaning. Here, it will be defined as a comprehensive set of capabilities based on technologies such as machine learning, reasoning, and decision technologies; natural language, speech, and vision technologies; human interface technologies; distributed and high-performance computing; and new computing architectures and devices. In a commercially available solution [Ref. for IBM Watson], the computation algorithm continuously learns from previous interactions, gaining in value and knowledge over time. The system can sense, create conclusions, and learn from experience. When integrated with traditional computing and data sources, these capabilities are designed to solve a wide range of problems, boost productivity, and foster new discoveries across many industries.

A typical distribution utility consists of two-thirds of the employees concentrating on field technical problems. The machine-learning-based algorithm could take verbal questions from the field tech and integrate information from current and recent SCADA information, past written maintenance reports about diagnosis of that equipment, and equipment supplier service documents (of course keeping current with all manufacturers' updates). Furthermore, each service call becomes another lesson to be learned. This now begins to attack the "aging workforce" problem of lost experience due to retirement. While two decades ago AI often involved programmer interviews with subject matter experts, now, algorithms can learn from past enterprise asset management system service records and build that experience knowledge base. It also helps on the other end—the recruitment of new field techs to whom the idea of working with a cognitive system may have a certain "Wired magazine" appeal.

The following is an example scenario of IBM Watson cognitive processing of big data in power systems, a service advisor for junior technicians.

For most customers, having faulty equipment accurately and reliably serviced one of the most important drivers of their satisfaction with a manufacturer. When these machines require servicing, customers rely on the expertise of a field technician to maintain the continuity of their business. However, the consistency and quality of maintenance expertise across an organization can prove difficult in an industry that is constantly experiencing changes and increasing complexity within its equipment.

With Watson for field service, field technicians can drastically minimize the time spent diagnosing a problem and searching for a proper solution to a service call. This will help them improve their first-time fix rate, allowing the customer to get back to running their equipment with fewer visits.

With a competitive environment that forces manufacturers to differentiate themselves with superior service and with so many variables that can go into resolving a work order, empowering field technicians to make the right decisions without needing to escalate the issue to higher levels has become a necessity, not a "nice to have."

Here is how a junior technician's experience can be impacted. First, the tech wants to start his day by getting a bird's eye view of orders that need to be fulfilled. After logging in, he sees a comprehensive list of all the orders that are still open and may decide to get a more detailed view on the 8:00 a.m. There, one can see the key pieces of info needed to get this job done: the issue, the parts and tools needed to bring, and even extra things that need to be done to fulfill this customer's unique level or expectation of support. In this case, the customer is at gold level, implying fast response. Upon arriving on site to fulfill the work order, he is prepared to address the service request. He can view a fully comprehensive page detailing the customer's order. Here, he can see the equipment model, exact error, customer complaint, and service history. Using information from the service manuals and the symptoms reported by the customer, Watson provides a recommendation on how to fix this order, along with an estimate of how long it will take to finish.

However, the tech may want to get an even more detailed view of the information Watson used to come to this conclusion. The system is an assistant, not a replacement, and human judgment is still important.

Here, the cognitive system can show the exact passages from the relevant service manual or other sources it used as evidence for its suggestion. Vendor manuals, which are often updated annually, are a large source of unstructured data. Such systems can understand not only the text in the manual but also the images and diagrams. A social paradigm is also included. One can see comments and ratings from other technicians who used this procedure to supplement the recommendation. The tech's input and rating for this procedure, along with that of other technicians, will help further train Watson to provide accurate suggestions.

If the tech is sure, this suggestion might not be the best action to take for the current problem, he can troubleshoot the issue to see the next best alternative. Watson provides a list of the next best options it thinks can be taken to solve the order, ranked by confidence. It uses cases from prior work orders that are relevant or similar to the issue being dealt with right now. Using the service history, another often untapped big data source, Watson behaves the same way that a senior engineer or expert-level technician would, taking all the information available and depth of "experience" to help inform the best route to resolving a problem.

However, in the deployment of innovative technologies, there needs to be a method of human interaction. To dig a little deeper into the problem, a natural language query can be done. Here, the tech can see more information related to the issue. Again, this information is derived from the manuals and procedures that have proved successful with this issue. However, this feature is used only when the prior two steps did not produce the proper solution, which is unlikely.

To obtain a second opinion, the tech can open up the chat log and view what has been said about the issue while providing his own input to peers. Here, all the chats between with other engineers who have worked on problems like are available. One can view what is already been said or can communicate with them directly if they are available. Even the inputs from the administrative and senior level are included.

Armed with these tools to navigate the sea of knowledge that goes into field service, every technician can be helped by an expert.

6   WEATHER, THE BIGGEST DATA TOPIC FOR POWER SYSTEMS

Weather historically has driven load demand, especially for residential consumption. Later, weather also became an essential element for modeling power outages. The third frontier for weather as a source of big data has become prediction of power output from renewable generation. Solar projects such as the DoE "SunShot" and wind turbine farm aggregate output, blade tip icing, and tower vibration and stress have been the latest beneficiaries of advanced big data and analytics for power systems.

To understand an application of weather as big data, consider what is essentially a CFD model of the atmosphere [15]. These are examples of big data computational science and statistical and measurement engineering. To attempt an understanding of the scope of "big" data here, one system handles 15–26 billion API requests each day (peak—340,000/s), many initiated from individual activation of a smartphone weather app. The system ingests over 100 Tb of third-party data and 300 Tb of proprietary data each day. It handles 60 different types of data from 150 different sources. Given that, the mean forecast creation time is about 10 ms, and the mean total delivery time is less than 300 ms. This content serves consumer forecast services and powers many distribution portals.

The platform is cloud-based, ingests large volumes and types of data, and delivers insights from precise weather data combined with location and other data sets. The result is enabling companies, from electric power to utilities to aviation, to embed weather insights into decision support platforms, in order to take a variety of responsive operational actions. The science of numerical weather prediction involves a mathematical model that describes the physics of the atmosphere. Weather begins as the sun adds energy and gases rise from the surface, resulting in convection. Unequal heating of the surface causes temperature and pressure differences, which drives winds.

To provide the precision for wind turbine and solar forecasting requires high spatial and temporal resolution, customized to the business needs. An example requirement would be a 1 km horizontal resolution, tens to hundreds of meters of vertical resolution, and output every 10 min. In addition, a diversity of input data is used from public and private organizations. To be useful to the consumer of the data, there is tailored dissemination and visualization. "Coupled" modeling is used to integrate this knowledge into decision-making, and outputs are customized to the geographic area and end user's weather sensitivity. There is also a massive amount of quasistatic information, which is often not readily available at the required resolutions. This means specific power system projects may have to capture the geographic characteristics that locally affect weather, horizontally, vertically, and temporally. Often for training and predictive validation, there is the capability for retrospective analysis of past impactful events, via hindcasts and reanalysis modes.

In power systems, associated data instrumentation for weather impacts include SCADA/telemetry; AMI, also known as smart meters, providing end-point power measurement, including no power indications; geographic information systems (GIS) for assets and infrastructure; electric network models from the utility's grid; and operational management systems, such as a utility's outage management system. Cognitive methods now can use anecdotal data, such as storm logs, social media, and thresholds for decision-making.

References

[1] http://resourcecenter.smartgrid.ieee.org/ (accessed 27.11.16).

[2] http://submissions.mirasmart.com/PESGM2016/Itinerary/TechnicalProgramDetail.asp?id=40 (accessed 27.11.16).

[3] https://www.ibm.com/us-en/marketplace/energy-analytics (accessed 27.11.16).

[4] https://www.ge.com/digital/predix (accessed 27.11.16).

[5] https://www.ibm.com/internet-of-things/iot-solutions/asset-management/ (accessed 27.11.16).

[6] http://new.abb.com/enterprise-software/asset-optimization-management (accessed 27.11.16).

[7] https://www.usef.energy/Home.aspx (accessed 27.11.16).

[8] http://www.pnwsmartgrid.org/ (accessed 27.11.16).

[9] http://gridoptics.pnnl.gov/ (accessed 27.11.16).

[10] http://www.research.ibm.com/client-programs/seri/ (accessed 27.11.16).

[11] C.A. Pickover (Ed.), IBM Journal of Research and Development, vol. 60, Issue 1. http://ieeexplore.ieee.org/xpl/tocresult.jsp?isnumber=7384400 (accessed 27.11.16).

[12] https://www.naspi.org/ (accessed 27.11.16).

[13] Pacific Northwest National Lab, July 9, 2015, Franny White (Battelle). http://www.pnnl.gov/news/release.aspx?id=4210 (accessed 27.11.16).

[14] M. Sinn, F. Dinuzzo, IBM Research Blog, May 14, 2015. https://www.ibm.com/blogs/research/2015/05/demand-forecasting-by-means-of-data-driven-technique/ (accessed 27.11.16).

[15] J. Davis, Information Week, May 2, 2016. http://www.informationweek.com/strategic-cio/the-weather-company-brings-together-forecasting-and-iot-/d/d-id/1325362 (accessed 27.11.16).

Further Reading

Cybersecurity in the Electric Power Industry

[1] N. Hitpas, George Mason University School of Business, February 8, 2015. http://business.gmu.edu/news/981-mason-ibm-nsf-partnership-produces-cybersecurity-report/ (accessed 27.11.16).

[2] N. Hitpas, George Mason University School of Business, November 16, 2015. http://business.gmu.edu/news/1104-mason-ibm-nsf-partnership-yields-second-cyber-report/ (accessed 27.11.16).

Cloud Computing for Electrical Power Utilities

[3] B. Ramsay, National Association of Regulatory Utility Commissioners, Final Resolutions Adopted at the 2016 Annual Meeting, November 16, 2016. http://pubs.naruc.org/pub/4FDD6D6B-F303-DE7B-5B46-7B25C04E6317 (accessed 27.11.16).

Analytics for Utilities

[4] http://www.utilityanalyticsweek.com/ (accessed 27.11.16).

[5] http://www.utilityanalyticssummit.com/ (accessed 27.11.16).

Big Data and Analytics

[6] D. Cnota, Infogix, November 16, 2016. http://www.infogix.com/press-releases/infogix-identifies-top-ten-transformative-data-trends-2017/ (accessed 26.12.16).

[7] Big Data University—Analytics, Big Data, and Data Science Courses. http://bigdatauniversity.com/ (accessed 04.01.17).

[8] IEEE Utility Big Data Workshop, http://2017isap.tamu.edu/ieee-utility-big-data-workshop/

CHAPTER 4

Frameworks for Big Data Integration, Warehousing, and Analytics

Feng Gao

Tsinghua University Energy Internet Research Institute, Beijing, China

CHAPTER OVERVIEW

Big data is a term for large and complex datasets that traditional processing approaches are not suitable to deal with them. Big data usually comes from the Internet, enterprise systems, Internet of Things, and other information systems. Data collection and preparation, storage management, data processing, data analysis, and knowledge presentation would generate new insights to support decision-making and business intelligent operation. Smart grid is a developing trend of electrical power energy industry. The core message is to implement the next generation of cyber-physical systems shaping future energy industry that is based on a deep merge of operational technology and internet information technology. The growth of smart grid is dependent on the availability of high performance computing (HPC) and analytics technology to process a massive amount of data set. Deployment of advanced technologies within smart grid and usage of state-of-the-art computing systems provide utility companies with innovative capabilities. These advances lead to unprecedented explosion of data volumes. As smart grid operations will leverage advanced metering infrastructure to drive more real time decision-making and operational activities, complex event processing and stream computing are needed for the modern smart grid. The chapter discusses one core technique that would support the growth of smart grid, big data with HPC, with a focus on the platform, data integration, warehousing, and analytics that are particularly adaptive to handle a variety of characteristics of energy industry data. Finally, the chapter summarizes and proposes a comprehensive, technical solution for smart grid platform with applications focusing on complementary operation of multiform energy system that supports all aspects within a data lifetime cycle, e.g., acquisition, storage, analytics, and visualization.

1 INTRODUCTION

In recent years, big data quickly becomes a "hot topic" in industry and academia. "Nature", "Science," and other magazines have published special issues to explore the challenges and opportunities of big data [1,2]. "Data have penetrated into every industry and business area today and have become an important factor in production. The discovery and use of big data signal a new wave of productivity growth and consumer surplus," said McKinsey & Company, a leading management consulting firm.

Big data is a term for large and complex datasets that traditional data processing approaches are not suitable to deal with them. Big data usually comes from the

Big Data Application in Power Systems. https://doi.org/10.1016/B978-0-12-811968-6.00004-8

Internet, enterprise systems, Internet of Things, and other information systems. Data collection and preparation, storage management, data processing, data analysis, and knowledge presentation [3] would generate new insights to support decision-making and business intelligent operation.

Smart grid is the direction and trend of the electrical power industry evolution. Smart grid is a new generation of physical information systems which integrate the advanced communication information technology and physical energy systems. The essence is to provide comprehensive energy services based on complementary operation of multienergy sources. The core is the energy production, transportation, distribution, and consumption based on distributed and renewable energy. At the same time, it can integrate heat, electric power, gas, and even water together [4]. Smart grid business model includes two aspects: integrated energy, that is, electric power, gas, and other complementary multiexchange; integrated services, such as energy security monitoring, energy efficiency management, operation and maintenance services, and green energy deployment services [4]. Smart grid is based on integrated, high-speed communication network, and makes full use of advanced sensing, measurement technology, optimization control technology, and decision support technology to realize the reliable, safe, economical, efficient, and environment-friendly goals. The intelligent operation of the smart grid is based on a high degree of "observability" and "controllability." The observability and controllability are based on the need to obtain real-time panoramic data that can reflect the operating status of the system [5]. Thus, the high-performance data analysis technology will become a significant support for the development of smart grid. One of the core technologies of supporting the development of smart grid will be big data analytics with high performance computing (HPC). We mainly focus on frameworks for big data integration, warehousing, and analytics.

2 FRAMEWORKS FOR BIG DATA PLATFORM

The electric power industry is facing unprecedented challenges caused by extremely high volume and high frequency measurement data. According to the Navigant Research Report, the estimated installed base of smart meters worldwide will surpass 1.1 billion by 2022 [6]. Advanced metering infrastructure (AMI) typically collects electricity usage data in the range of 15 min to 1 h. This is up to a 3000-fold increase in the amount of data utilities would have processed in the past [7]. Meanwhile, synchrophasor is being deployed around the global that collects a large volume of low-latency, real-time streaming measurement data. Phasor measure unit (PMU) can measure AC waveforms (voltages and currents) typically at a rate of 48 samples per cycle (2880 samples per second for 60 Hz systems) [8]. Just one phasor data concentrator collecting data from 100 PMUs of 20 measurements each at 30-Hz sampling rate generates over 50 GB of data 1 day [9].

Big data analytics provides a suite of techniques for the utility industry that are deemed to resolve these challenges. The in-memory calculation engine and parallel computing framework, Hadoop/MapReduce and Spark, are ready for handling an extremely large scale of dataset; on the other hand, the stream processing engine, Storm, Streams, and Spark Streaming are built to analyze data in motion and act on information as it is happening.

Consequently, big data could be applied to improve both power system short-term operations and long-term planning processes. The promising applications for big data analytics include detection of energy theft, strategic adoption for electric vehicle and rooftop solar integration, fine granularity load forecast and renewable generation forecast, distribution system topology identification, online asset risk assessment, distribution system voltage and var. optimization, customer segmentation & targeting, and revenue protection, etc. [6].

The architecture of big data platform includes data integration, warehousing, analytics, and combines the demand of smart grid to put forward a set of frameworks which have excellent computing ability and can adapt to various business requirements. The popular big data framework is based on the Apache Hadoop ecosystem, the Hadoop distributed file system (HDFS) is used as the underlying file system. Yet Another Resource Negotiator (YARN), as resource management and task scheduling tool, uniformly distributes Hadoop MapReduce and Spark to support multiple business scenarios and business requirements. The specific architecture is shown in Fig. 1.

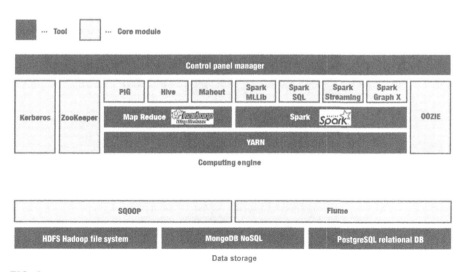

FIG. 1
Frameworks of big data platform.

A system performing data science requires some formats of data storage. For decades, the term "database" has been more or less synonymous with relational database management systems, which are widely used in power system. Recently there has been increased adoption of Not only SQL (NoSQL) databases with the primary motivation being horizontal scaling to handle big data [10].

The Lambda architecture is designed to handle massive quantities of data by taking advantage of both batch- and stream-processing methods. It attempts to balance latency, throughput, and fault tolerance by using batch processing to provide comprehensive and accurate precomputed views, while simultaneously using real-time stream processing to provide dynamic views. The Lambda architecture has three major components: the batch layer, the serving layer, and the speed layer. The batch layer manages the immutable master dataset, and precomputes batch views, whereas the serving layer indexes batch views and loads them as needed, and the speed layer handles new data and updated real-time views.

2.1   Architecture

The new generation of big data frameworks features in support of semistructured, unstructured data processing, support data visualization, also with higher processing performance and easier operation and maintenance features. Lambda architecture is the reference model for the next generation of big data frameworks, which was first proposed by Twitter. As shown in Fig. 2, Lambda architecture contains three layers, batch layer, speed layer, and serving layer [11].

Ideally, any data access procedure can start with a single argument input and a function result output, but if the data reaches a significant level (such as

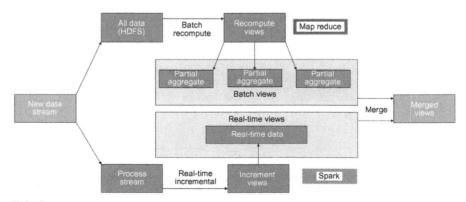

FIG. 2
Lambda architecture.

PB (petabyte)) and needs to support real-time queries, it can be very expensive. Lambda architecture introduces the concept of batch view, which preevaluates part of the query results so that when the query needs to be executed, the results can be read from the batch view. A precomputed view can be indexed, thus enabling fast random reading.

The batch view is generated from the batch layer, the use of batch layer is to make the large data small, to use computing resources to improve real-time query performance effectively.

Speed layer uses the same data processing logic as the batch layer, but the main difference is that the speed layer deals with real-time streaming data and batch layer deals with bulk off-line data. Another difference is that to meet the minimum computing delay, the speed layer does not read all new data at the same time; on the contrary, it will receive one piece of new data and update the speed view. Speed layer is an incremental calculation, but not reoperation.

2.2 Storage

Energy data are of multiple sources and heterogeneous features [12], for structure data storage, open-source database MySQL and PostgreSQL are the better choices. For unstructured data storage, NoSQL databases such as HBase, MongoDB, and Cassandra are the most appropriate.

The raw data can be stored in MongoDB (cluster mode), and the data schema of the cloud computing platform is shared. When Spark/MapReduce calculation engine is ready, the data will be automatically imported into HDFS through Kafka. The results will be exported back to MongoDB via Kafka, as well as to other data sources already deployed in the system, such as PostgreSQL or HDFS.

2.3 Security

Hadoop's security certification is based on the Kerberos. Kerberos is a network authentication protocol. Users only need to enter authentication information to verify that they can access multiple Kerberos-based services by obtaining Kerberos ticket. Single sign-in of the machine can also be done based on this protocol. Hadoop itself does not create user accounts but uses the Kerberos protocol for user authentication.

3 BIG DATA WITH HPC

HPC is the use of parallel processing techniques for running advanced application programs efficiently, reliably, and quickly. HPC is a critical component for frameworks for big data integration, warehousing, and analytics.

3.1 System Architecture (Fig. 3)

(1) Monolithic architecture

Monolithic architecture is a very common form of computer software architecture, which is often layered by functions. Common layers are the presentation layer, business logic layer, and data layer [13]. The business logic layer can be modularized to components by different business responsibilities and functions. The "monolithic architecture" is a "single block" at the physical deployment architecture level. It is usually compiled, packaged, deployed, and maintained as a single application.

(2) Microservice architecture

The microservice architecture is a new concept, which divides the application into a series of small services. Each service focuses on a single function, runs in a separate process, and has a clear boundary between services. Services can communicate with each other with light protocols (such as HTTP/RESTful), to achieve a complete application that meets business and user demands [14].

3.2 Service Type

HPC services currently have a variety of ways, which are recognized by IaaS, PaaS, SaaS, BaaS, etc. [15].

(1) IaaS

IasS is short for "Infrastructure-as-a-Service," usually as a hardware server rental service. IaaS service providers provide off-site server,

A monolithic application puts all its functionality into a single process…

A microservices architecture puts each element of functionality into a separate service…

… and scales by replicating the monolith on multiple servers

… and scales by distributing these services across servers, replicating as needed.

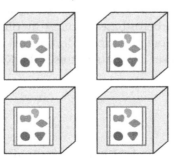
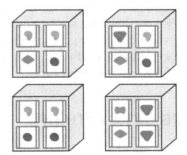

FIG. 3
Monolithic architecture vs microservice architecture.

virtualization, storage, and network hardware for users to rent. It can save maintenance costs and office space for users. Some of the successful IaaS products include Amazon's AWS cloud service, Microsoft's Azure, Alibaba's Aliyun, and Tencent's Qcloud.

(2) PaaS

PaaS is short for "Platform-as-a-Service." PaaS provide a complete application runtime environment, middleware, database, and so on. Developers only need to upload and deploy the code; the application can run up. PaaS not only reduces the cost of IT operation and maintenance but also eliminates a lot of development workload. PaaS platform includes Google's App Engine, Baidu's App Engine (BAE), and Sina's App Engine (SAE).

(3) SaaS

SaaS is short for "Software-as-a-Service." Usually, it provides an entirely web-based application, the user does not need to consider any software development, deployment, and do not have to worry about server hardware, bandwidth, just directly purchase and the software is ready to go.

(4) BaaS

BaaS is short for "Backend-as-a-Service." It is a new PaaS-based cloud service. It is designed to provide backend cloud services for mobile and web applications, including cloud data, storage, account management, messaging, social media integration, big data analysis interfaces, and other services [9]. The BaaS platform provides users with a variety of core components and microservices (middleware) to complete specific functions and meet individual demands.

3.3 Internet of Things

The Internet of Things is to let devices sense, communicate, interact, and collaborate within a network. From the perspective of monolithic architecture, the traditional IoT cloud computing platform consists of three layers [16], as shown in Fig. 4.

The protocol layer acts as a gateway and accesses the IoT devices through MQTT (Message Queue Telemetry Transport), XMPP (Extensible Messaging and Presence Protocol), or other proprietary protocols, and provides APIs (Application Programming Interfaces) and corresponding SDKs (Software Development Kits) for other Internet devices and applications [17–19]. IoT devices can also directly connect to the gateway through WiFi, 3G, 4G, and other network communication module, and complete the data transfer. To ensure the stability of data access and push services in the high-concurrency environment, load balancing mechanism is usually implemented, and the number of gateways is extended according to the traffic [18,20].

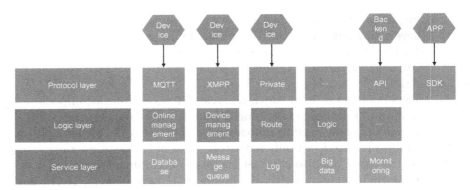

FIG. 4

IoT high-performance computing architecture.

The business logic layer is mainly based on the user's request parameters, matching the target hardware, the business data transfer distribution, the completion of the business system data subscription and distribution, to achieve intelligent device control and monitoring [19,20].

Core services layer provides all the essential services, including data storage database, message queue, log management, big data platform docking, and monitoring warning services [21].

3.4 HPC Platform for Smart Grid

The HPC platform abstracts and unifies backend system's requirements in most scenarios, and implements a common, powerful, customizable BaaS (Backend as a Service) service through cloud engine.

Cloud engine is the hosting service launched by the smart grid HPC platform. The engine is based on runtime environment and able to run backend applications. The engine supports multiinstance service load balancing and scaling.

(1) Cloud functions, hooks, and background tasks
Smart grid HPC platform provides cloud function, function hook, and background tasks. Through the cloud function, users can easily customize the business logic based on data interaction and can be exposed through the routing interface. Function hook allows the user to define a function that will be called when target event happened, such as new object saved or object modified. Background tasks enable you to set nonblocking tasks and perform tasks such as big data processing.

(2) Real-time data streaming
HTTP protocol is well known as the "request/response" model, but this model does have a lot of limitations. In many scenarios, we want the

server to actively send notifications to the browser, such as notifying the browser to refresh the data display in real time when receiving new meter data. HTML5's WebSocket protocol allows the server to send packets to the client directly, so it can make real-time data streaming possible.

To ensure the communication connectivity and service scalability of WebSocket in the multiinstance cloud engine environment, we take the necessary measures to prevent WebSocket from becoming the bottleneck of the whole service. If the user's service has high throughput, the single instance of the cloud engine may cause the packets blocking. To avoid such situation, we use Redis as a global message queue to handle all messages' distribution and subscription. This will not only ensure the reliability and efficiency but also let the whole cloud engine scalable.

The collection, storage, and analytical processing of relevant data sets are a key strategic objective of nearly every modern business entity, and the utility sector is one which may see large improvements in operational efficiency and planning by having situational awareness of the smart grid through proper management of data.

The collection and organized storage of data is managed by a well-developed collection of online transaction processing (OLTP) database systems, which are capable of reliably organizing vast quantities of data when deployed as a storage cluster. These systems are typically oriented around the goal of individual read/write operations to the database, while ensuring accuracy and consistency within the stored data.

For developing analyses and insights from a collected data set, the data access pattern is likely to vary from an OLTP system significantly, which calls for the use of an alternative system for querying the data set, typically referred to as a data warehouse. For an online analytics processing (OLAP) platform in a data warehouse [16], rather than the database interaction consisting of many individual read/write operations, commonly there will be a small number of users of the OLAP platform performing large batch-read operations for analyzing aggregated data over the dimensions of time, location, etc. The commonly queried data will often be organized into "cubes" of preaggregated results, and any query which cannot be addressed by the preaggregated results of the OLAP cube will be served by a partially denormalized data structure called a star schema.

4 BIG DATA WITH COMPLEX EVENT PROCESSING

For modern big data platform systems, the ability for the timely reaction to the occurrence of real-world situations in the system environment has become a fundamental requirement. This applies to many different applications, e.g., in smart grid, automatic stock trading, logistics, and production control. For example, in a smart grid scenario, the detection of a divergence between the energy consumption and the energy production can enable the rapid deployment of an intelligent Demand Response system, adapting the energy demand of intelligent appliances to the energy production, which reduces the demand for operating reserve provided in expensive supplemental power plants. In such applications, incoming data streams of low level information arrive from heterogeneous sources at high rates, and need to be processed in real time in order to detect more complex situations. Those data streams can be busy, and their rates fluctuate tremendously.

To tackle the problem, the paradigm of real-time complex event processing (CEP) has emerged as a favorable approach. CEP is a new big data integration, warehousing, analytics technique which takes data as unordered series of events coming from different sources. CEP has found wide uses in industries such as financial systems, homeland security, and sensor data processing. In each of these cases, the common element is that data from edge devices must be processed "on the fly," whether it comes in streams or asynchronous bursts. The CEP technology is capable of applying complex queries to multiple data streams simultaneously to detect specified conditions (events), thus triggering appropriate actions in real time.

CEP can be used in a variety of utility business functions. These include meter data management, demand response, fault detection, outage management, billing, and remote equipment monitoring, etc. CEP is a flexible tool, and when included in an overall data management strategy and architecture, it can tremendously improve the flexibility needed to implement the data management solutions for smart grid.

Smart grid is inherently complex due to the dynamic nature of power generation equipment, the use of complex technologies, the long distance of electric power transportation, and the instantaneous balance of production and consumption. Electric power networks are among the world's most complex human-made systems. The problems existing in power system fit naturally within the paradigm of CEP.

CEP is the use of technologies to track streams of data from multiple sources, to analyze trends, patterns, and events in real time in order to respond to them as quickly as possible. It can be further leveraged to monitor diverse and disparate data sources or events, brining organizations enhanced situational knowledge and increased business agility. CEP allows users to access events that happened in the past and use them in any order. The event can come from various sources and may occur over a long period of time. CEP requires sophisticated event interpreters, event pattern definition and matching along with correlation techniques.

With CEP, incoming data is being continuously monitored and acted upon using declarative conditions. Moreover, the data monitoring and processing works at a near-zero latency. Different events may come from different sources, and the CEP system can assemble a complex event to internally model the components as one object. A CEP system is aimed at solving the velocity problem of big data, while data comes as a stream of predefined events. The sliding window approach used by CEP systems ensures that only a portion of actual data simultaneously passes into the main memory, whereas the old events may be discarded or archived. This way, all the data does not have to fit into system memory, but still the most recent events can be efficiently analyzed.

In a smart grid, the data sources may include PMUs (typically 2880 samples per second), SCADA (typically acquiring data every 2–5 s), AMI (typically acquiring data every 1–15 min), weather data, and third-party data, etc. Since data is transferred at a steady high-speed rate, the input data may be treated as streams. The data is continuously evaluated by queries. Fig. 5 describes the system architecture of CEP.

The state-of-the-art CEP takes advantage of stream computing to handle unstructured data and large numbers of business events per second. CEP pushes data through highly sophisticated analytics processes to deliver real-time analytics results on data in motion, to help increase responsiveness when dealing with high levels of input. It enables both descriptive (simple) and predictive (sophisticated) analytics to support real-time decisions. Essentially, CEP enables organizations to capture and analyze as much data as they can handle at any time and in near-real time.

Apache Spark is a fast and general engine for large-scale data processing. It is a unified platform combining Spark SQL, Spark Streaming, and MLLib for machine learning and GraphX. Spark now boasts the ability to not only process streams of data at scale but also to "query" that data at scale using SQL-like syntax. This ability makes Spark a viable alternative to established CEP platforms and provides advantages over other open-source stream processing systems. Especially with regard to the former, Spark will now allow for the creation of "rules" that can run within stream "windows" of time and make decisions with the ease of SQL queries. This is a remarkably powerful combination.

Spark Streaming is an interesting extension to Spark that adds support for continuous stream processing. All the strengths of Spark's unified programming model apply to Spark Streaming, which is particularly relevant for real-time analytics that combine historical data with newly collected data. Spark Streaming ingests data from any source, including file systems such as S3 and HDFS. Users can express sophisticated algorithms easily using high-level functions to process the data streams. The core innovation behind Spark Streaming is to treat streaming computations as a series of deterministic microbatch

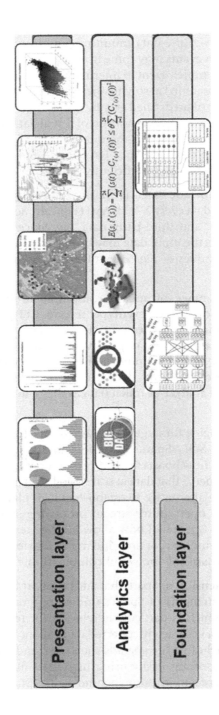

FIG. 5

Big data complex event processing architecture.

computations on small time intervals, executed using Spark's distributed data processing framework. Microbatching unifies the programming model of streaming with that of batch use cases and enables strong fault recovery guarantees while retaining high performance. The processed data can then be stored in any file system (including HDFS), database (including Hbase), or live dashboards.

5 APPLICATION OF BIG DATA TECHNIQUES FOR POWER SYSTEMS

A major goal of developing big data techniques in smart grid is to promote complementary operation of multiform energy system that will achieve energy cascading utilization and boosts the energy efficiency. Distributed combined cooling heating and power (CCHP) system develops as a novel energy supply way in recent years, which is regarded as environment-friendly, economical, and reliable option for the future [17]. CCHP is one of the key components for successful operation of multiform energy system.

The typical primary movers for CCHP include steam turbines, gas turbines, reciprocating internal combustion engines, and microturbines, etc., which are integrated with thermally activated cooling technologies. Due to its unique characteristics and complexities, the new research needs to be done on system design, control, operation, and planning. Consequently, optimization and economic evaluation techniques based on thermo-dynamics and thermo-economic principles are critical to the successful deployment of CCHP system. A data-driven method for better operating multiform energy system, with a consideration of uncertainty on price and demand, would be beneficial for directing practical system operation.

The multiform energy system employs more than one thermos-dynamic cycle. For example, a CCHP refers to the combination of gas turbine generators with exhaust waste heat boilers and thermally activated chilling machines for the production of electric power, heating, and cooling energy together [18]. The overall efficiency of CCHP can reach around 80%–90% due to energy cascading utilization [19]. The CCHP is particularly fit for multiform energy system where distributed solar, wind, energy storage, and load are systematically integrated. The CCHP can provide most of heating and cooling energy, and partially fulfill electricity demand. The outstanding electricity demand is compensated by power grid (Fig. 6). The overall reliability of energy supply is improved by a combination of distributed and centralized models. Our platform implements a general methodology to optimize CCHP performance and introduces a dynamic model to evaluate CCHP operating cost and combine power network analysis with heating/cooling energy dispatch [20]. The data-driven method is

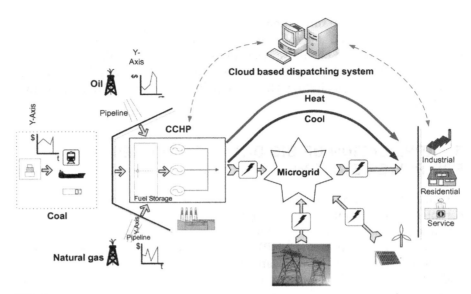

FIG. 6
CCHP based multienergy form system.

demonstrated on a simulated data set. The reduction of total cost is about 20% compared to the old design of electricity, heating, and cooling energy being produced separately.

Fig. 7 demonstrates mechanism of a CCHP system. Quality fuel (e.g., natural gas, oil, or gasified coal) is fed into furnace together with compressed air. The burning of fuel produces high temperature and high pressure air that is propelling the turbine and generate electricity. The waste heat is fed into furnace to drive chilling machine and produce cooling energy in summer, heating energy in winter. There are also electric-powered heating or cooling devices to produce the rest of energy. The outstanding electricity demand is compensated by municipal power grid.

Fig. 8 shows electricity, heating, and cooling production by CCHP based on data-driven analytics. During the price peak spots, the CCHP reduces production and microgrid uses more electricity from power grid. The reduction of total cost is about 20% compared to the old design of electricity, heating, and cooling energy being provided separately.

The demonstration system is connected to a variety of power consumption and electricity generation device located on the campus of Tsinghua University, including teaching buildings, dormitories, and other functional buildings. The data-driven techniques are based on a general methodology called Collocation Algorithm to optimize CCHP performance and apply a dynamic model

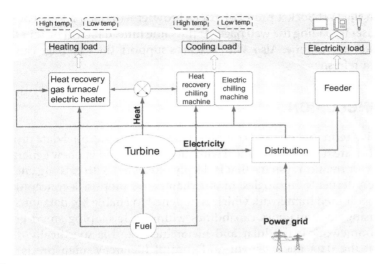

FIG. 7
Heating, cooling, and electricity provided by CCHP.

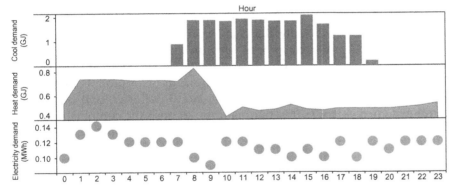

FIG. 8
Electricity, heating, cooling energy production by CCHP.

to evaluate CCHP operating cost and combine power network analysis with heating/cooling energy dispatch [21–24].

The data is updated every hour. The updated data includes current energy cost and price. Python is used to implement a simulator as data upload tool, the simulator upload data from cloud engine's open API. Once the data is stored in the database, the cloud engine will start a background task to let the big data platform perform data calculation and analysis. The results will be automatically stored in the corresponding database and ready for the frontend page. Plotly.js, leaflet, and other web components are used for data visualization.

Based on the WebSocket protocol, user's browser will open a long connection when a user is loading the web page. At the same time, data will be sent through WebSocket in real time. Also, with react.js's support, the web page can update without a refresh.

6 CONCLUSION

Big data is a term for large and complex datasets that traditional data processing approaches are not suitable to deal with them. Smart grid is a new generation of physical information systems that is deeply integrated with existing energy systems and internet technologies. In this chapter, we propose a conceptual data-driven distributed framework which is designed to enable big data integration, warehousing, and analytics capabilities within a developing smart grid. The chosen framework is modular and hierarchical, and is specifically designed such that the data management and control hierarchy align precisely with the typical hierarchy of smart grid. We also propose a new big data processing platform for a smart grid, based on the Lambda architecture, where a real-time CEP engine is embedded in the speed layer. We discuss the advantages for big data platform, HPC, CEP, and complementary operation of multiform energy system. The development of smart grid counts on big data integration, warehousing, and analytics. Finally, this chapter proposes a case to optimize CCHP performance by combining power network analysis with heating/cooling energy dispatch. The reduction of total cost is about 20% compared to the old design of electricity, heating, and cooling energy being produced separately. In future, we will integrate solar, battery, electric vehicles, and other green distributed energy devices into our platform, to push the combination of big data, cloud computing, and smart grid technology development.

Acknowledgment

The author would like to thank my colleagues, particularly Rong Zeng, Rui Fu, Jun Hu, Wendong Zhu, and Chris Saunders, for their contributions.

References

[1] G. David, Big data, Nature 455 (7209) (2008) 1–136.

[2] L. Wouter, W. John, Dealing with big data, Science 331 (6018) (2011) 639–806.

[3] J. Han, M. Kamber, J. Pei, Data Mining: Concepts and Techniques, The Morgan Kaufmann Series in Data Management Systems, third ed., Morgan Kaufmann, 2011.

[4] Z. Dong, J. Zhao, F. Wen, Y. Xue, From smart grid to energy internet: basic concept and research framework, Autom. Electr. Power Syst. 38 (15) (2014) 1–11.

[5] Y. Song, G. Zhou, Y. Zhu, Present status and challenges of big data processing in smart grid, Power System Technology 37 (4) (2013) 927–935.

[6] Navigant Research, Smart electric meters, Advanced Metering Infrastructure, and Meter Communications: Global Market Analysis and Forecasts, Available from: www.navigantresearch.com/research/smart-meters, 2013.

[7] N. Yu, S. Shah, R. Johnson, R. Sherick, M. Hong, K. Loparo, Big data analytics in power distribution systems, IEEE Innovative Smart Grid Technologies Conference, Washington, DC, 2015.

[8] Wikipedia, Available from: https://en.wikipedia.org/wiki/Phasor_measurement_unit.

[9] L. Xie, Y. Chen, P.R. Kumar, Dimensionality reduction of synchrophasor data for early anomaly detection: linearized analysis, IEEE Trans. Power Syst. 29 (6) (2014) 2784–2794.

[10] N. Marz, Big Data Lambda Architecture [EB/OL]. (2016–08-03)[2012–09-05], http://www.databasetube.com/database/big-data-lambda-architecture/.

[11] D. Li, S. Geng, J. Zheng, Development tendency of power big data in energy internet circumstances, Mod. Electr. Pow. 32 (5) (2015) 10–14.

[12] J.-Z. Luo., et al., Cloud computing: architecture and key technologies, J. China Inst. Commun. 32 (7) (2011) 3–21.

[13] P. Sareen, Cloud computing: types, architecture, applications, concerns, virtualization and role of IT governance in cloud, Int. J. Adv. Res. Comput. Sci. Softw. Eng. 3 (3) (2013) 533–538.

[14] R. Machado, R. El-Khoury, Monolithic Architecture, Prestel Publishing, Munich, 1995.

[15] S. Newman, Building Microservices, O'Reilly Media, Sebastopol, CA, 2015.

[16] J. Wen, Y. Shi, Research on the cloud-based computing service platform for the mega eyes, Telecommun. Sci. 6 (2010) 48–52.

[17] Q. Liu, L. Cui, H. Chen, Key technologies and applications of internet of things, Comput. Sci. 37 (6) (2010) 1–10.

[18] H. Huang, J. Deng, Discussion on the technology and application of IOT gateway, Telecommun. Sci. 4 (2010) 20–24.

[19] Z. Qian, Y. Wang, IoT technology and application, Acta Electron. Sin. 40 (5) (2012) 1023–1029.

[20] Q. Sun, J. Liu, S. Li, Internet of things: summarize on concepts, architecture and key technology problem, J. Beijing Univ. Posts Telecommun. 33 (3) (2010) 1–9.

[21] D. Wu, "Multiple Objective Thermodynamic Optimization and Application Study of Distributed Combined Cooling Heating and Power System (PhD dissertation), Shanghai Jiao tong University, 2008.

[22] H. Hui, C. Yu, F. Gao, Combined cycle resource scheduling in ERCOT nodal market, in: 2011 IEEE Power Energy Society General Meeting Proceeding, Detroit, MI, 2011.

[23] Y. Liu, "Optimal Design of Distributed Combined Cooling, Heating, and Power System (MS thesis), North China Electric University, 2012.

[24] F. Gao, Integration of Wind Generation With Storage Techniques (MS thesis), Department of Economics, Iowa State University, 2008.

Further Reading

[1] J. Cao, M. Yang, D. Zhang, Energy internet: an infrastructure for cyber-energy integration, South. Power Syst. Technol. 8 (4) (2014) 1–10.

Section

Harness the Power of Big data

CHAPTER 5

Moving Toward Agile Machine Learning for Data Analytics in Power Systems

Yuxun Zhou, and Reza Arghandeh

UC Berkeley and Florida State University, Tallahassee, FL, United States

CHAPTER OVERVIEW

This chapter starts with a brief discussion on classical supervised and unsupervised learning paradigms. The focus is not to give an extensive review of the field, which is impossible due to its many ramifications, but rather to equip the readers with fundamental ideas and popular approaches for regression, classification, dimension reduction, etc. The chapter moves on with a discussion on the important issue of model selection (hyperparameter tuning), which is pivotal to the performance of those off-the-shelf machine learning (ML) tools. Because all ML tools are "garbage in garbage out," the next section of this chapter is devoted to the problem of feature selection (FS), in which existing FS methods and their recent variants are presented in some depth. The rest of this chapter is devoted to the introduction of recent schemes of ML that seem promising for power system data analysis applications. The topics include but are not limited to semisupervised learning, multitask learning, transfer learning, multiview learning, etc. Overall this chapter aims at providing power system practitioners with basic knowledge of ML tools and their proper usage, as well as motivating researchers to develop new models and methods by combining their expertise from both ML and power system fields.

1 INTRODUCTION

The recent developments in monitoring, sensor networks, and advanced metering infrastructure dramatically increase the variety, volume, and velocity of measurement data in electricity transmission and distribution networks. For a concrete example, the advent of phasor measurement unit (PMU) has granted researchers and operators the access to high quality system information that would otherwise be unobservable using traditional technologies. The PMUs provide real-time measurement of three-phase voltage and current magnitude and phase angle with a high accuracy and a refined time resolution [1]). The GPS time stamped PMU data enables time-synchronized observability of the system that would otherwise be unavailable with traditional measurement technologies. Topology detection [2], phase labeling [3], event detection [4], and linear state estimation [5] are among applications of PMU data that are explored so far. Fig. 1 illustrates the integration of PMU network, machine learning (ML)-based event detection tools, and traditional power system infrastructures.

Big Data Application in Power Systems. https://doi.org/10.1016/B978-0-12-811968-6.00005-X

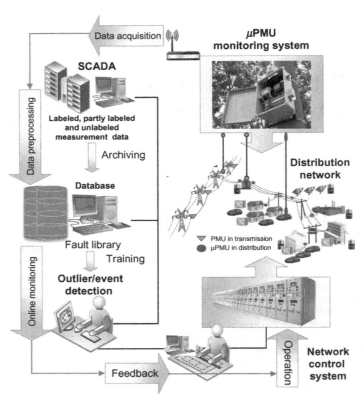

FIG. 1
The integration of ML-based event detection tool with PMU monitoring and SCADA system.

In general, modern power systems feature a combination of networking, system dynamics, measurement technology, and computational techniques (such as ML and control), and are tightly integrated with the demand and behavior of their users. It is foreseeable that in the near future, the incorporation of those intelligent computations, ubiquitously connected usage patterns, and increased efficiency and reliability demands will shift the traditional concept of power system from a pure physical system to a cyber-physical system (CPS), which requires fundamentally new modeling, control, monitoring, design, and diagnosis approaches to enhance adaptability, autonomy, efficiency, functionality, reliability, safety, and usability.

Meanwhile, the progress in artificial intelligence, advanced statistics, ML, data base, and data-mining methodologies has transformed many research fields, including computer vision, natural language processing, speech recognition, robot control, etc., in a significant manner [6]. Being considered as a CPS that incorporates intelligent computation, modern power system can also benefit from the increasing availability of open source platforms and existing tools in

ML to transform the area and utilities into data-driven enterprises. This chapter particularly discusses ML and data-mining algorithms, methods, and techniques that are adaptable for exploratory data analysis, regression, and pattern recognition in power systems. The objectives are two folds. On the one hand, both classical and status quo ML paradigms are reviewed and discussed, motivating the proper usage of traditional supervised/unsupervised learning tools and recent development of semisupervised learning, multitask, multiview learning, sparse representation, deep learning, etc., for various tasks in power systems. The hope is that the dramatic progress in ML can be fully harnessed to reform the solution of power system state estimation, load forecasting, event detection, structure identification, etc. On the other hand, the reverse direction (i.e., the challenges and new problems brought by power system data to ML) is discussed. Similar to the impact of computer vision, natural language processing, speech recognition, or robot control on the advancement of ML, it is expected that the complexity of the interconnected system, the behavior-related data-generating process, as well as the unique sensing and measurement techniques in power systems would inspire novel theoretical and methodological results for ML.

2 CLASSIC SUPERVISED AND UNSUPERVISED LEARNING

2.1 Supervised Learning Overview

Traditionally, there have been two fundamentally different paradigms of ML. The first one is the supervised learning, with the goal of learning a mapping from some input x to output y. Usually the observations (x_i, y_i), $i = 1, ..., n$ are called samples, $x_i \in R^d$ are referred to as features of sample i, and $y_i \in \mathcal{Y}$ are called labels or targets. To find the "optimal" mapping f from a function class $\mathcal{H}$, a wide variety of methodologies have been proposed from very different perspectives. One of the commonly used strategies is to formulate the learning task, that is, the identification of f, as a "regularized empirical risk minimization" problem [7], shown in the following

$$\min_{f \in \mathcal{H}} \frac{1}{n} \sum_{i=1}^{n} L(y_i, f(x_i)) + \frac{\lambda}{2} \|f\|_{\mathcal{H}} \tag{1}$$

in which the first term measures the "goodness of fit" of the classifier f with some loss function L, and the second regularization term penalizes some norm of the mapping in the functional space $\mathcal{H}$. The first term is essentially the averaged loss (empirical risk) aggregated over the training samples, and the second term controls the complexity of the f to avoid over-fitting.[1] Hence overall, the learning problem can be thought of as a complexity constrained functional

---

[1]In some context the regularization can also help alleviate ill-posed problem and induce sparsity.

fitting. Note that although some ML formulations do not have the regularization, such as the classic linear regression or simple neural network, the inclusion of the term for complexity control is crucial to improve the performance of a learning algorithm [8]. This is known as the Occam's razor principle, which favors simpler models (i.e., the learned mapping f) for generalization (testing on unseen data) purposes. By tuning the "hyperparameter" c, which is also called "model selection" in the jargon of ML, one is able to balance training fitness and model complexity, hence finding the optimal classifier that generalizes well to unseen dataset. The issue of model selection in ML is an important one, and will be discussed later.

2.1.1 Examples of Supervised Learning as a Regularization Empirical Risk Minimization

Depending on different choices of the function class $\mathcal{H}$, the loss function L, and the norm $\| \cdot \|_{\mathcal{H}}$, different learning problems can be formulated accordingly to accomplish various tasks. Following are some examples of well-known ML paradigms that fit in this category.

Linear/Ridge Regression

Perhaps the simplest realization of the general formulation (1) is to assume a linear function space, that is,

$$\mathcal{H} = \left\{ f(x) = w^T x | w \in R^d \right\} \tag{2}$$

a squared lose function $L = (\cdot)^2$ together with a squared L_2 norm regularization on R^d. Then the empirical risk minimization reads

$$\min_{w} \frac{1}{n} \sum_{i=1}^{n} \left(y_i - w^T x_i \right)^2 + \frac{\lambda}{2} \| w \|^2 \tag{3}$$

This realization gives rise to the learning problem of ridge regression, and with $\lambda = 0$ it reduces to the classic linear regression. Since the problem can be written in a more compact matrix form as

$$\min_{w} \| Y - Xw \|^2 + \frac{\lambda n}{2} \| w \|^2 \tag{4}$$

the solution can be obtained explicitly with

$$w^* = \left(X^T X + \lambda nI \right)^{-1} X^T Y \tag{5}$$

where X is called the design matrix, each row of which contains a sample x_i, and Y is the vector of the target variable. By working with a simplified case with orthogonal features, one can derive that the regularization term essentially "shrinks" the value of w,[2] thus controls the complexity of the linear model

---

[2]For general cases with nonorthogonal features one can perform the singular value decomposition (SVD) first to analyze the shrinkage effect.

by favoring "light" coefficients. Yet another popular choice of the regularization is the L_1 norm, which leads to the well-known LASSO learning formulation. An advantage using the modified regularization is that the solution w can shrink directly to zero, hence achieving sparsity or feature selection (FS) in the process of modeling learning. This will be revisited in the FS part. Note that the previous derivation is only one perspective for linear/ridge regression. Commonly linear/ridge regressions are proposed within a Bayesian framework for statistical analysis, as will be seen shortly.

Logistic Regression

To cope with classification problems in which the target variable is categorical (e.g., $y \in \{0, 1\}$), logistic regression again assumes a linear functional space. However, it uses the following transformation to model the conditional probability:

$$P(y=1|x) = \frac{1}{1 + e^{-w^T x}} \tag{6}$$

In the binary case the probability $P(y = 0|x)$ is simply one minus the term earlier,

$$P(y=1|x) = 1 - \frac{1}{1 + e^{-w^T x}} = \frac{e^{-w^T x}}{1 + e^{-w^T x}} \tag{7}$$

Using the negative of the log likelihood (or cross entropy) as the loss function, the empirical risk minimization problem reads

$$\min_w -\frac{1}{n} \sum_{i=1}^{n} \log\left(P(y_i|x_i)\right) + \frac{\lambda}{2} \|w\|^2 \Rightarrow \frac{1}{n} \sum_{i=1}^{n} \left[-y_i w^T x_i + \log\left(1 + e^{-w^T x_i}\right) \right] + \frac{\lambda}{2} \|w\|^2 \tag{8}$$

which brings about the L_2 regularized logistic regression. Unlike linear/ridge regression, the previous optimization problem does not allow an explicit solution, but it is still convex and is usually solved by second-order numerical optimization methods such as the Newton-Raphson algorithm.

Support Vector Machine

In the original proposition of support vector machine (SVM), the classifier was constructed with considerations of "large margin separation" [9]. Although that construction is intuitively satisfactory and mathematically inspirational, the SVM learning method can also be understood within the regularized empirical risk minimization framework. To begin with, the classifier is still assumed to be a linear function of features and L_2 regularization is adopted, but a hinge loss is used to penalize the misclassification errors, that is,

$$L_{\text{hinge}} = [1 - y_i w^T x_i]_+ \tag{9}$$

where

$$[t]_+ = \begin{cases} 0 & \text{if } t \leq 0 \\ t & \text{if } t > 0 \end{cases} \tag{10}$$

Intuitively, the loss is zero if the fitted label is correct (i.e., $y_i w^T x_i \geq 1$), otherwise the loss is positive and becomes larger if the fitting is away from the true value (i.e., $y_i w^T x_i \ll 1$). With those terms the overall learning task can be written as

$$\min_{w} \frac{1}{n}\sum_{i=1}^{n}\left[1 - y_i w^T x_i\right]_{+} + \frac{\lambda}{2}\|w\|^2 \tag{11}$$

An important property of SVM is that it allows the usage of the so-called "kernel trick" to perform classification in a nonlinear functional space. More specifically, the Lagrangian dual of the optimization (11) reads,

$$\min_{\alpha} \frac{1}{2}\alpha^T Q \alpha - \mathbf{1}^T \alpha$$
$$\text{s.t. } 0 \leq \alpha_i \leq \frac{1}{\lambda n} \quad \forall i \in \{1,\dots,n\} \tag{12}$$

The matrix involved can be expressed in the following

$$Q = XX^T \circ YY^T$$

where X is still the design matrix containing all feature samples, and Y the target (column) vector. The symbol $\circ$ is used for component-wise multiplication of the two $n \times n$ matrices. Due to strong duality, solving the primal SVM problem in Eq. (11) is equivalent to solving its dual form (12). However, a crucial property of the dual is that it only involves the inner product of feature samples. In particular, the (i, j)th entry of the matrix XX^T is simply $x_i^T x_j$. If a nonlinear feature transformation $x \rightarrow \phi(x)$ is applied to all samples, then the inner product becomes

$$\langle \phi(x_i), \phi(x_j)\rangle \triangleq k(x_i, x_j) \tag{13}$$

which avoids the explicit calculation of the feature transformation through a kernel function. Commonly used kernel functions are Mercer's kernels [10] such as the RBF Gaussian kernel, polynomial kernel, which ensure that the matrix Q is positive semidefinite. In that case the dual problem is a convex quadratic programming.

Nonparametric Regression

The so-called nonparametric method covers a class of learning paradigms in which the number of model parameters grows as the amount of training data increases. One classic example of the nonparametric method is the simple k-nearest neighbor algorithm for classification or regression. It is demonstrated here that the nonparametric regression method, namely the natural spline method, can be derived under the regularized empirical minimization framework.

In the context of nonparametric regression, the function space is taken to be all functions f that have continuous second-order derivatives, i.e., f'' exists and is continuous. Consider minimizing the penalized residual sum of squares:

$$\min_{f \in L^{(2)}} (y_i - f(x_i))^2 + \frac{\lambda}{2} \int [f''(t)]^2 dt \qquad (14)$$

Intuitively the regularization term imposes the smoothness of the function, hence favors "simple" solutions. Remarkably, one can show that the problem (14) admits an explicit, unique, and finite dimensional solution in the form of natural cubic splines [11]. Hence, the regressor can be written as

$$f(x) = \sum_{i=1}^{n} w_i S_i(x) \qquad (15)$$

where $S_i(x)$ are basis functions for the family of n-dimensional natural splines. It is also seen that the number of model parameters w_i increases with the amount of the training data.

Decision Tree

Yet another example is the decision tree method. Consider the following functional class:

$$f(x) = \sum_{m=1}^{M} c_m I(x \in R_m) \qquad (16)$$

If the criterion of empirical risk minimization is adopted as the sum of squared loss $(y_i - f(x_i))^2$, and a greedy heuristic is applied on rectangle regions R_m, then the learning process reduces to a decision tree for regression. Note that the complexity of the model can be directly controlled by specifying the maximal depth of the search tree in the algorithm.

2.1.2 *Bayesian Perspectives*

Certainly, the viewpoint of regularized empirical risk minimization is not (and should not be) the only perspective to understand and derive supervised ML paradigms. In effect, traditional ML literature or text books usually proceed from a probabilistic perspective that tries to describe the data generation process and to learn the parameters of that process for future inference purposes. In contrast to the previous section that considers various learning tasks as a frequentist function fitting problem, this perspective is focused on data modeling and is oftentimes called the Bayesian framework. A comprehensive survey of the Bayesian ML deserves an entire book and is beyond the scope of this chapter. Here two examples are given to equip the readers with some basic concepts.

Linear/Ridge Regression Revisited

Instead of directly minimize the empirical risk, it is assumed that the training samples $\{x_i, y_i\}$ are generated with the following process

$$y_i = w^T x_i + e_i \quad \forall i \tag{17}$$

where the noise e_i are drawn independently from a Normal distribution $\mathcal{N}(0, \sigma^2)$. Then the complete likelihood under the above probabilistic assumption is

$$J(w) = \prod_{i=1}^{n} \frac{1}{\sqrt{2\pi}\sigma} \exp\left\{ -\frac{[y_i - w^T x_i]^2}{2\sigma^2} \right\} \tag{18}$$

The model estimation can be done via likelihood maximization, which, after taking negative log and ignore constant terms, yields the minimization of $\sum_{i=1}^{n} (y_i - w^T x_i)^2$. Hence the classic linear regression is derived. Note that the solution is usually called the OLS estimator, and based on the probabilistic models one can derive diverse statistical properties of the estimator, such as the Gaussian-Markov theorem, hypothesis testing for model parameters, variance estimation, etc. Now consider incorporating a "prior knowledge" about the model parameter w by assuming that it is centered around a zeros vector and each dimension follows a normal distribution $\mathcal{N}(0, \tau^2)$. Maximizing the posterior likelihood, one arrives at the following optimization problem:

$$\max_{w} \prod_{i=1}^{n} \frac{1}{\sqrt{2\pi}\sigma} \exp\left\{ -\frac{[y_i - w^T x_i]^2}{2\sigma^2} \right\} \frac{1}{\sqrt{2\pi}\tau^d} \exp\left\{ -\frac{\|w\|^2}{2\tau^2} \right\} \tag{19}$$

which is equivalent to

$$\min_{w} \sum_{i=1}^{n} (y_i - w^T x_i)^2 + \frac{\sigma^2}{\tau^2} \|w\|^2 \tag{20}$$

So far the learning formulation of ridge regression is rediscovered. Under the Bayesian framework, many ML problems can be solved with two steps: (1) modeling the data-generating process and (2) likelihood maximizing or $K - L$ divergence minimization. In the following, a more comprehensive example is provided to give the readers a flavor about how to develop new ML paradigms under this framework.

Example: HMM With Autoregressive Emissions

This is a variant of the classic Hidden Markov Model (HMM) that incorporates observation auto-correlations. Readers who are not familiar with HMM and the expectation-maximization algorithm are referred to Chapter 14 of [12] for background knowledge, or they can simply ignore this part and save it for future reading. To start with, consider an HMM where $(q_1, q_2, \ldots, q_T)$ is the state sequence and where $(y_1, y_2, \ldots, y_T)$ is the observation sequence. The following modifications are made to include the observation auto-correlations:

- Instead of letting y_t depend solely on q_t as in a standard HMM, we let y_t also depend on y_{t-1}.
- In particular, let each y_t be a univariate Gaussian random variable, and distributed as $N(\mu_{q_t} + \beta y_{t-1}, \sigma^2)$.

With such data-generating process, the learning problem is reduced to estimating the transitional matrix, denoted by $A_{k, l}$, and the parameters μ_k and β of the Gaussian distribution. The graphical model of the above generating process is illustrated in Fig. 2. Now that the model has been established, the likelihood of the training data is maximized with the EM algorithm, which in a sense can be regarded as an alternating optimization procedure. More specifically, the following steps are conducted:

Step 1: Write out the complete log-likelihood. Let $\theta = (\pi, A, \mu, \beta, \sigma^2)$

$$
\begin{aligned}
\log p(Y, q | \theta) &= \log\left\{ \prod_{l=1}^{m} \pi_{q_1} \prod_{t=2}^{T} A_{q_{t-1}, q_t} \prod_{t=1}^{T} N(\mu_{q_t} + \beta y_{t-1}, \sigma^2) \right\} \\
&= \sum_{l=1}^{m} I(q_1 = l) \log \pi_l + \sum_{t=2}^{T} \sum_{k=1}^{m} \sum_{l=1}^{m} I(q_{t-1} = k, q_t = l) \log A_{k,l} \\
&\quad - \frac{1}{2\sigma^2} \sum_{t=1}^{T} \sum_{l=1}^{m} I(q_t = l)(y_t - u_l - \beta y_{t-1})^2 - \frac{T}{2} \log(2\pi\sigma^2)
\end{aligned}
$$

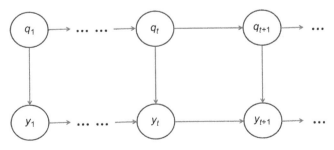

FIG. 2
Graphical model for hidden Markov model with autoregressive emissions.

Step 2: Find objective function by taken expectation conditioning on Y, θ^{old}.

$$
\begin{aligned}
E_{q|Y,\theta^{\text{old}}}[\log p(Y,q|\theta)] &= \sum_{l=1}^{m} p(q_1 = l|Y,\theta^{\text{old}}) \log \pi_l \\
&+ \sum_{t=2}^{T} \sum_{k=1}^{m} \sum_{l=1}^{m} p(q_{t-1} = k, q_t = l|Y,\theta^{\text{old}}) \log A_{k,l} \\
&- \frac{1}{2\sigma^2} \sum_{t=1}^{T} \sum_{l=1}^{m} p(q_t = l|Y,\theta^{\text{old}})(y_t - u_l - \beta y_{t-1})^2 \\
&- \frac{T}{2} \log(2\pi\sigma^2)
\end{aligned}
$$

This gives rise to the EM algorithm for learning the HMM with autoregressive emissions.

E-step: Compute conditional expectations (conditional probabilities) based on observation and current estimation of parameter.

$$
p(q_t = l|Y,\theta^{\text{old}}) = \frac{\alpha(q_t)\beta(q_t)}{p(Y)} \triangleq \gamma(q_t) \, p(q_{t-1} = k, q_t = l|Y,\theta^{\text{old}}) = \xi(q_t, q_{t+1})
$$

Each terms can be updated recursively with the following formula:

$$
\begin{aligned}
\alpha(q_{t+1}) &= p(y_1, \dots, y_{t+1}, q_{t+1}) \\
&= \sum_{q_t} p(y_1, \dots, y_{t+1}, q_t, q_{t+1}) \\
&= \sum_{q_t} p(y_1, \dots, y_{t-1}|q_t, q_{t+1}, y_t, y_{t+1}) p(q_t, q_{t+1}, y_t, y_{t+1}) \\
&= \sum_{q_t} p(y_1, \dots, y_{t-1}|q_t, y_t) p(y_{t+1}|y_t, q_{t+1}) p(q_{t+1}|q_t) p(y_t, q_t) \\
&= \sum_{q_t} \alpha(q_t) A_{q_{t+1}, q_t} p(y_{t+1}|y_t, q_{t+1})
\end{aligned}
$$

$$
\begin{aligned}
\beta(q_t) &= \sum_{q_{t+1}} p(y_{t+1}, \dots, y_T, q_{t+1}|q_t, y_t) \\
&= \sum_{q_{t+1}} p(y_{t+2}, \dots, y_T|q_t, y_t, q_{t+1}, y_{t+1}) p(q_{t+1}, y_{t+1}|q_t, y_t) \\
&= \sum_{q_{t+1}} p(y_{t+2}, \dots, y_T|q_{t+1}, y_{t+1}) p(y_{t+1}|y_t, q_{t+1}) p(q_{t+1}|q_t) \\
&= \sum_{q_{t+1}} \beta(q_{t+1}) A_{q_{t+1}, q_t} p(y_{t+1}|y_t, q_{t+1})
\end{aligned}
$$

$$
\begin{aligned}
\xi(q_t, q_{t+1}) &\triangleq p(q_t, q_{t+1}|Y) \\
&= \frac{p(y_1, \dots, y_{t-1}|q_t, y_t) p(y_{t+1}, \dots, y_T, q_{t+1}|q_t, y_t) p(q_t, y_t)}{p(Y)} \\
&= \frac{\alpha(q_t) p(y_{t+2}, \dots, y_T|q_{t+1}, y_{t+1}) p(y_{t+1}|y_t, q_{t+1}) p(q_{t+1}|q_t)}{p(Y)} \\
&= \frac{\alpha(q_t) \beta(q_{t+1}) A_{q_{t+1}, q_t} p(y_{t+1}|y_t, q_{t+1})}{p(Y)}
\end{aligned}
$$

M-step: Based on current expectation of hidden variable, maximize $E_{q|Y,\theta^{old}}[\log p(Y,q|\theta)]$ by make the gradient vanish. The following updating formula can be obtained:

$$\hat{\pi}_l = \gamma_1^l$$

$$\hat{A}_{kl} = \frac{\sum_{t=2}^{T} \xi_{t-1,t}^{kl}}{\sum_{t=2}^{T} \gamma_t^k}$$

$$\hat{\sigma}^2 = \frac{1}{T}\sum_{t=1}^{T}\sum_{l=1}^{m} \gamma_t^l (y_t - \hat{\mu}_l - \hat{\beta}y_{t-1})^2$$

$$\hat{\mu}_l = \frac{\sum_{t=1}^{T} \gamma_t^l (y_t - \hat{\beta}y_{t-1})}{\sum_{t=1}^{T} \gamma_t^l}$$

$$\hat{\beta} = \frac{\sum_{t=1}^{T}\sum_{l=1}^{m} \gamma_t^l (y_t - \hat{\mu}_l)y_{t-1}}{\sum_{t=1}^{T} y_{t-1}^2}$$

Note that the update only involves sufficient statistics, and at this step the sub-problem is concave because complete distribution is in exponential family.

2.2 Unsupervised Learning Overview

The second task of ML is the unsupervised learning [13]. Under this setting, only the unlabeled observations $X = \{x_1, \ldots, x_n\}$ are given. Typically, the goal of unsupervised learning is to identify interesting structures in the data X, such as clusters, quantiles, support, low-dimensional embedding, or more generally the patterns related with the distribution of the data. Note again that this part is not a comprehensive survey of existing unsupervised learning method, but rather to provide readers with the intuition and the linear algebra or optimization techniques behind those ML tools. In particularly, this part discusses two popular unsupervised ML tasks.

Principle Component Analysis

Given a data set with large number of features, such as the PMU measurement from power networks, it is often desirable or even necessary to find a low-dimensional representation of the data. Dimension reduction (or manifold learning) is one of the most important problems in unsupervised learning. The argument is that many high-dimensional data sets are in effect noisy versions of a low-dimensional embedding. Hence by finding their low-dimensional structure, one is able to achieve efficient information representation, computation, denoising, feature extraction, as well as for visualization purposes.

Ideally, the transformed data in the reduced space should maintain key properties of the original data set. This implies some loss functions to minimize. Let $X \in R^{n \times d}$ be a centered data (design) matrix (i.e., each row contains a data sample, and each column is centered to have zero mean). The classic principle component analysis (PCA) aims at finding a linear subspace of the d-dimensional data, such that the reconstruction error is minimized. More formally, let the rank of the subspace be k, then PCA essentially solves

$$\min_{P \in \mathcal{P}^k} \| X - XP \|_{\mathcal{F}}^2 \tag{21}$$

where $\mathcal{P}^k$ is a $d \times d$ orthogonal projection matrix with rank k, and $\| \cdot \|_{\mathcal{F}}$ is the Frobenius norm of a matrix. The optimization problem shown previously seems intractable at first glance; however, one can show that its solution is simply the projection of each data sample onto the top k singular vectors of the empirical covariance matrix. In effect, we have

$$\| X - XP \|_{\mathcal{F}}^2 = Tr\left[(X - XP)(X - XP)^T \right] = -Tr[XPX^T] + Tr[XX^T]$$

Hence minimizing the reconstruction error is equivalent to maximize $Tr[XPX^T]$. Since P is an orthogonal projection matrix of rank k, it can be written as $P = UU^T$ for some $U \in R^{d \times k}$. Then using the cyclic permutation property of the trace operator, one arrives at

$$Tr[XPX^T] = U^T X^T X U = \sum_{j=1}^{k} u_j^T X^T X u_j$$

Given the previous equivalence, it is clear that the top k singular vectors of X maximize the final objective. Hence PCA is closely related with the SVD and is usually computed by perform exact or approximate SVD on the data matrix. Moreover, PCA is also amendable for the usage of "kernel trick," since it only involves the inner products. Please refer to [14] for an application of kernel PCA for abnormal event detection in power systems.

k-Means Clustering

Another popular unsupervised learning task is the clustering, which tries to group a collection of data samples into several subsets or clusters. Ideally, the sample points within the same cluster are "close" or "similar" to each other, while samples in different groups or clusters are different. Hence, clustering is also called data segmentation and is closely related with the NP-hard set partition problem. From the objective, one immediate realizes that the definition of closeness and similarity is the key for the success of any clustering algorithms. In practice, the naive choice of the similarity metric (e.g., common distance metrics, such as L_2, Hamming distance, etc.) may not produce meaningful clustering. Hence, it is advisable to select the distance/similarity measure with expert knowledge or using metric learning algorithms.

The following provide a brief review of the classic k-mean algorithm again from an optimization viewpoint. Consider the cost function

$$J(z,\mu) = \sum_{i=1}^{n}\sum_{j=1}^{K} z_i^j \, \| x_i - \mu_j \|^2$$

where x_i and μ_j are vectors in $\mathfrak{R}^d$ and each vector $z_i = (z_i^1, ..., z_i^K)$ is constrained to have one component equal to 1 and the others 0.

- When the cluster centers $\mu_1, ..., \mu_k$ are fixed, minimizing the objective w.r.t z decomposes as minimizing each z_i^j separately, that is, for each i

$$z_i^j = \mathrm{argmin} J(z,\overline{\mu}) = \underset{z_i^j \in \{0,1\}; \sum_j z_i^j = 1}{\mathrm{argmin}} \left\{ \sum_{k=1}^{K} z_i^j \, \| x_i - \overline{\mu}_k \|^2 \right\} = 1\left[j = \arg \min_k \| x_i - \mu_k \| \right]$$

as the minimum is achieved simply by assigning $z_i^j = 1$ to the minimum value of $\| x_i - \mu_k \|$.
- When the group indicators z_i's are fixed, minimizing the objective w.r.t μ also decomposes as minimizing each μ_j independently (i.e., for each $j \in \{1, ..., K\}$),

$$\mu_j = \mathrm{argmin} J(\overline{z}, \mu) = \mathrm{argmin}_c \left\{ \sum_{i=1}^{n} z_i^j \, \| x_i - c \|^2 \right\}$$

which is a quadratic form of the vector c, taking derivative and setting it to zero we get

$$\mu_j = \frac{\sum_i z_i^j x_i}{\sum_i z_i^j}$$

In both of the previous two steps, one set of decision variables is fixed and the overall objective is decreased by minimizing w.r.t another set of variables. Hence, the k-means algorithm is indeed a special case of coordinate descent with z, μ as coordinates, or alternating optimization in general. Its convergence is guaranteed as a bounded objective is decreased in each iteration; however, the problem is not convex and it does not necessarily converge to the global minimum. Some efforts have been made to alleviate this issue, such as the well-known k-means++ which systematically find good initial points. Other clustering techniques include spectral clustering, hierarchical clustering, self-organized map, etc. Although the readers are encourage to try out exiting packages for those algorithms, it is always helpful to keep in mind that a proper definition of distance/similarity is the key to the success of those unsupervised ML methods.

3   MODEL SELECTION (HYPERPARAMETER SELECTION)

ML algorithms for regression, pattern recognition, or classification depend on multiple hyperparameters. In the earlier example concerning soft margin SVMs, the parameter C allows the tuning of the margin error penalty, which implicitly determines the number of support vectors or the complexity of the model. The kernel function and its corresponding hyperparameter maps the original feature space to a high-dimensional Hilbert space. In the decision tree example the maximal depth of the tree directly determines the complexity of the splitting process. It has also been seen that the penalty parameter λ of ridge regression reflects our prior knowledge of the model and has an effect to "shrink" the learned model coefficients. Given available training data set D_t and an ML model characterized by a set of hyperparameters $\boldsymbol{\theta}$,[3] the model selection essentially tries to solve:

$$\boldsymbol{\theta}^* = \mathrm{argmin}_{\boldsymbol{\theta}} \mathbb{E}_{D_t, D_u} \Psi \left(\hat{f}(D_t, \boldsymbol{\theta}), D_u \right) \tag{22}$$

where $\hat{f}(D_t, \boldsymbol{\theta})$ is the learned ML model with the training data D_t and hyperparameter $\boldsymbol{\theta}$. D_u denotes some unseen data set, and Ψ is certain loss function that evaluates the goodness of the ML model on that testing data. Since the true distribution of the training and testing data is unknown, the expectation has to be approximated with an estimation, yielding the following problem:

$$\boldsymbol{\theta}^* = \mathrm{argmin}_{\boldsymbol{\theta}} \hat{\Psi} \left(\hat{f}(D_t, \boldsymbol{\theta}), D_u \right) \tag{23}$$

In general, the tuning of ML models through those hyperparameters can greatly impact the complexity, flexibility, cost sensitiveness, noise-resilience, and many other performance metrics of the ML method. An optimal classifier $f^* \in \mathcal{H}$ resulted from the optimal set of hyperparameters is then expected to minimize the generalization cost on some unseen data, and the process of selecting those

---

[3]Note the difference between model hyperparameters and model parameters, which is denoted by w throughout this chapter. w reflects the way we model the data and is determined by the training process (e.g., empirical risk minimization), while the hyperparameters θ are introduced and selected by the users.

hyperparameters is called model (or hyperparameter) selection. This section is devoted to the discussion of commonly used techniques for model selection. Some theoretical foundations about generalization bound in the probably approximately correct (PAC) learning framework are first introduced, then various model selection strategies are presented for practitioners.

3.1 Theoretical Intuition

Generalization Bound

A natural question arises when an ML model, say f, is learned from data, that is, "How will f perform (in terms of mean squared error, classification error, etc.) on unseen data in the future?" This question cannot be answered in general without further assumptions, because the source of the future unseen data could be arbitrary even adversary. However, this question can be answered when some mild conditions are imposed on the true data generating process/model. In the terminology of ML, the previous question is addressed with *generalization performance* (also known as the out-of-sample performance) analysis. One popular way to conduct such analysis is through a framework called PAC learning, which stands for PAC learning. Within the PAC learning framework, a concept class[4] is PAC learnable if there exists a tractable learning algorithm A such that for any $c \in C$, any distribution $\mathcal{D}$ and some $\epsilon > 0$, $\delta > 0$,

$$\Pr_{D \sim \mathcal{D}}\{\text{err}[f(D)] \le \epsilon\} \ge 1 - \delta \tag{24}$$

Although no assumptions are made for $\mathcal{D}$, the learning framework do require that both training and testing data are from the same distribution. It is called "probably" as the error is controlled in a probabilistic sense with confidence $1 - \delta$, and the name "approximately" is used to emphasize that a soft control of the accuracy, $1 - \epsilon$, is adopted. Despite of being abstract and largely mathematical, the PAC learning analysis is able to reveal the generalization performance of a learning method as a function of model complexity, training sample size, etc., and is quite helpful for model selection purposes. As an example, in the case of binary classification with i.i.d. (independent identically distributed) samples, one can show that [8] with probability at least $1 - \delta$,

$$\underbrace{\Pr(y \ne f(x))}_{\text{testing error}} \le \underbrace{\hat{\Pr}_n(y \ne f(x))}_{\text{training error}} + \underbrace{\frac{1}{2}R_n(F)}_{\text{model complexity}} + \underbrace{\sqrt{\frac{\ln(1/\delta)}{2n}}}_{\text{vanishing term}} \tag{25}$$

It can be seen that the LHS, the testing error, is controlled by the three terms in the RHS, which are (1) the empirical error on the training data set, (2) the

---

[4]Roughly speaking, the notion "concept class" C is an abstraction of the target function one would like to learn from data.

model complexity measured by the Rademacher complexity of the function class F, and (3) a term that vanishes as the training sample size n increases. The above generalization bound reveals an important trade-off: A complex model will induce smaller training error in the first term; however, it will also increase the value of the second term. Therefore, to find the optimal model $f \in F$, one has to balance the fitness on training data and the complexity of the ML model. As a side note for the second term, many other complexity measures, such as VC-dimension, metric entropy, can be used for similar analysis. Interested readers are referred to the first few chapters of [15] for details. Moreover, the complexity term often can be expressed explicitly as a function of hyperparameters for particular ML models, such as the kernel SVM and its variants [16]. This property can be exploited to perform model selection in practice to avoid over fitting.

Bias-Variance Trade-Off

Yet another way to reveal the effect of model complexity on generalization performance is through a simple bias-variance analysis. For example, in a regression setting let us assume that $y = f(x) + \varepsilon$ where $E(\varepsilon) =$ and $Var(\varepsilon) = \sigma^2$, then the expected prediction squared error for any new input x' can be written as:

$$
\begin{aligned}
L(x') &= E\left[(y - \hat{f}(x'))^2 | x' \right] \\
&= \underbrace{\sigma^2}_{\text{irreducible error}} + \underbrace{\left(E[\hat{f}(x')] - f(x') \right)^2}_{\text{biase}^2} + \underbrace{E\left[\hat{f}(x') - E[\hat{f}(x')] \right]^2}_{\text{variance}}
\end{aligned}
$$

Typically, complex models f fit the data well and incur a small bias, while the induced variance is high. On the contrary simple models produce high bias but low variance. Hence, the fitness and complexity trade-off is revealed again. To illustrate the above points, Fig. 3 demonstrates the training, validation (which will be discussed later), and testing error as a function of model complexity. One observes that as model complexity increase, the training error decreases constantly, while the testing error first decreases and then increases due to over-fitting. Hence, model selection tries to estimate testing error via validation error and choose the "best" model (the circle point in the figure) for future use.

3.2   Practical Model Selection Techniques

Cross-Validation

In practice, the most well-known technique for hyperparameters selection is the cross-validation (CV), in which classifiers induced by different sets of parameters are applied to a validation sets. The parameters that minimize the average validation cost are then selected for future use. CV is particularly useful if we are in a data-rich situation: the available data at hand is randomly divided into two parts: a training set and a validation set. Here the testing set is assumed to be

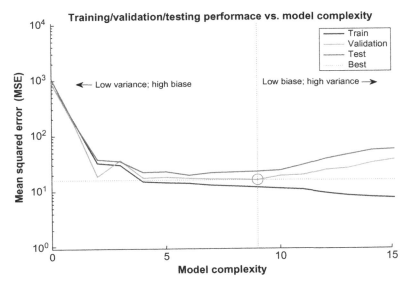

FIG. 3
The effect of model complexity on training, validation, and testing performance.

unavailable, but in practice to evaluate the final ML model one can simply "hold out" a subset of the data by not including it in the model selection process. The training set is used to train the models, and the validation set is used to estimate generalization cost for model selection, that is, in problem (23) the estimator $\hat{\Psi}$ is taken to be the cost on the validation data set. With that, a search is conducted in the hyperparameter space for different combinations of θ, and the one inducing minimal validation cost is chosen. Finally, the model obtained from the optimal hyperparameters is used on the unseen (or the held out) testing set for assessment.

Two problems arises. The first one is how to split the available data. Although the answer depends largely on the size, signal-to-noise ratio, dimensionality, etc., of the data at hand, in practice one can simply adopt the K-folds CV strategy, shown in Fig. 4. Note that when K is small, the estimation of the generalization cost has low variance but high bias, while as K increases the estimation incurs low bias but high variance, and is computationally more costly.[5] The second problem is about how to search within the hyperparameter space. This is essentially an optimization problem, although the function/objective to be optimized is random and hardly bears any nice properties (e.g., convexity, smoothness, differentiability) for optimization purposes. For low-dimensional θ, an exhaustive search

---

[5]Although for some ML models, such as linear/ridge regression, the leave-one-out CV $(n-1)$-folds CV can be calculated explicitly.

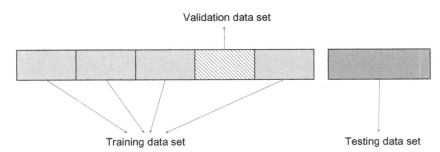

FIG. 4
Illustration of *K*-folds CV.

(or coarse to fine grid search) can be performed; however, when the number of hyperparameters is large, a resort to more advanced techniques, such as the parametric gradient [17], or black-box function optimization, is necessary.

Being a Bayesian

In the Bayesian perspective, the model selection problem can be formulated as picking the model largest marginal likelihood, that is,

$$M_\theta^* = \text{argmax}_m \Pr(D|m_\theta) \qquad (26)$$

Together with additional assumptions on the prior distribution of the hyperparameters, and techniques such as Gaussian approximation, the above maximization problem can either be solved with various sampling methods, or lead to some simple criterion for model selection. Among others, one such example is the Bayesian information criterion. Interested readers are referred to Chapter 11 of [12] for more discussion.

Regularization Path Algorithms

It is worth pointing out that the model selection problem for certain popular ML methods, such as LASSO or SVM, has been extensively studied in literature, which yields both theoretical guarantees and algorithmic methods. In particular, the so-called regularization path algorithm tries to characterize the learned model as a function of hyperparameters. In this way, the computational cost of CV can be greatly reduced. For instance, it can be shown that the solution α^* of the SVM dual Eq. (12) is a piece-wise linear function of $1/\lambda$, and an explicit form of the dependence can be derived at each intervals [17].

Bayesian Optimization

For complex nonprobabilistic ML models having large number of hyperparameters, such a deep neural network, all of the above methods may not be applicable. CV with an exhaustive search over the hyperparameter space is intractable, as the number of training-validations scales exponentially to the number of hyperparameters. Also, it seems unlikely that one can derive a simple regularization path for such complicated models. To cope with this problem, recently the ML community has been treating the model selection problem as a black-box optimization problem with a computational constraint. The input of the objective function is the settings of hyperparameters, and the output is an estimation of the generalization performance, such as the testing error in the CV framework. The optimization is then solved by leveraging the advancements of black-box optimization theory and algorithms. One optimization technique that demonstrates remarkable success is the Bayesian optimization, which builds Gaussian process models based on observed function value, and decides the next point to evaluation by balance the exploration-exploitation trade-off. Readers are referred to [18, 19] for theoretical details and examples.

4 FEATURE SELECTION

4.1 Overview

Identifying the most informative patterns or features from the observed data (i.e., FS) is one of the underpinnings for the success of any data analytic methods. Especially with the recent advancement in instrumentation and measurement technologies, power system researchers now have access to rich and real-time system, loads, and user data. Often, the raw datasets present themselves in very different forms, such as multiple coevolving time series, independent records, survey, etc., and need to be preprocessed to extract tentative patterns/features for pattern recognition or predictive applications. Among the vast quantity of information generated by such process, some features are correlated with the target application while others may be less relevant or redundant. For most ML methods, incorporating irrelevant or redundant patterns as input will not help but instead will deteriorate their performance. As such, the goal of FS is to identify the most informative feature set from the observed data. The overall information processing procedure for the input of ML is called feature engineering in the jargon of data mining. As can be seen from the earlier, feature engineering actually involves two steps, that is, (1) feature extraction and (2) FS. Because feature extraction is application specific and requires domain knowledge such as physics, networking, and signal processing, this section will only provide an example in power system event detection. The mainly focus, on the other hand, will be techniques for FS (Fig. 5).

In the literature, FS methods are classified into three categories, filter method, wrapper method, and embedded method. A more comprehensive survey can be found in [20].

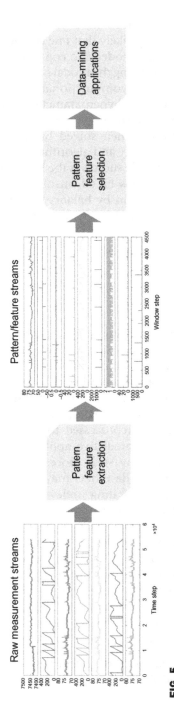

FIG. 5

Raw measurement data preprocessing and FS for data-mining applications.

Filter Method

This method selects informative features and suppresses the least interesting ones regardless of the underlying model assumption. Some informativeness metrics (i.e., a score), which indicate the usefulness of a feature subset, are adopted as the objective for selection purposes. Commonly used metrics include mutual information, t-statistics, signal-to-noise ratio, Hilbert-Schmidt Independence Criterion (HSIC), etc. Then the selection is done via optimizing the metric through combinatorial search or greedy approximation. Since the definition of the metric is not bounded by a particular ML model and is easy to compute from data, Filter method can be used for general and efficient FS. However, except for a few metrics, most filter methods are heuristics that lack theoretical justification.

Wrapper Method

The wrapper method blends in a learner (e.g., classifier or predictor) with the straightforward goal to minimize the classification or prediction error. In other words, a candidate feature set is evaluated by training a particular ML model and the estimated generalization performance (with CV for example) is used as the selection criterion. The selection procedure is usually done in a stage-wise or greedy manner to avoid combinatorial search. Features selected by the wrapper method can yield high accuracy for the particular learner but are not always suitable for others. Besides, the wrapper method becomes computationally intensive when the number of candidate patterns is large, as the wrapped ML model needs to be trained each time a feature set is evaluated.

Embedded Method

This method implicitly selects feature subsets by introducing sparsity in the learning model construction. The core idea, as is briefly introduced in the previous section, is to incorporate sparsity constraints or penalties in the risk minimization. Exemplary models include the LASSO, L_1-SVM, etc., which exploit the "truncated shrinkage effect" of the L_1 norm. The embedded methods are still model-specific, and their selection consistency is an issue in both theory and practice.

Some examples of the earlier FS methods are summarized in Table 1.

4.2 An Practical Example of FS Using Information Theory Criterion

In this section an example is provided to demonstrate both the feature extract and selection steps in the context of power system event detection with μPMU [30]. Fig. 6 shows a snapshot of the μPMU measurement data. Facing

Table 1 Examples of Various FS Methods and Comparison

Method	Type	Dependence	Optimization	Target	Scalable?
Centroid [21]	Filter	Linear	Com/Greedy	Classification	Yes
t-score	Filter	Linear	Com/Greedy	Classification	Yes
B-statistic [22]	Filter	Linear	Com/Greedy	Classification	Yes
Correlation	Filter	Linear	Com/Greedy	Regression	Yes
mRmR [23]	Filter	Nonlinear	Greedy	Both	Yes
IGFF [24]	Filter	Nonlinear	Greedy	Both	Yes
LASSO	Embedded	Linear	Convex	Both	Yes
LDFS [25]	Embedded	Linear	Convex	Classification	No
FVM [26]	Embedded	Nonlinear	Nonconvex	Both	No
HSIC [27]	Filter	Nonlinear	Greedy	Both	Yes
QPFS [28]	Filter	Nonlinear	Nonconvex	Both	No
SpAM [29]	Embedded	Nonlinear	Convex	Both	Yes

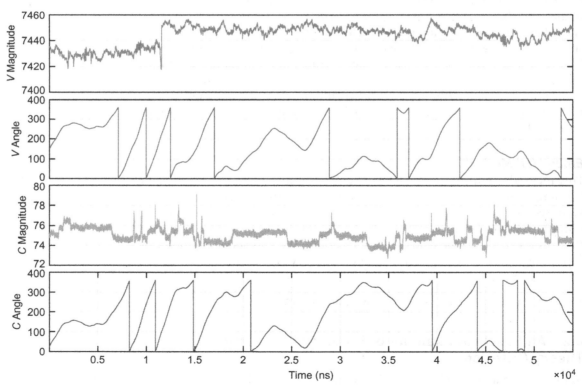

FIG. 6

Raw μPMU data from a real-world power distribution network.

milliseconds μPMU data that has not been explored before, limited prior knowledge is available on the effectiveness of different feature extraction methods. To use the data in an effective way, firstly all plausible ways of feature extraction are conducted, then the minimum-redundancy-maximum-relevance (mRmR) criterion is used to selected most informative ones for each type of events.

Notation-wise, the multistream time series μPMU data are written as $\{X_1, \ldots, X_T\}$. Each X_t is an $M \times C$-dimensional vector where M is the number of μPMUs and C is the number of channels of each μPMU. Because the raw data are in millisecond's resolution and almost all practical events happen at a larger time scale, one can safely use a sliding window to extract useful information. The window size L should be chosen according to the time scale of the event of interest. For example, in order to detect certain transient event in 0.1s scale, one takes $L = 12$ and processes the data in each window. For ease of notation let $w_t^i \triangleq \{x_t^i, \ldots, x_{t+L}^i\}$ be the tth window of stream i.

Now diverse techniques are considered to construct feature candidates. Intuitively, some events, such as voltage sag or voltage disturbance, could be revealed by investigating single streams (voltage magnitude or phase) fluctuations, while other events, such as high impedance fault and voltage oscillation, might be more obvious by analyzing the interbehavior/dependence of multiple voltage and current streams. For the purpose of detecting different types of events, both single stream and interstream feature extracts should be included with a variety of metrics.

Single Stream Features Extraction
- Classic statistics: Including mean, variance, and range of voltage/current magnitude in each window. These features capture the average voltage/current values as well as their fluctuations in the time slot. The median is also included as it is a more "robust" metric of average value from a statistical viewpoint. To further characterize the variations of magnitude in each window, the distributional features, including entropy, and histogram are calculated.
- First-order difference: We compute $x_{t+1}^i - x_t^i$ for each stream and take the corresponding mean and variance in each window. The intuition is that some transient events may exhibit significant "jumps" in voltage and current magnitude, which can be well captured by "spikes" in the first-order difference. As for streams associated with phase information, the average difference is an indicator of voltage/current frequency and is also an important indicator of system stability.
- Transformation: Notice that many distribution side events, such as ON/OFF of reactive loads, usually lead to oscillations in both magnitude

and phase measurement, we propose to use fast Fourier transform (FFT) to capture this frequency domain information. Also, wavelet transformation is adopted to capture local fluctuations and abrupt changes.

Interstream Features

- Deviation: The difference between any two of the three phases, for both voltage and current. The resulted time sequences are processed as single streams in each window with classic statistics. In this way, we incorporate information for the events that exhibit phase imbalance.
- Correlation between any two of the three phases, for both voltage and current. The correlation constitutes a metric of dependence for these time series, and is also helpful in providing information related with interphase behavior.

A summary of feature extraction candidates are given in Table 2. Note that the interstream features for different nodes (hence from different μPMUs) should be very interesting for subsystems width event detection, for which one can include not only correlation as dependence metric, but also causal information [31] that pinpoints the propagation of the event.

With the presented feature extraction procedure, a total number of 260 features have been pooled together. Obviously, some of them may be redundant as there are significant similarities among extracted features, for example, when the three phases are balanced, their single stream mean, variation, etc., are almost the same. For another instance, the first-order difference and wavelet transformation of one specific stream might have very similar pattern as they both reflect the sudden change of the same time series. From an ML point of view, adding redundant features does not help detection/classification, but instead would introduce extra learning noise and cause computational difficulties.

Table 2 Extracted Features Candidates

Single stream	Statistics	$\mathrm{mean}(w_t^i), \mathrm{var}(w_t^i), \mathrm{range}(w_t^i)$ $\mathrm{median}(w_t^i), \mathrm{entropy}(w_t^i), \mathrm{hist}(w_t^i)$
	Difference	$u_t^i = \mathrm{Diff}(x_t^i);$ statistics
	Transformation	$\mathrm{fft}(w_t^i), \mathrm{wavelet}(w_t^i)$
Interstream	Deviation	$x^i - x^j \;\; \forall i, \forall j \in \mathcal{N}(i)$
	Correlation	$\mathrm{corr}(x^i, x^j) \;\; \forall i, \forall j \in \mathcal{N}(i)$

More importantly, for a particular event, or type of events, in practice only subsets of the calculated features are relevant, as it is mentioned earlier when those feature extraction techniques are proposed. After all, it is always beneficial to find out the "fingerprint" of each types of the event, not only for algorithmic concern, but also for system diagnosis purposes.

Here mRmR is used for FS. The procedure uses mutual information as the metric of goodness of a candidate feature set, and resolve the trade-off between relevancy and redundancy. To be specific, let $I(X;Y)$ be the mutual information between random variable X and Y, the first part of FS objective is to maximize the average dependence of selected feature set S on the target label c, i.e.

$$\max_{S \in \mathcal{X}} D(S) \triangleq \frac{1}{|S|} \sum_{x_i \in S} I(x_i; c) \tag{27}$$

where we have denoted S as the set variable for a collection of features, and $\mathcal{X} = \{x^1, \cdot, x^d\}$ as the set of all candidate features. Considering that features selected only according to Max-Relevance criterion could have rich redundancy, a second objective, a penalty on average first-order redundancy is introduced

$$\min_{S \in \mathcal{X}} R(S) \triangleq \frac{1}{|S|^2} \sum_{x_i, x_j \in S} I(x_i, x_j) \tag{28}$$

Combining the earlier two consideration yields the mRmR FS objective

$$\max_{S \in \mathcal{X}} \{D(S) - R(S)\} \tag{29}$$

which is approximately solved with a greedy heuristic: suppose we already have S_{m-1}, the feature set with $m - 1$ features, then the next feature is found by optimizing the following one variable problem:

$$\max_{x_j \in \mathcal{X} \setminus S_{m-1}} \left[I(x_j; c) - \frac{1}{m-1} \sum_{x_i \in S_{m-1}} I(x_j; x_i) \right] \tag{30}$$

For a theoretical analysis of the greedy algorithm, please refer to [32]. The mRmR implementation is available online for C++, R, and MATLAB, with a multilayer discretization technique for mutual information estimation. For each event, the selection method can be performed to choose the most informative feature set for event classification or detection. The top four selected features for high impedance fault are shown in Fig. 7.

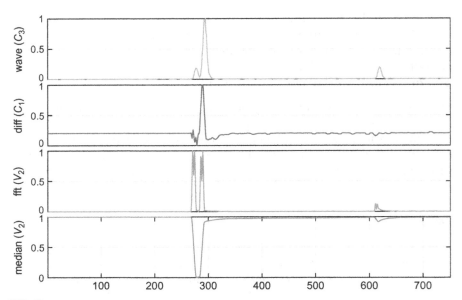

FIG. 7
Selected features for high impedance fault.

5   OTHER PROMISING RESEARCH DIRECTIONS

Leverage Unlabeled Data With Semisupervised Learning

The presence of both labeled and unlabeled data motivates the so-called semi-supervised learning [33]. The hope is that, by combining both types of available data sets, semisupervised learning could find better models and reduce the cost of expert engagement [34]. In the context of power system applications, data with detailed labels are precious but scant—power system experts are needed to inspect the measurement and provide insights. On the other hand, partial information may be obtained less costly by using unsupervised learning methods. Moreover, unlabeled data can be acquired in large quantity simply by collecting sensor measurement. Therefore, one possible (and may be necessary) direction for big data analytic in power systems is the development of specialized semisupervised learning methods for various tasks like prediction, state estimation, load forecast, event detection, fault localization, etc. Some recent work (e.g., [30]) has shown promising results in this direction.

Knowledge Transfer With Multitask Learning

The learning tasks in power system applications are often related with each other. For a simple example, the load forecast problem at one particular node of the distribution network bears much similarity to the forecast problems at other locations. This is because they are all driven by similar user behavior, weather condition, etc. Recent development of multitask learning [35–37] and transfer learning [38] allows the sharing of information and knowledge from one or several learning tasks to related tasks. The advantages are many folds. For instance, it might be the case that the training data for some learning task are not sufficient, and by information sharing among similar tasks it has an effect to "increase the training data size." In addition, recent multitask learning tools are able to automatically identify the dependence structure among tasks. This provides interesting "structure" information for different processes that are coevolving in power systems. Last but not least, the multitask learning setup can help the learning of "system or network level" events that would otherwise be unobservable with single-task learning methods. See Chapter 3 of [39] for an example that builds multitask time series model for distribution network event detection.

Information Fusion With Multiview Learning

The problem of combining information from different sources has been addressed by diverse control and system engineering literature, usually under the framework of Kalman filtering or its variants like particle filtering. From an ML perspective, those fusion methods amount to assuming a model with additive noise and then including side information using Bayes' rule. Yet another growing subarea of ML, namely multiview learning, is able to explore the consistency and complementary properties of different information sources or feature sets, and produce more effective and more promising models with better generalization ability than traditional single-view learning models [40–42]. The techniques used for multiview learning include cotraining, multiple kernel learning, and subspace learning, and all of them have well-established theoretical guarantees and algorithmic implementations. For power system applications, multiview learning seems promising to combine time domain system dynamics, spectral analysis, topological information, as well as external covariates such as weather, transportation, user mobility, etc., in an unified ML framework for a more general and automatic information fusion.

Other Promising Directions

Among many others, the following advancements in ML can be exploited for power system applications:

- Deep learning for end-to-end classification/regression and automatic information representation (see Chapter 7).
- Compress sensing for load/resource prediction and system structure mining (see Chapter 8).

- Manifold learning and interaction learning to model high-dimensional power system state.
- Tensor analysis and modeling for the data generated from multichannel sensors equipped in smart grid.

References

[1] A. Von Meier, D. Culler, A. McEachern, R. Arghandeh, Micro-synchrophasors for distribution systems, in: Innovative Smart Grid Technologies Conference (ISGT), IEEE, 2014, pp. 1–5.

[2] R. Arghandeh, M. Gahr, A. von Meier, G. Cavraro, M. Ruh, G. Andersson, Topology detection in microgrids with micro-synchrophasors, in: Power & Energy Society General Meeting, IEEE, 2015.

[3] M. Wen, R. Arghandeh, A. von Meier, K. Poolla, V. Li, Phase identification in distribution networks with micro-synchrophasors, in: Power & Energy Society General Meeting, IEEE, 2015.

[4] Y. Zhou, R. Arghandeh, C.J. Spanos, Partial knowledge data-driven event detection for power distribution networks, IEEE Trans. Smart Grid (2017).

[5] L. Schenato, G. Barchi, D. Macii, R. Arghandeh, K. Poolla, A. Von Meier, Bayesian linear state estimation using smart meters and PMUS measurements in distribution grids, in: International Conference on Smart Grid Communications (SmartGridComm), IEEE, 2014, pp. 572–577.

[6] M.I. Jordan, T.M. Mitchell, Machine learning: trends, perspectives, and prospects, Science 349 (6245) (2015), 255–260.

[7] V.N. Vapnik, V. Vapnik, Statistical Learning Theory, vol. 1 Wiley, New York, 1998.

[8] P.L. Bartlett, S. Mendelson, Rademacher and Gaussian complexities: risk bounds and structural results, J. Mach. Learn. Res. 3 (2002) 463–482.

[9] B.E. Boser, I.M. Guyon, V.N. Vapnik, A training algorithm for optimal margin classifiers, in: Proceedings of the Fifth Annual Workshop on Computational Learning Theory, ACM, 1992, pp. 144–152.

[10] J. Shawe-Taylor, N. Cristianini, Kernel Methods for Pattern Analysis, Cambridge University Press, Cambridge, 2004.

[11] F. Harrell, Regression Modeling Strategies: With Applications to Linear Models, Logistic and Ordinal Regression, and Survival Analysis, Springer, Dordrecht, 2015.

[12] K.P. Murphy, Machine Learning: A Probabilistic Perspective, MIT Press, Cambridge, MA, 2012.

[13] T. Hastie, R. Tibshirani, J. Friedman, Unsupervised learning, in: The Elements of Statistical Learning, Springer, New York, 2009, pp. 485–585.

[14] Y. Zhou, R. Arghandeh, I. Konstantakopoulos, S. Abdullah, A. von Meier, C.J. Spanos, Abnormal event detection with high resolution micro-PMU data, in: Power Systems Computation Conference (PSCC), IEEE, 2016, pp. 1–7.

[15] M.J. Kearns, U.V. Vazirani, An Introduction to Computational Learning Theory, MIT Press, Cambridge, MA, 1994.

[16] Y. Zhou, N. Hu, C.J. Spanos, Veto-consensus multiple kernel learning, in: Thirtieth AAAI Conference on Artificial Intelligence, 2016.

[17] Y. Zhou, J.Y. Baek, D. Li, C.J. Spanos, Optimal training and efficient model selection for parameterized large margin learning, in: Pacific-Asia Conference on Knowledge Discovery and Data Mining, Springer, 2016, pp. 52–64.

[18] G. Malkomes, C. Schaff, R. Garnett, Bayesian optimization for automated model selection, in: Adv. Neural Inf. Process. Syst., 2016, pp. 2892–2900.

[19] E. Brochu, V.M. Cora, N. De Freitas, A tutorial on Bayesian optimization of expensive cost functions, with application to active user modeling and hierarchical reinforcement learning. 2010 (arXiv preprint arXiv:1012.2599).

[20] G. Chandrashekar, F. Sahin, A survey on feature selection methods, Comput. Electr. Eng. 40 (1) (2014) 16–28.

[21] J. Bedo, C. Sanderson, A. Kowalczyk, An efficient alternative to SVM based recursive feature elimination with applications in natural language processing and bioinformatics, in: Australasian Joint Conference on Artificial Intelligence, Springer, 2006, pp. 170–180.

[22] G.K. Smyth, et al., Linear models and empirical Bayes methods for assessing differential expression in microarray experiments, Stat. Appl. Genet. Mol. Biol. 3 (1) (2004) 3.

[23] H. Peng, F. Long, C. Ding, Feature selection based on mutual information criteria of max-dependency, max-relevance, and min-redundancy, IEEE Trans. Pattern Anal. Mach. Intell. 27 (8) (2005) 1226–1238.

[24] D. Li, Y. Zhou, G. Hu, C.J. Spanos, Optimal sensor configuration and feature selection for AHU fault detection and diagnosis, in: IEEE Trans. Ind. Inf., 2016.

[25] M. Masaeli, J.G. Dy, G.M. Fung, From transformation-based dimensionality reduction to feature selection, in: Proceedings of the 27th International Conference on Machine Learning (ICML-10), 2010, pp. 751–758.

[26] F. Li, Y. Yang, E. Xing, From Lasso regression to feature vector machine, in: NIPS, 2005, pp. 779–786.

[27] L. Song, A. Smola, A. Gretton, J. Bedo, K. Borgwardt, Feature selection via dependence maximization, J. Mach. Learn. Res. 13 (2012) 1393–1434.

[28] I. Rodriguez-Lujan, R. Huerta, C. Elkan, C.S. Cruz, Quadratic programming feature selection, J. Mach. Learn. Res. 11 (2010) 1491–1516.

[29] P. Ravikumar, H. Liu, J. Lafferty, L. Wasserman, Spam: sparse additive models, in: Proceedings of the 20th International Conference on Neural Information Processing Systems, Curran Associates Inc., 2007, pp. 1201–1208.

[30] Y. Zhou, R. Arghandeh, I. Konstantakopoulos, S. Abdullah, C.J. Spanos, Data-driven event detection with partial knowledge: a hidden structure semi-supervised learning method, in: American Control Conference (ACC), IEEE, 2016, pp. 5962–5968.

[31] Y. Zhou, Z. Kang, L. Zhang, C. Spanos, Causal analysis for non-stationary time series in sensor-rich smart buildings, in: 2013 IEEE International Conference on Automation Science and Engineering (CASE), IEEE, 2013, pp. 593–598.

[32] Y. Zhou, C.J. Spanos, Causal meets submodular: subset selection with directed information, in: Adv. Neural Inf. Process. Syst., 2016, pp. 2649–2657.

[33] O. Chapelle, B. Scholkopf, A. Zien, Semi-supervised learning (Chapelle, O. et al., eds.; 2006) [Book Reviews], IEEE Trans. Neural Netw. 20 (3) (2009) 542.

[34] X. Zhu, A.B. Goldberg, Introduction to semi-supervised learning, in: Synthesis Lectures on Artificial Intelligence and Machine Learning, vol. 3(1), Morgan & Claypool Publishers, 2009, pp. 1–130.

[35] T. Evgeniou, M. Pontil, Regularized multi-task learning, in: Proceedings of the Tenth ACM SIGKDD International Conference on Knowledge Discovery and Data Mining, ACM, 2004, pp. 109–117.

[36] L. Jacob, J.P. Vert, F.R. Bach, Clustered multi-task learning: a convex formulation, in: Adv. Neural Inf. Process. Syst., 2009, pp. 745–752.

[37] A. Kumar, H. Daume III, Learning task grouping and overlap in multi-task learning, (2012) (arXiv preprint arXiv:1206.6417).

[38] S.J. Pan, Q. Yang, A survey on transfer learning, IEEE Trans. Knowl. Data Eng. 22 (10) (2010) 1345–1359.

[39] Y. Zhou, Statistical Learning for Sparse Sensing and Agile Operation (Ph.D. thesis), EECS Department, University of California, Berkeley, 2017, Available from: http://www2.eecs. berkeley.edu/Pubs/TechRpts/2017/EECS-2017-39.html (Accessed 12 May 2017).

[40] C. Xu, D. Tao, C. Xu, A survey on multi-view learning, 2013 (arXiv preprint arXiv:1304.5634).

[41] S. Sun, A survey of multi-view machine learning, Neural Comput. Applic. 23 (7–8) (2013) 2031–2038.

[42] Z. Zhang, Z. Zhai, L. Li, Uniform projection for multi-view learning, in: IEEE Trans. Pattern Anal. Mach. Intell., 2016.

CHAPTER 6

Unsupervised Learning Methods for Power System Data Analysis

Thierry Zufferey*, Andreas Ulbig*,†, Stephan Koch*,†, Gabriela Hug*

*ETH Zurich, Power Systems Laboratory, Zurich, Switzerland, †Adaptricity AG, c/o ETH Zurich, Power Systems Laboratory, Zurich, Switzerland

ABSTRACT

This chapter focuses on the use of the K-Means clustering algorithm for an enhanced visibility of the electrical distribution system which can be provided by advanced metering infrastructure and supported by big data technologies and parallel cloud computing environments such as Spark and H2O. Based on smart meter data of more than 30,000 loads in the City of Basel, Switzerland, and thanks to an appropriate cluster analysis, it is shown that useful knowledge of the grid state can be gained without any further information concerning the type of consumer and their habits. Once energy data is judiciously prepared, the features extraction is an important step. A graphical user interface is presented which illustrates the potentially great flexibility in the choice of features according to the needs of distribution system operators (DSOs). For example, the distribution of the various types of customers across the power system is of interest to DSOs. This chapter presents thus some pertinent examples of clustering outcomes that are visualized on the map of Basel, which notably enables to easily identify heating and cooling demand or gain insight into the energy consumption throughout the day for different neighborhoods.

1 INTRODUCTION

For the last few years, we have observed a rapid rollout of new sensor elements, e.g., smart meters, in the electrical distribution grid. This enables accurate high-resolution measurements on both the spatial scale (on a household level) and the temporal scale (between 1 min and 1 h) for parts of the distribution grid for which previously only spatially aggregated measurements on the substation and transformer level have been available. At first glance, the main motivation of distribution system operators (DSOs) to install electricity meters is the efficient integration of billing data into the existing billing systems by avoiding manual data gathering. It also facilitates the tracking in case of customers moving to a different property or changing their electricity supplier. This digitalization of electricity consumers is additionally an excellent opportunity for a better operation and planning of active distribution grids. Indeed, DSOs used to monitor load flows on a medium-voltage level for an ensemble of consumers whereas the low-voltage grid was considered as a black box, but the

Big Data Application in Power Systems. https://doi.org/10.1016/B978-0-12-811968-6.00006-1

development of smart meters that gather energy data in single households changes the situation. According to their habits, single consumers or a couple of consumers can have a significant influence on the state of the distribution system, especially on the voltage, which might induce unexpected stresses in the nearby grid components. Instead of overdimensioning the grid infrastructure, DSOs have now the possibility to rely on high-resolution measurements for an enhanced visibility of their system state. This is all the more important given the increasing share of renewable energy feed-in in the distribution system, which imposes further grid operation challenges and would require the active participation of all entities connected to the grid, including consumers.

In this chapter, a comprehensive data-driven clustering approach is presented, going from a proper data preparation to the visualization of results in an intuitive way. The analysis is part of the project *Optimized Distribution Grid Operation by Utilization of Smart Metering Data* funded by the Swiss Commission for Technology and Innovation (CTI [1]) and carried out at the ETH Zurich. This is based on smart meter data that has been gathered by "Industrielle Werke Basel" (IWB [2]), the public utility of the City of Basel, and processed by the ETH spin-off company Adaptricity [3] that develops simulation and optimization software tools for adapting electric distribution grids for the transition toward renewable energies. Notice that, in this case study, data is pseudonymized for privacy reasons such that no additional information on the type or habits of individual consumers is available, except the load profiles. Nevertheless, the distribution grid of the City of Basel is equipped with so-called data concentrators (DCs) that collect measurements from a manageable number of smart meters within a defined neighborhood before forwarding the data to the central server of IWB. The address of each DC as well as the assignment of smart meters to their DC is known, which gives an indication of the approximate location of each consumer inside the city's neighborhoods.

The remainder of this chapter is organized as follows: Section 2 describes more precisely the smart meter data used in this case study and related preprocessing tasks to obtain a tidy dataset. Section 3 provides the necessary theoretical basis concerning K-Means, one of the most widely used clustering algorithm. Section 4 details an approach to provide DSOs with useful knowledge based on the clustering of smart meter data. Section 5 finally summarizes the ideas developed in this chapter and enhances the benefits of unsupervised learning methods for power system data analysis.

2   SMART METER DATA PREPARATION

Good quality data is a necessary condition to obtain meaningful outcomes from a learning algorithm in terms of accuracy and interpretability. Generally, two facets of data preprocessing can be distinguished. On the one hand, real measurements are usually gathered from different sources and contain inconsistencies, noise, or data gaps. After consistent data integration, suitable preparation techniques such as an anomaly detection, data cleaning and missing values imputation allow to enhance the data quality. On the other hand, learning methods that give actual value to measurements generally do not process the entire dataset but work on specific features that need to be selected from the dataset. A detailed features extraction process for clustering purposes is presented in Section 4.

IWB is currently deploying the largest smart meter infrastructure in all of the German speaking countries with more than 50,000 equipped customers so far, where the active energy consumption is recorded every 15 min. The dataset used for analysis spans from April 2014 to September 2016. As mentioned above, it is important to tidy raw data by properly dealing with anomalies and missing values. Corrupt measurement devices must be discarded while preserving consumers with an uncommon but still correct profile. It also makes sense to work only with load profiles that can reasonably capture the habits of the consumer and that are relevant for the subsequent analysis. Therefore, from all equipped customers, only the ones whose electricity consumption has been recorded for at least 1 year and is greater than a threshold of 100 kWh/year remain in the dataset. In addition, energy time-series often contain missing values, due to both sporadic failures of individual meters and connection issues for a group of devices. Smart meters showing more than 10% of missing values or exhibiting data gaps larger than 2 weeks are simply discarded whereas the remaining ones are subject to a missing values imputation. Data gaps smaller than 1 h are filled by linear regression. Bigger data gaps are imputed by the average of a couple of values at the same instant and on the same weekday in the surrounding weeks. To complete this anomaly detection and data cleaning process, load profiles with more than 30% of zero values are removed from the dataset since a large amount of zero values in a consumption profile is a typical characteristic of corrupt meters. Consequently, the final dataset consists of more than 30,000 clean time-series (15 min sampling period) with a duration going from 12 to 30 months (i.e., 35,040–87,744 values per time-series). This overall dataset is about 16 GB large, which is considered as a Big Data problem in the context of power systems. Nevertheless, features extracted for clustering purposes represent a considerably smaller amount of data (a few MB). This is relatively small in comparison with the necessary data for forecasting that can be dozens of times larger than the original dataset.

This data preparation phase and the creation of following charts are based on the software tool R [4] due to the ease of programming and its wide variety of statistical and graphical methods. Although the processing of large dataset can be time-intensive, its compatibility with most databases as well as parallel processing tools like APACHE SPARK [5] makes R a comprehensive and powerful platform in the field of big data analysis.

2.1   Statistical Analysis

At this point, a statistical analysis of the dataset can already give some insight into the variety of metered consumers. Figs. 1 and 2 represent their distribution according to the mean energy consumption and to the correlation with the outside temperature. Notice that both histograms exhibit a Gaussian shape when the logarithm of energy is considered.

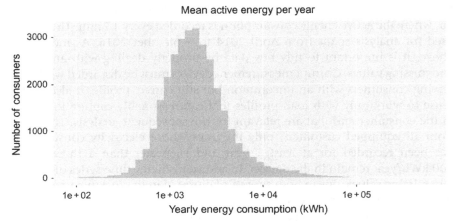

FIG. 1

Histogram of the active load consumption per year.

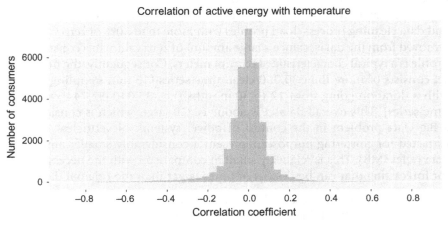

FIG. 2

Histogram of the correlation coefficient of load profiles with the outside temperature.

First, although most of the loads are residential customers (e.g., households) and consume a few MWh per year (mean of 3 MWh/year), some commercial and industrial customers whose yearly consumption can reach several hundred MWh are also part of the dataset. Special attention should be paid to them, due to their significant contribution to changes in the grid state. Second, one can observe a considerable diversity concerning their dependence to the temperature even if the majority is barely influenced by the meteorological conditions. Positively correlated load profiles mainly characterize customers which are generally active in the daytime or in summer. Conversely, negatively correlated profiles suggest the presence of electrical heating appliances. DSOs could then benefit from this knowledge on a local scale and better anticipate the demand of temperature-sensitive customers.

These two examples illustrate how a simple statistical analysis can help to provide a first evaluation of the characteristic features of consumers. However, it is somewhat difficult for DSOs to directly enhance the grid operation based on simple statistics which need to be properly interpreted. This chapter addresses therefore machine learning methods such as clustering techniques that can give a deeper understanding of the different types of electricity consumers in an intuitive way. Especially, these methods focus on the marginal customers that might not be perceptible on an aggregate level but still play a role locally in the distribution grid.

3   CLUSTERING ALGORITHM

An unsupervised learning method is an algorithm that is trained to discover some hidden structure in a certain dataset. Unlike supervised learning methods that classify instances or build a regression function, features in unsupervised learning are not associated with a label, i.e., the outcome cannot be compared with a supposedly good answer. Unsupervised learning includes various approaches such as Hidden Markov Models (HMM), dimensionality reduction, or anomaly detection. One of the most popular methods is nevertheless the K-Means clustering algorithm which is detailed in this section and applied to the smart meter data described above. Notice that most of the numerical computing tools (R, MATLAB) and machine learning environments (H_2O, WEKA) as well as libraries in Java or Python contain an implementation of the K-Means algorithm.

The objective of clustering is to group together similar instances, also called measurements or observations in the data mining literature. In this case, it enables to detect multiple types of consumers among the heterogeneous population. Initially, features need to be extracted from a tidy dataset to serve as benchmarks for the formation of clusters. The number of clusters K also has to be defined depending on the data diversity, on the number and type of features, and on the clustering purpose. A couple of clusters might be sufficient if the main types of loads are of interest. However, several dozen groups enable the capture of more subtle differences among the clusters and can reveal uncommon consumers. More detail about the appropriate number of clusters and the selection of suitable features for power system analysis is provided in the following section.

Fig. 3 illustrates how the K-Means clustering algorithm works to build three clusters based on the Fisher's Iris dataset. This famous dataset is often used in statistical classification and consists of the features of 150 flowers belonging to three iris species. Here, feature 1 corresponds to the sepal length and feature 2 is the sepal width, and both are normalized. In order to create three clusters, three points called "cluster centroids" first have to be initialized. For example, $K = 3$ examples in the training set can be arbitrarily picked as the first centroids. "K-Means++" is another algorithm for the selection of initial values, where only

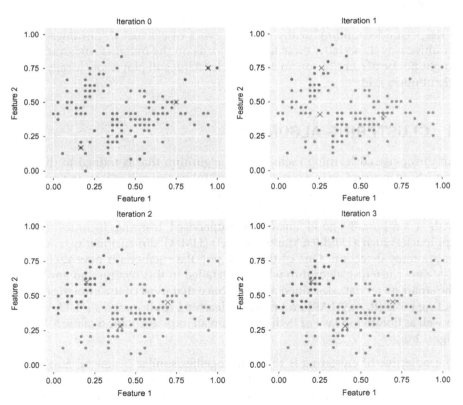

FIG. 3
Iterative process of the K-Means algorithm to build three clusters out of the Iris dataset.

one centroid is chosen totally randomly. Then, one computes the Euclidean norm between that point and remaining instances, which is used to define a weighted probability distribution from which the next centroid is picked randomly. This process is repeated until all centroids are chosen. The choice of these first centroids is crucial since it influences the convergence of the method and the final cluster formation, and K-Means++ usually outperforms the random initialization. After the initialization phase, K-Means is based on an iterative process with basically two tasks. First, a cluster assignment step goes through each of the examples and assigns them to the nearest centroid in terms of Euclidean distance. Second, a cluster update step moves the K cluster centroids to the average of all data points assigned to each of them. These two steps are then iterated until centroids stabilize, which implies that the algorithm has converged. Thus, three iterations of K-Means are necessary to cluster the Iris dataset.

4 CLUSTERING APPROACH AND VISUALIZATION

4.1 Features Extraction

According to the type of partitioning that is of interest among consumers, diverse features must be extracted from the smart meter profiles. It is particularly important to gather the right amount of information which is representative of the consumer habits without overwhelming the clustering algorithm with unnecessary data. Fig. 4 shows a MATLAB-based graphical user interface (GUI) that gives an overview of the wide variety on features which can be selected for the cluster analysis. Besides the usual statistical measures like the mean energy consumption, the standard deviation and the maximum energy value, some more advanced features require the extraction of multiple values. For example, a typical pattern consists of computing the average profile over a predefined period like a day or a week, and normalizing it. Based on a 15 min sampling period, this would result in the creation of 96 features for a typical daily pattern. Furthermore, before the installation of smart meters in households, the only information available to DSOs concerning their customers was the total consumption per billing period. Currently, one can easily get insight into daily, weekly, or seasonal fluctuations for each single metered load. For example, the energy requirements in a household, an office, or a shop during working days typically differ from the needs at the weekend. The peak demand can notably be inferred from high-resolution measurements with the aim of detecting the critical time periods for individual components in the distribution grid. Additionally, based on the autocorrelation coefficient with a lag of 1 day or 1 week, customers with an irregular stochastic load demand and therefore probably harder to predict are more easily identifiable thanks to the clustering process. Finally, the effect of exogenous variables such

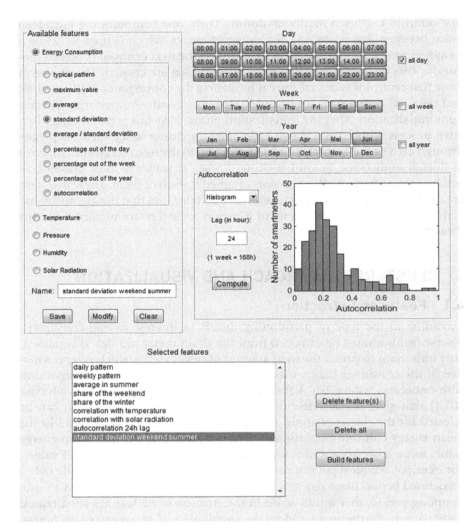

FIG. 4
User interface for the selection of clustering features.

as meteorological data on the electricity consumption (clearly visible during extreme winter days when electricity price peaks are reached) can be inferred. The ability of a clustering algorithm to automatically point out weather-sensitive loads can contribute to the decision support tools for the grid operation. Of course, all these features can be combined, which gives an additional value in comparison to raw statistics when they are processed by clustering. Notice also that extracting features from the entire measurement period may not be necessary. A good analysis tool should give the possibility to focus on specific hours or specific days that are more problematic from the point of view of the DSO and quickly identify the customers who have the largest impact on grid operation, e.g., who may lead to voltage band violations or the overloading of lines and transformers.

4.2   Typical Daily Patterns

In this section, two different clustering outcomes are presented and discussed based on their load profiles in a typical day. For the sake of clarity, the average profile in each cluster is indicated by a black curve. The first example, illustrated in Fig. 5, relies on normalized typical daily patterns. Considering a 15 min granularity, 96 features for each consumer have been extracted based on the entire measurement duration and supplied to the clustering algorithm. Notice that features directly correspond to the plotted data points in this particular case, providing relatively homogeneous profiles in each category. The black profile represents precisely the cluster centroids. As a rule of thumb, the number of clusters has been fixed to 25 in order to better detect unusual loads while having a sufficient amount of consumers per cluster. However, only the six most distinctive clusters are shown and analyzed in the following.

According to the significant peaks during lunchtime and in the evening (i.e., outside regular working hours), it is probable that Cluster 4 mainly consists of restaurants, cafeterias, and maybe a few households. In contrast, customers from Cluster 9 have a relatively constant power consumption, typically like 24 h active industrial loads. Nevertheless, it must be kept in mind that profiles are an average over a large number of days and do not necessarily reflect the behavior on a single day. However, they can still give insight into the profile of an aggregation of multiple similar loads on a daily basis. For example, typical daily profiles of electrical heating systems can appear fairly constant due to the averaging effect. In Cluster 10, energy is mainly consumed between 8 a.m. and 8 p.m., which is indicative of the needs of a shop or a department store where variations in load during opening hours can depend, to a limited degree, on the number of visitors. Cluster 19 contains the largest number of consumers in the presented subset and likely consists of households as illustrated by their mean profile. Indeed, a slightly higher demand appears around 6 a.m. and at noon, indicating a more energy-intensive activity like cooking. Nevertheless, most of the demand occurs after working hours since inhabitants probably come back from work, use cooking devices, turn on the lights, and watch television. Notice that a majority of clusters that are not shown here also exhibit this type of profiles, which suggests that most smart meters are installed at residential customer sites. Furthermore, offices seem to be assembled in Cluster 23, considering the characteristic drop at lunchtime and a very low consumption outside of regular business hours. Finally, Cluster 24 mostly consists of ripple-controlled units

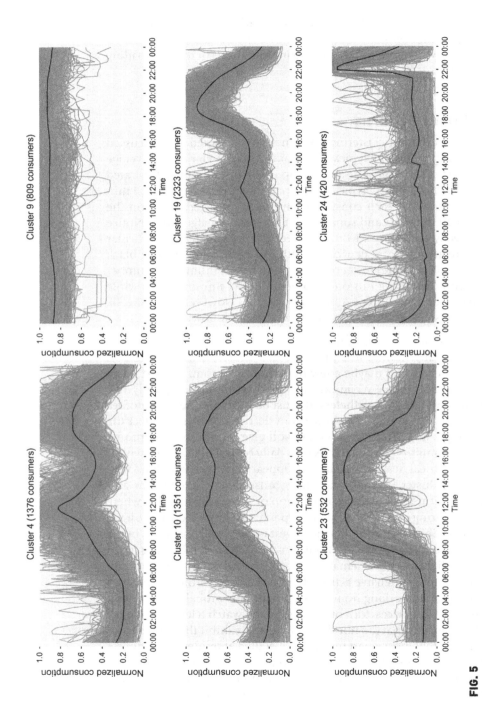

FIG. 5

Clustering of typical daily load profiles.

such as boilers that are programmed to start working based on the time of the day. This is particularly clear at 10 p.m. where the electricity price switches from high tariff to low tariff regime. Customer segmentation is one of the main applications of clustering with metered energy data and some DSOs make use of this information to implement dynamic pricing strategies with the aim of smoothening the demand by encouraging consumers to shift some of their activities to less critical hours. This cluster analysis can obviously be extended to an entire week, where a considerable energy drop will be visible at the weekend for some of the categories.

As shown in Fig. 6, a further interesting analysis is the evolution of the typical profile according to the mean energy consumption. In this case, there is only one feature and 15 clusters have been created. On average, small consumers exhibit a typical household pattern even if their individual profile can be of any shape. Nevertheless, the more energy is consumed, the more rectangular the mean load profile looks like while the usual household evening peak tends to vanish. This can certainly be explained by the higher share of commercial and industrial loads that are mainly active during regular working hours with a fairly constant energy demand. This is furthermore confirmed if one considers the weekly profile, where small consumers show a higher activity at the weekend, which gradually decreases with rising power consumption. As it has already been observed in Fig. 1, the size of clusters drastically decreases while the power consumption per customer increases. Finally, the last cluster presented here indicates that no specific trend can be expected for very large consumers. Again, this type of data mining can easily be combined with other measures or focus on a smaller set of customers, e.g., living in the same neighborhood. Notice, though, the added value provided by an unsupervised learning algorithm compared to a simple histogram although both analysis methods rely on the same metric. In this way, a DSO can potentially get a very quick intuition of a customer's behavior on the sole basis of its electricity bill, keeping in mind that it does not replace a more thorough and specific analysis.

4.3   Visualization Tool

It becomes clear that the application of clustering techniques to energy time-series provides interesting insights for the grid operator. However, it is also important to present this knowledge to the grid operators in an intuitive and useful way. Therefore, this section shows an interactive Leaflet [6] based visualization tool developed at Adaptricity [3]. In addition to standard statistical analysis and charts of typical patterns, clustering outcomes are displayed in the form of pie charts based on the smart meter location. This provides a good overview of the load variety across the grid under consideration. In the case of the City of Basel, the exact smart meter location is not available; however

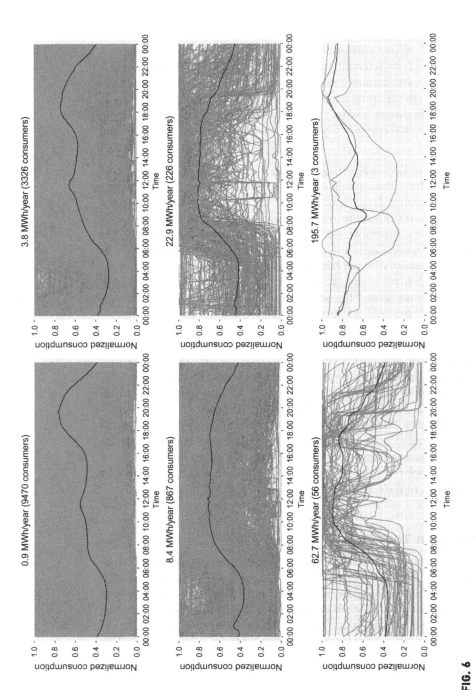

FIG. 6

Typical daily profiles of metered loads clustered according to their mean energy demand.

approximate DC-based addresses are sufficiently precise to obtain an accurate picture of consumers spread over the city.

According to the selected features, one can observe the penetration of a certain load category in each area of a distribution system where smart metering devices are installed. Let us start with the example shown in Fig. 7. As suggested by Silipo et al. [7] a day has been divided into five periods, i.e., early morning (7 a.m.–9 a.m.), morning (9 a.m.–1 p.m.), afternoon (1 p.m.– p.m.), evening (5 p.m.–9 p.m.), and night (9 p.m.–7 a.m.). For each time window, a feature represents the percentage of energy used, yielding five features that sum up to 100% for every load. Notice that this clustering process is similar to the

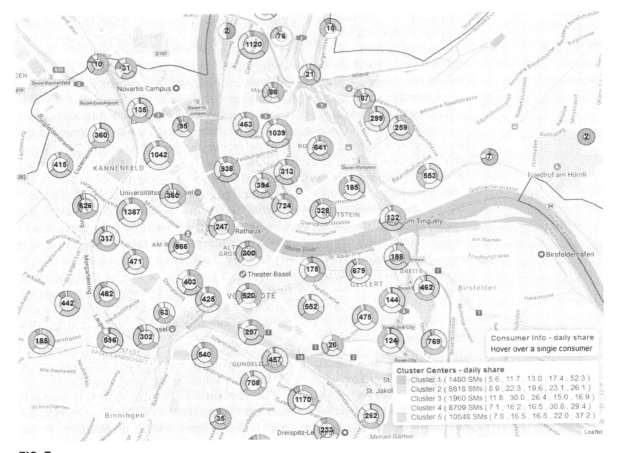

FIG. 7
Visualization of metered loads clustered according to their share of energy consumed at different representative periods of the day (early morning, morning, afternoon, evening, night).

typical daily load profile presented in the previous section with the exception that a much lower number of features are used for the clustering. Based on this, five clusters are created with K-Means. A small number of clusters are preferable for the sake of simplicity in a visualization tool. The box at the bottom right corner of the figure summarizes the cluster characteristics, including the number of related consumers and the cluster centroids (i.e., mean feature values). In this case, the smallest group is Cluster 1 (red) with less than 5% of metered consumers and consists of so-called night owls that mainly consume overnight. A large part of them are situated in areas with a majority of apartment buildings, like in the most easterly neighborhood of Basel, and their typical daily profile reveals that they are configured to be active especially from 10 p.m. on (off peak tariff). This gives a good indication of the buildings equipped with electric boilers, which implies higher power flows at night in these areas. Since these loads appear to be particularly price sensitive, they would be most likely to participate if the DSO introduces demand side participation. Conversely, consumers of Cluster 3 (yellow) clearly have their highest activity during business hours and are mainly concentrated in the old city center where shops, museums, offices, and restaurants are located, which accounts for 6% of the total number of customers. A high share of these loads also characterizes a shopping mall in the north of Basel. The other three clusters encompass the vast majority of the IWB customers and are well represented across the entire distribution grid. Cluster 2 (blue) is rather active between 9 a.m. and 7 p.m. and might include offices, restaurants, and a few households. Clusters 4 (green) and 5 (orange) seem to contain a great number of residential loads and probably restaurants since the corresponding consumers are characterized by a relatively high activity in the evening. To sum up, the tool facilitates the visualization of the energy requirements in the different parts of the grid and at different periods of a typical day. This concept can then be adapted to longer periods like a whole year instead of a day in order to compare the various levels of consumption according to the seasons. In addition, a similar analysis where only the data over a specific period (e.g., working days or winter) is considered during the features extraction process provides complementary information and should be more representative of the system state during this specific period.

As previously mentioned, it is important for a system operator to monitor temperature-sensitive consumers and this visualization tool enables the easy identification of such consumers if the correlation of load profiles with the temperature profile is used as a clustering feature. Fig. 8 exhibits the exact same loads but clustered according to their dependency on the temperature. While the statistical analysis presented in Section 2.1 shows a quasi-perfect normal distribution around zero, K-Means could build distinct meaningful groups based on this unique feature. Notice that in Switzerland, households are

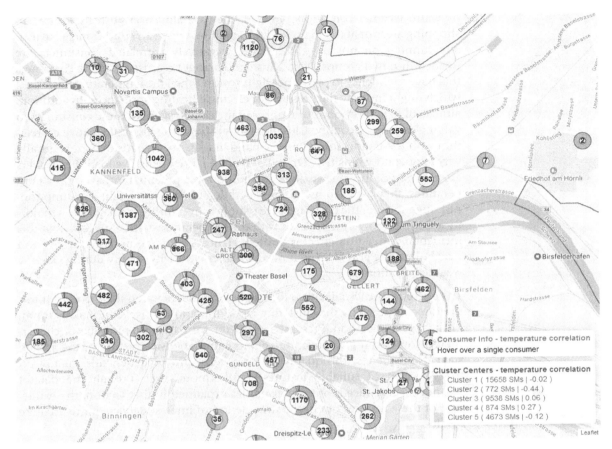

FIG. 8

Visualization of metered loads clustered according to their correlation with the outside temperature.

usually not equipped with air conditioners and electrical heating systems are less common than gas-fired ones. Hence, half of the consumers belong to the same cluster (red) and are barely influenced by the temperature. Most probably, they are not equipped with an electrical heating system. Cluster 3 (yellow) is the second largest group with 30% of all customers and the load profile of its members is slightly correlated with the temperature. However, this can be explained to a large extent from the fact that they consume mainly during the daytime, i.e., when temperatures are naturally higher. Furthermore, although the number of customers in Cluster 4 (green) is relatively small, it is an interesting category since they have the largest positive correlation and they are situated in characteristic areas in the city. Even if these consumers have a higher demand in electricity during the daytime, they are definitely influenced by warm weather conditions. Indeed, they are almost exclusively located in shopping areas or at the main football stadium, where air conditioners are usually running on hot days. Moreover, negatively correlated consumers are divided into two groups. First, Cluster 5 (orange) consists of many customers that have a tendency to consume slightly more when the temperature decreases, which might indicate the presence of electrical heating appliances. In addition, almost all neighborhoods of Basel contain a small share of these consumers to varying degrees. Secondly, Cluster 2 (blue) shows the highest negative correlation with the temperature. On one hand, it consists of loads that are naturally very temperature sensitive. On the other hand, price-sensitive loads like boilers are also part of this cluster since, in addition to a potential impact of low temperatures, they get even more correlated due to their overnight consumption. Although most of the customers are mainly active during the daytime, a majority are still negatively correlated with the temperature, which suggests that this incidental correlation effect is limited. Hence, DSOs can notably gain a good insight into the grid areas that need a higher electricity supply during extreme weather conditions.

To conclude, the visualization tool makes the link between the pieces of information computed by the clustering algorithm and the decision makers for an optimal grid operation. Presented under the form of cluster pie charts, clustering outcomes can give an intuitive and comprehensive picture of the system state. Nevertheless, the quality of knowledge provided depends mainly on the selection of suitable features. Furthermore, even though the dataset has not been reduced for both examples explained in this section, the extraction of these features out of the time period of interest delivers more accurate results.

5   CONCLUSIONS

To summarize, a huge amount of data has been collected for the last few years in the low-voltage grid by smart meters, which allows to leverage additional benefits for the grid operation. Nevertheless, it is important to develop suitable methods that can deal with the quantity of data and turn it into knowledge which must be quickly and easily interpretable by DSOs and be able to support them in the grid operation. While the high spatial and temporal resolution provided by smart meters enables a previously unattainable degree of detail in state estimation and other grid functionalities, unsupervised learning techniques such as clustering add value by linking and putting into perspective the multiple consumers.

In this chapter, a complete clustering approach has been presented on the basis of large sets of measurement data gathered by IWB, the DSO of the City of Basel, and enhanced by the ETH spin-off Adaptricity. First of all, it is necessary to correctly prepare the available raw data in order not to process useless or unreliable information. This notably includes proper data integration, the removal of large anomalies, and the imputation of missing values. Secondly, the extraction of clustering features is the key element of a successful cluster analysis since they define the points of similarity between energy consumers in order to build clusters. They can consist of standard statistical metrics, a combination of these metrics, the dependence of load profiles on weather variables, or more advanced features like a typical profile. Thirdly, a number of clusters have to be chosen and extracted features based on all consumers are used to train the K-Means clustering algorithm. As soon as the clusters are computed, each metered consumer is assigned to one of them. Finally, these clustering outcomes can be visualized on a map of the city in combination with the postal address of consumers or of the DC, depending on anonymization requirements.

Apart from the data preparation phase, clustering is not a time-intensive method compared to other learning algorithms such as forecasting and adds a significant value to the measurement data. Especially, a cluster analysis is often of interest to detect the most uncommon loads. Their behavior with respect to certain characteristics can be considerably different from the consumers of the main clusters and, as it has been shown thanks to the visualization tool, they are usually located in specific areas of the city. Therefore, a good knowledge of these unusual loads can notably support the DSO to rapidly cope with critical states of the low-voltage grid. For example, a large concentration of temperature-sensitive consumers at a certain branch can heavily load the nearby grid components in case of extreme weather conditions. The extent of this issue is not necessarily visible on an aggregation level and can go undetected if no suitable tools for analyzing and visualizing smart meter data are available. Furthermore, the customer segmentation obtained by clustering can set the basis for the implementation of demand side participation and dynamic pricing. In this way, the distribution grid can get more flexible and, for example, better integrate renewable energy if the DSO were able to reasonably estimate where this surplus of energy can flow and be immediately absorbed.

Finally, smart metering is a recent source of measurement data in power systems and such a large amount of information on the end consumer has not previously been available. Hence, new challenges arise from the software perspective in order to get the most value out of it. More data implies more knowledge if, and only if, suitable methods and tools for data processing, analysis, and visualization are available. Although unsupervised learning techniques are usually combined with Big Data technologies such as parallel computing tools, the application to power system data analysis is a very new field of research and opens up great opportunities.

References

[1] Commission for Technology and Innovation, https://www.kti.admin.ch/kti/en/home.html.

[2] Industrielle Werke Basel, https://www.iwb.ch.

[3] Adaptricity AG, https://www.adaptricity.com.

[4] The R Project for Statistical Computing, https://www.r-project.org.

[5] Apache Spark, https://spark.apache.org.

[6] Leaflet, https://leafletjs.com.

[7] R. Silipo, P. Winters, Big Data, Smart Energy, and Predictive Analytics, 2013. KNIME, Technical Report.

CHAPTER 7

Deep Learning for Power System Data Analysis

Elena Mocanu, Phuong H. Nguyen, Madeleine Gibescu
Eindhoven University of Technology, Eindhoven, The Netherlands

CHAPTER OVERVIEW

Unprecedented high volumes of data are available in the smart grid context, facilitated by the growth of home energy management systems and advanced metering infrastructure. In order to automatically extract knowledge from, and take advantage of this useful information to improve grid operation, recently developed machine learning techniques can be used, in both supervised and unsupervised ways. The proposed chapter will focus on deep learning methods and will be structured as follows: Firstly, as a starting point with respect to the state of the art, the most known deep learning concepts, such as deep belief networks and high-order restricted Boltzmann machine (i.e., conditional restricted Boltzmann machine, factored conditional restricted Boltzmann machine, four-way conditional restricted Boltzmann machine), are presented. Both, their theoretical advantages and limitations are discussed, such as computational requirements, convergence, and stability. Consequently, two applications for building energy prediction using supervised and unsupervised deep learning methods will be presented. The chapter concludes with a glimpse into the future trends highlighting some open questions as well as new possible applications, which are expected to bring benefits toward better planning and operation of the smart grid, by helping customers to adopt energy conserving behaviors and their transition from a passive to an active role.

1 INTRODUCTION

1.1 From Neural Network Towards Deep Learning

The power to predict an uncertain event makes the science community to continuously search for more and more accurate methods. In an attempt to determine which approaches are the most popular, and to integrate deep learning methods in the existing literature, a short bibliometric analysis of the collections of publications related with the electricity prediction problem is performed by using specialized queries on Scopus database. With focus on energy prediction there are 6613 publications in the last decade, from which in 2015 there are 839 publications. On the one hand, in Fig. 1 is reported the distribution of this existing literature based on the type of publications (conferences, articles, reviews, and others). On the other hand, using specified queries we classified these papers into machine learning and nonmachine learning methods.

125

Big Data Application in Power Systems. https://doi.org/10.1016/B978-0-12-811968-6.00007-3

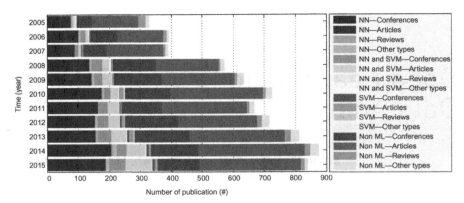

FIG. 1

Electricity prediction—a summary of the Scopus-indexed publications with focus on electricity prediction in the years 2005–15.

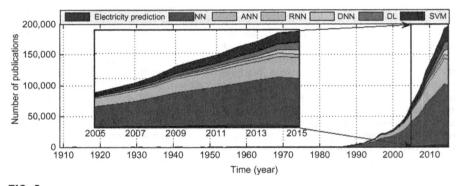

FIG. 2

Prediction—a summary of the Scopus-indexed publications with focus on prediction over the last century (i.e.,1909–2015), including a zoom over the years 2005–15.

Specifically, the overall count for Neural Networks is 2380 publications and 820 for Support Vector Machine, while a number of 703 publications are using both methods simultaneously. From a more general perspective, in Fig. 2 a short overview over the evolution of the machine learning methods applied to prediction is shown. A zoom over the last decade is added. This is a starting point with respect to the state of the art, and a glimpse into the future trends in prediction with a focus on electricity prediction.

Overall, perhaps the most investigated machine learning (ML) methods are based on neural networks (NNs) and their variations (e.g., artificial neural networks (ANNs), recurrent neural networks (RNNs), deep neural networks (DNNs), or deep belief networks (DBNs)). It is worth mentioning that up to 2016, the collection of publications indexed by Scopus and related with

NNs counts for more than one million. From this, we can observe that Deep Learning models (including DNNs) represent the most important trend in the last years.

1.2 Deep Learning Methods

Since its conception, deep learning [1] is widely studied and applied, from pure academic research to large-scale industrial applications, due to its success in different real-world machine learning problems such as audio recognition [2], reinforcement learning (RL) [3], transfer learning [4], and activity recognition [5]. Deep learning models are ANNs with multiple layers of hidden neurons, which have connections only among neurons belonging to consecutive layers, but have no connections within the same layer. In general, these models are composed by basic building blocks, such as restricted Boltzmann machines (RBMs) [6]. In turn, RBMs have proven to be successfully not just providing good initialization weights in deep architectures (in both supervised and unsupervised learning) but also as standalone models in other types of applications. Examples are density estimation to model human choice [7], collaborative filtering [8], information retrieval [9], or multiclass classification [10]. Thus, an important research direction is to improve the performance of RBMs on any component, e.g., computational time, generative and discriminative capabilities [11].

So, we start by putting in the context of the recent developments our proposed methods, and we argue that a unified topological investigation of this deep learning architecture based on the order of tensor factorization, as depicted in Fig. 3, could highlight two general directions of research. By increasing the number of hidden layers in the well-known ANNs, nowadays we are referring them as deep networks and it is the principal direction in deep learning (Fig. 3, vertical shadowed area). Nevertheless, based on the order of the multiplicative interactions (tensor connection) between various layers of RBMs and their derivatives a second direction using higher order tensor factorization arises. Therefore, the primary contribution of this chapter is to extend the red shadowed area, while in the remaining sections both areas are covered.

2 SUPERVISED ENERGY PREDICTION USING DEEP LEARNING

Commercial and industrial buildings represent a tremendous amount of the global energy used. A future energy ecosystem is emerging, which connects green buildings with a smart power grid to optimize energy flows between them. This requires prediction of energy consumption in a wide range of time horizons. It is important to predict not only at the aggregated level but to go deep into the individual building level so distributed generation resources can be deployed based on the local forecast. Decomposition of demand

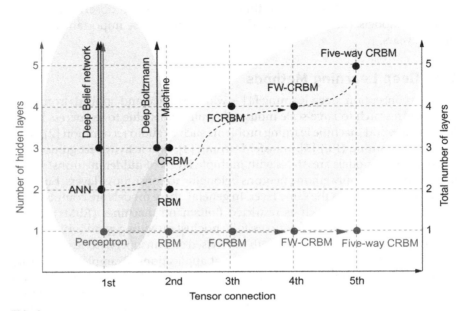

FIG. 3

Deep Learning—a unified schematic representation of high-order restricted Boltzmann machine architectures and their corresponding factorization.

forecasting helps analyze energy consumption patterns and identify the prime targets for energy conservation. Moreover, prediction of temporal energy consumption enables building managers to plan out the energy usage over time, shift energy usage to off-peak periods, and make more effective energy purchase plans.

The complexity of building energy behavior and the uncertainty of the influencing factors, such as more fluctuations in demand, make energy prediction a hard problem. These fluctuations are given by weather conditions, the building construction and thermal properties of the physical materials used, the occupants and their behavior, sublevel system components lighting or HVAC (heating, ventilating, and air conditioning). Many approaches have been proposed aiming at accurate and robust prediction of the energy consumption. In general, they can be divided into two types. The first type of models is based on physical principles to calculate thermal dynamics and energy behavior at the building level. Some of them include models of space systems, natural ventilation, air conditioning system, passive solar, photovoltaic systems, financial issue, occupants' behavior, climate environment, and so on. Overall, the numerous approaches depend on the type of building and the number of parameter used. The second type is based on statistical methods. These methods are used to predict building energy consumption by correlating energy consumption with influencing variables such as weather and energy cost. We refer to Krarti [12] and Dounis [13] for a more comprehensive discussion about building energy systems, and more recently reviews [14,15]. Moreover, to shape the evolution of future buildings systems there are also some hybrid approaches which combine some of the above models to optimize predictive performance, such as [16–18]. Actually, the most widely used machine learning methods for energy prediction are ANNs and Support Vector Machines [19]. Hidden Markov model (HMM) [20] is other popular stochastic model for time-series analyses. This model shows good results in different fields, from bio-informatics to stock market and it was not so much investigated in the context of building energy prediction [21].

This section focuses especially on deep learning methods for energy prediction, by the characterization of load profiles on measured data. Due to the fact that energy consumption can be seen as a time-series problem, we proposed the use of conditional restricted Boltzmann machine (CRBM) [8] and factored conditional restricted Boltzmann machines (FCRBMs), recently introduced stochastic machine learning methods which were used successfully until now to model highly nonlinear time series (e.g., human motion style, structured output prediction) [22–24]. As a secondary contribution, we adapt the FCRBM architecture for energy prediction problems by merging the style and feature labels into one, and by rewriting the equations and the derivatives of the learning rules according to the new configuration of the model.

2.1 Conditional Restricted Boltzmann Machine

CRBMs [22] are an extension over RBMs [6] used to model time-series data and human activities [25]. They are energy-based models for unsupervised learning. These models are probabilistic, with stochastic nodes and layers, which make them less vulnerable to local minima. Further, due to their multiple layers and their neural configurations, RBMs possess excellent generalization capabilities [1]. Formally, a RBM consists of visible and hidden binary layers. The visible layer represents the data, while the hidden increases the learning capacity by enlarging the class of distributions that can be represented to an arbitrary complexity [25]. In CRBMs models [22] the RBMs is extended by including a conditional history layer. The general architecture of this model is depicted in Fig. 4 and the total energy function is calculated considering all possible interactions between neurons and weights/biases, such as

$$E(v, h, u; W) = -v^T W^{vh} h - v^T b^v - u^T W^{uv} v - u^T W^{uh} h - h^T b^h \tag{1}$$

where $\mathbf{u} = [u_1, \ldots, u_{n_u}]$ represents a real valued vector with all history neurons, with n_u being the index of the last history neuron (input), $\mathbf{v} = [v_1, \ldots, v_{n_v}]$ is a

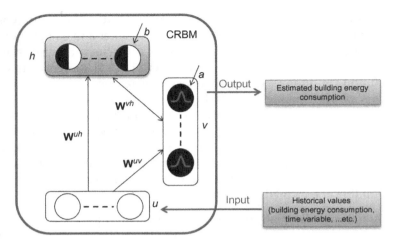

FIG. 4

The general architecture of conditional restricted Boltzmann machines, where *u* is the conditional history layer (input), *h* is the hidden layer, and *v* is the visible layer (output); the hidden layer has binary neurons and input layer represents the real values and the others are Gaussian values.

real valued vector collecting all visible units v_i, and n_v is the index of the last visible neuron (output), $\mathbf{h} = [h_1, \ldots, h_{n_h}]$ is a binary vector collecting all the hidden units h_j, with n_h being the index of the last hidden neuron. $\mathbf{W}^{vh} \in \mathbb{R}^{n_h \times n_v}$ represents the matrix of all weights connecting $\mathbf{v}$ and $\mathbf{h}$, $\mathbf{W}^{uv} \in \mathbb{R}^{n_u \times n_v}$ represents the matrix of all weights connecting $\mathbf{u}$ and $\mathbf{v}$, and $\mathbf{W}^{uh} \in \mathbb{R}^{n_u \times n_h}$ represents the matrix of all weights connecting $\mathbf{u}$ and $\mathbf{h}$. The biases for hidden neurons are given by $\mathbf{b}^h \in \mathbb{R}^{n_h}$ and the biases for visible neurons $\mathbf{b}^v \in \mathbb{R}^{n_v}$.

It is worth mentioning that in comparison with ANNs, the weights in CRBMs can be bidirectional. More exactly, $\mathbf{W}^{vh}$ is bidirectional. The other weight matrices $\mathbf{W}^{uv}$ and $\mathbf{W}^{uh}$ are unidirectional.

2.1.1   Inference in CRBM

In CRBMs probabilistic inference means determining two conditional distributions. The first is the probability of the hidden layer conditioned on all the other layers, i.e., $p(\mathbf{h}|\mathbf{v}, \mathbf{u})$ while the second is the probability of the present layer conditioned on the others, such as $p(\mathbf{v}|\mathbf{h}, \mathbf{u})$. Since there are no connections between the neurons in the same layer, inference can be done in parallel for each unit type, leading to

$$p(h = 1 | \mathbf{u}, \mathbf{v}) = \mathrm{sig}\left(\mathbf{u}^T \mathbf{W}^{uh} + \mathbf{v}^T \mathbf{W}^{vh} + \mathbf{b}^h\right) \qquad (2)$$

where $\mathrm{sig}(x) = 1/1 + \exp(-x)$, and

$$p(v|\mathbf{h}, \mathbf{u}) = \mathcal{N}\left(\mathbf{W}^{uv^T}\mathbf{u} + \mathbf{W}^{vh}\mathbf{h} + \mathbf{b}^v, \sigma^2\right) \tag{3}$$

where for convenience σ is chosen to be 1. Probability of the hidden neurons is given by a sigmoidal function evaluated on the total input to each hidden unit and probability of the visible neurons is given by a Gaussian distribution over the total input to each visible unit.

2.1.2 Learning for CRBM Using Contrastive Divergence

Parameters are fitted by maximizing the likelihood function. In order to maximize the likelihood of the model, the gradients of the energy function with respect to the weights have to be calculated. Because of the difficulty of computing the derivative of the log-likelihood gradients, Hinton [26] proposed an approximation method called contrastive divergence (CD). In maximum likelihood, the learning phase actually minimizes the Kullback-Leibler (KL) measure between the input data distribution and the approximate model. In CD, learning follows the gradient of:

$$CD_n \propto D_{KL}(p_0(\mathbf{x})||p_\infty(\mathbf{x})) - D_{KL}(p_n(\mathbf{x})||p_\infty(\mathbf{x}))) \tag{4}$$

where, $p_n(\cdot)$ is the distribution of a Markov chain running for n steps. The update rules for each of the weight matrices and biases can be computed by deriving the energy function with respect to each of these variables (i.e., the visible weights). Formally, this can be written as:

$$\frac{\partial E(v, h, u)}{\partial W^{uh}} = -uh^T; \frac{\partial E(v, h, u)}{\partial W^{uv}} = -uv^T; \frac{\partial E(v, h, u)}{\partial W^{vh}} = -vh^T \tag{5}$$

The update equation for the biases of each of the layers is

$$\frac{\partial E(v, h, u)}{\partial b^v} = -v \text{ and } \frac{\partial E(v, h, u)}{\partial b^h} = -h \tag{6}$$

Since the visible units are conditionally independent given the hidden units and vice versa, learning can be performed using one-step Gibbs sampling, which is carried in two half steps: (1) update all the hidden units, and (2) update all the visible units. Thus, in CD_n the weight updates are done as follows:

$$\mathbf{W}^{uh}_{\tau+1} = \mathbf{W}^{uh}_{\tau} + \alpha\left(\langle\mathbf{uh}^T\rangle_{data} - \langle\mathbf{uh}^T\rangle_{recon}\right) \tag{7}$$

$$\mathbf{W}^{uv}_{\tau+1} = \mathbf{W}^{uv}_{\tau} + \alpha\left(\langle\mathbf{uv}^T\rangle_{data} - \langle\mathbf{uv}^T\rangle_{recon}\right) \tag{8}$$

$$\mathbf{W}^{vh}_{\tau+1} = \mathbf{W}^{vh}_{\tau} + \alpha\left(\langle\mathbf{vh}^T\rangle_{data} - \langle\mathbf{vh}^T\rangle_{recon}\right) \tag{9}$$

and the biases updates are

$$\mathbf{b}_{\tau+1}^{v} = \mathbf{b}_{\tau}^{v} + \alpha\big(\langle\mathbf{v}\rangle_{\text{data}} - \langle\mathbf{v}\rangle_{\text{recon}}\big) \tag{10}$$

$$\mathbf{b}_{\tau+1}^{h} = \mathbf{b}_{\tau}^{h} + \alpha\big(\langle\mathbf{h}\rangle_{\text{data}} - \langle\mathbf{h}\rangle_{\text{recon}}\big) \tag{11}$$

where τ is the iteration and α is the learning rate.

2.2 Factored Conditional Restricted Boltzmann Machine

This section introduces FCRBMs, shown in Fig. 5. For this method an intuition describing the model as well as the configuration is discussed. Secondly, FCRBMs' mathematical details including the energy function, probabilistic inference, and learning/update rules are detailed. To formalize FCRBMs, three main ingredients are required. Firstly, an energy function providing scalar values for a given configuration of the network is essential. Secondly, probabilistic inference (i.e., the procedure in which conditionals are calculated) needs to be detailed. Finally, the update/learning rules required for fitting free parameters have to be derived.

Taylor et al. [25] introduced the FCRBMs where they add styles and the concept of factored, multiplicative, three-way interactions to predict multiple styles of human motion. Originally, FCRBMs consist of the previous three layers from CRBM and two new introduced layers for styles and features. However, to fit our needs we reduced the style and features layers to one and we used it to represent different parameters useful for prediction. More exactly, after the aforementioned reduction has been done, FCRBM consists of: (1) a real valued

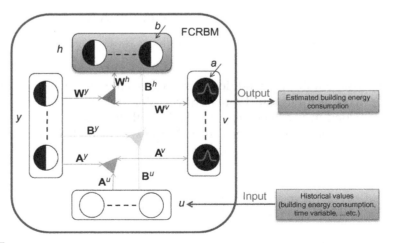

FIG. 5

The general architecture of factored conditional restricted Boltzmann machines, where u is the conditional history layer (input), h is the hidden layer, and v is the visible layer (output).

visible layer $\mathbf{v}$, (2) a real valued history layer $v_{<t}$ (i.e., $v_{<t} = v_{<t,t-N:t-1}$, where $N \in \mathbb{N}$, (3) a binary hidden layer $\mathbf{h}$, and (4) a style layer $\mathbf{y}$. Each of the above layers is essential for the success of FCRBMs. The visible layer encodes the current values of a time series which needs to be predicted. The history of the time sequence, being the basis of such predictions, is encoded on the history layer. The hidden layer guarantees the discovery of important features essential for the analysis of the time sequence, while the style layer encodes different parameters useful in the prediction. To learn the inherent relations between these layers, undirected or directed weights and factors, as shown in Fig. 5, are used as connections.

More formally, FCRBM defines a joint probability distribution over the visible $\mathbf{v}$, and hidden $\mathbf{h}$, neurons. The joint distribution is conditioned on the past N observations $\mathbf{v}_{<t}$, model parameters $\boldsymbol{\theta}$ (i.e., $\mathbf{W}^h$, $\mathbf{W}^v$, $\mathbf{W}^y$, $\mathbf{A}^{v<t}$, $\mathbf{A}^v$, $\mathbf{A}^y$, $\mathbf{B}^y$, $\mathbf{B}^v$, $\mathbf{B}^h$) and the style layer $\mathbf{y}$. Similar to CRBM, FCRBM assumes binary stochastic hidden units and real-valued visible units with additive, Gaussian noise. For notational ease, as in the original paper (Taylor and Hinton, 2011), we assume $\sigma_i = 1$.

2.2.1 Total Energy for FCRBM

The total energy function, $\mathbf{E}(\mathbf{v}_t, \mathbf{h}_t. | \mathbf{v}_{<t}, \mathbf{y}_t)$ for FCRBM, is computed as the sum of the first- and third-order energy terms as follows:

$$\mathbf{E} = \underbrace{\frac{1}{2} \sum_{i=1}^{n_1} (v_{i,t} - \hat{a}_{i,t})^2 - \sum_{j=1}^{n_2} \hat{b}_{j,t} h_{j,t}}_{E_I} \underbrace{- \sum_{f=1}^{F} \left[\sum_{i=1}^{n_1} W_{if}^v v_{i,t} \sum_{j=1}^{n_2} W_{jf}^h h_{j,t} \sum_{p=1}^{n_3} W_{pf}^y y_{p,t} \right]}_{E_{III}} \tag{12}$$

where E_{III} is defined as:

$$E_{III} = -\sum_{f=1}^{F} \left[\sum_{i=1}^{n_1} \left[\sum_{j=1}^{n_2} \left[\sum_{p=1}^{n_3} \left[W_{if}^v W_{jf}^h W_{pf}^y v_{i,t} h_{j,t} y_{p,t} \right] \right] \right] \right] \tag{13}$$

where F, n_1, n_2, and n_3 represent the total number of factors and the number of units in each of the visible, hidden, and label layers, respectively. The terms $\hat{a}_{i,t}$ and $\hat{b}_{j,t}$ are called dynamic biases, which are defined as:

$$\hat{a}_{i,t} = a_i + \sum_i A_{i,m}^v \sum_k A_{k,m}^{v<t} v_{k,<t} \sum_p A_{p,m}^y y_{p,t} \tag{14}$$

$$\hat{b}_{j,t} = b_j + \sum_n B_{j,n}^h \sum_k B_{k,n}^{v<t} v_{k,<t} \sum_p B_{p,n}^y y_{p,t} \tag{15}$$

with $A_{i,m}^v$, $A_{k,m}^{v<t}$, $A_{p,m}^y$, $B_{j,n}^h$, $B_{k,n}^{v<t}$, and $B_{p,n}^y$ as dynamic biases of each of the layers. These as well as the weights connections are free parameters that need to be trained as detailed further.

2.2.2 Inference in FCRBM

Inference in FCRBM is conducted in parallel, since there are no connections between the neurons in the same layer. Specifically, this means determining two conditional distributions. Firstly, the conditional probability distribution of the hidden neurons, $(h_{j,t} = 1 \mid v_t v_{<t} \gamma_t)$, is given by a sigmoidal function evaluated on the total input to each hidden unit, $h_{j,t}^* = \sum_f W_{if}^h \sum_i W_{if}^v v_{i,t} \sum_p W_{pf}^\gamma \gamma_{p,t}$, via the factors. Secondly, the probability of the visible neurons, $p(v_{i,t} = 1 \mid h_t v_{<t} \gamma_t)$, is given by a Gaussian distribution over the total input, $v_{i,j}^* = \sum_f W_{if}^v \sum_j W_{jf}^h h_{j,t} \sum_p W_{pf}^\gamma \gamma_{p,t}$, to each visible unit via the factors. Therefore, for each of the jth hidden and ith visible unit, inference is performed using:

$$p(h_{j,t} = 1 \mid v_t v_{<t} \gamma_t) = sigmoid\left(\hat{b}_{j,t} + h_{j,t}^*\right) \tag{16}$$

$$p(v_{i,t} = 1 \mid h_t v_{<t} \gamma_t) = \mathcal{N}\left(\hat{a}_{i,t} + v_{i,j}^*, \sigma_i^2\right) \tag{17}$$

where $\mathcal{N}\left(\mu, \sigma_i^2\right)$ denotes the Gaussian probability density function with mean μ and variance σ_i^2.

2.2.3 Learning and Update Rules for FCRBMs

The general update rule for all the hyperparameters $\boldsymbol{\theta}$ is given by

$$\boldsymbol{\theta}_{\tau+1} = \boldsymbol{\theta}\tau + \rho \Delta \boldsymbol{\theta}_\tau + \alpha(\Delta \boldsymbol{\theta}_{\tau+1} - \gamma \boldsymbol{\theta}_\tau) \tag{18}$$

where τ, ρ, α, and γ represent the update number, momentum, learning rate, and weights decay, respectively. More details regarding the choice of these parameters are described in Hinton [27]. The update rules for each of the weight matrices and biases can be computed by deriving the energy function from Eq. (12) with respect to each of these variables (i.e., the factored visible weights, factored label weights, factored hidden weights, and the biases of each of the layers), yielding:

Weights update: Three update rules corresponding to each of $\mathbf{W}^v$, $\mathbf{W}^h$, and $\mathbf{W}^\gamma$ need to be derived. Firstly, the factored visible weight W_{if}^v is computed by derivating the total energy function, provided in Eq. (12), with respect to W_{if}^v is

$$\frac{\partial \mathbf{E}\left(\mathbf{v}_t, \mathbf{h}_t \mid \mathbf{v}_{<t}, \mathbf{y}_t\right)}{\partial W_{if}^v} = -v_{i,t} \sum_{j=1}^{n_2} W_{jf}^h h_{j,t} \sum_{p=1}^{n_3} W_{pf}^\gamma \gamma_{p,t} \tag{19}$$

Secondly, the factored hidden weights W_{jf}^h are updated. Following the same reasoning we obtain:

$$\frac{\partial \mathbf{E}\left(\mathbf{v}_t, \mathbf{h}_t \mid \mathbf{v}_{<t}, \mathbf{y}_t\right)}{\partial W_{jf}^h} = -h_{j,t} \sum_{i=1}^{n_1} W_{if}^v v_{i,t} \sum_{p=1}^{n_3} W_{pf}^\gamma \gamma_{p,t} \tag{20}$$

Thirdly, by deriving the total energy function with respect to W_{pf}^y we obtain the update rule for the factored label weights:

$$\frac{\partial E\left(\mathbf{v}_t, \mathbf{h}_t \mid \mathbf{v}_{<t}, \mathbf{y}_t\right)}{\partial W_{pf}^y} = -y_{p,t} \sum_{i=1}^{n_1} W_{if}^v v_{i,t} \sum_{j=1}^{n_2} W_{jf}^h h_{j,t} \tag{21}$$

Biases update: The derivatives of Eq. (12) to find the update rules for the parameters which compose the dynamic biases of the present layer (i.e., A_{im}^v, $A_{k,m}^{v_{<t}}$, $A_{p,m}^y$) are

$$\frac{\partial E\left(\mathbf{v}_t, \mathbf{h}_t \mid \mathbf{v}_{<t}, \mathbf{y}_t\right)}{\partial A_{i,m}^v} = v_{i,t} \sum_k A_{k,m}^{v_{<t}} v_{k,<t} \sum_p A_{p,m}^y y_{p,t} \tag{22}$$

$$\frac{\partial E\left(\mathbf{v}_t, \mathbf{h}_t \mid \mathbf{v}_{<t}, \mathbf{y}_t\right)}{\partial A_{k,m}^{v_{<t}}} = v_{k,<t} \sum_i A_{i,m}^v \sum_p A_{p,m}^y y_{p,t} \tag{23}$$

$$\frac{\partial E\left(\mathbf{v}_t, \mathbf{h}_t \mid \mathbf{v}_{<t}, \mathbf{y}_t\right)}{\partial A_{p,m}^y} = y_{p,t} \sum_i A_{i,m}^v \sum_k A_{k,m}^{v_{<t}} v_{k,<t} \tag{24}$$

The derivatives to find the update rules for the parameters which compose the dynamic biases of the hidden layer (i.e., $B_{j,n}^h$, $B_{k,n}^{v_{<t}}$, $B_{p,n}^y$) are calculated using a similar procedure. The model free parameters (i.e., dynamical biases and weights) are learned using CD. Using the energy derivative of the hyperparameters and the CD expression shown in Eq. (4), we can calculate the Δ rule leading to

$$\Delta W \propto \left\langle \frac{\partial E}{\partial W} \right\rangle_0 - \left\langle \frac{\partial E}{\partial W} \right\rangle_k \tag{25}$$

$$\Delta A \propto \left\langle \frac{\partial E}{\partial A} \right\rangle_0 - \left\langle \frac{\partial E}{\partial A} \right\rangle_k \tag{26}$$

$$\Delta B \propto \left\langle \frac{\partial E}{\partial B} \right\rangle_0 - \left\langle \frac{\partial E}{\partial B} \right\rangle_k \tag{27}$$

with k being a Markov chain step running for a total number of K steps and starting at the original data distribution.

2.3 Experiments and Results

To achieve the goal of energy prediction the ANN, SVM, RNN, CRBM, and FCRBM models are evaluated and compared using a set of measured data. In this given set the collected data highlights the evolution in time of the electric power consumption, and other different electrical quantities, in one household with a one-minute sampling rate over a period of almost 4 years. Specifically, the dataset [28] contains 2,075,259 measurements gathered between December 2006 and November 2010 (47 months). In all the experiments performed in this section, we have used the first 3 years of data to train the models and the 4th year to test them. Furthermore, in our prediction experiments we used the following attributes information from the database:

1. Aggregated active power—Household global minute-averaged active power. It represents the active energy consumed every minute (in watt-hour) in the household by electrical equipment not measured in submetering 1, 2, and 3.
2. Energy submetering 1—It corresponds to the kitchen, containing mainly a dishwasher, an oven, and a microwave (in watt-hour of active energy).
3. Energy submetering 2—It corresponds to the laundry room, containing a washing machine, a tumble dryer, a refrigerator, and a light (in watt-hour of active energy).
4. Energy submetering 3—It corresponds to an electric water heater and an air conditioner (in watt-hour of active energy).

Predicting of different electrical quantities in an individual household is directly influenced by human behavior along with many other factors. All of these factors lead to a nonlinear time series. Different scenarios have been created to assess the performance of the proposed models. They are briefly described in Table 1.

In the remaining of this section, the results are presented for each of these scenarios in terms of root mean square error (RMSE) while in Mocanu et al. [29] the metrics for prediction assessment include also the correlation coefficient and a P-value computation in order to gain insights on statistical significance of the results (Fig. 6 and Table 2).

For all scenarios, it can be observed that the energy prediction for submetering 3, independently of the methods used, in general shows the most inaccurate results. This can be explained by the not easy predictable behavior of the users regarding the use of the electric water heater. This is also reflected in the

Table 1 Summary of the Experiments

	Notation	Time Horizon	Resolution
Scenario 1	S1	15 min	1 min
Scenario 2	S2	1 h	1 min
Scenario 3	S3	1 day	1 min
Scenario 4	S4	1 day	15 min average
Scenario 5	S5	1 week	15 min average
Scenario 6	S6	1 week	1 h average
Scenario 7	S7	1 year	1 week average

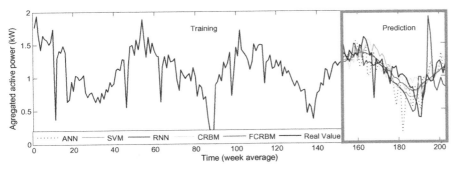

FIG. 6
Scenario 7—aggregated active power prediction for a year, with weekly average data, using ANN, SVM, RNN, CRBM, and FCRBM, versus the true values.

Table 2 Prediction Error in Terms of RMSE for All Seven Scenarios Using ANN, SVM, RNN, CRBM, and FCRBM

	Methods	S1	S2	S3	S4	S5	S6	S7
Aggregated active power (kW)	ANN	0.703	0.731	2.529	0.907	1.867	0.784	0.246
	SVM	0.649	1.995	1.881	1.344	1.559	0.790	0.188
	RNN	0.565	0.939	1.889	1.009	2.807	0.915	0.457
	CRBM	0.638	0.903	1.075	1.030	0.951	0.690	0.182
	FCRBM	0.621	0.666	0.828	0.899	0.797	0.663	0.170
Energy submetering 1 (Wh)	ANN	7.491	3.100	7.171	9.147	15.189	3.078	0.951
	SVM	3.548	2.651	5.056	7.115	5.122	3.443	0.508
	RNN	9.680	6.055	10.515	19.625	21.058	9.093	1.278
	CRBM	3.068	2.746	6.244	5.103	4.634	3.286	0.493
	FCRBM	3.241	2.605	6.193	4.996	4.565	3.128	0.462
Energy submetering 2 (Wh)	ANN	1.852	3.697	4.440	7.284	7.858	8.681	0.500
	SVM	0.761	0.948	9.897	5.321	4.883	3.635	0.468
	RNN	5.022	7.978	12.301	15.098	14.977	8.848	2.011
	CRBM	0.886	1.238	4.840	4.162	4.334	3.481	0.526
	FCRBM	0.687	1.105	4.047	3.790	4.260	3.318	0.436
Energy submetering 3 (Wh)	ANN	10.032	12.451	12.236	10.457	10.289	7.469	2.078
	SVM	9.637	12.013	10.262	8.469	8.535	6.729	2.001
	RNN	10.433	10.417	10.229	11.814	11.412	15.083	6.080
	CRBM	11.796	11.509	11.229	7.301	8.299	9.423	2.025
	FCRBM	8.971	8.140	8.556	6.837	7.709	6.648	1.579

standard deviation of 8.4347 for the energy consumption of submetering 3, being the biggest values from all the electrical quantities analyzed.

Overall, we observed that predicting outliers from the submetering data decreases the predictive performance for all methods significantly. In addition,

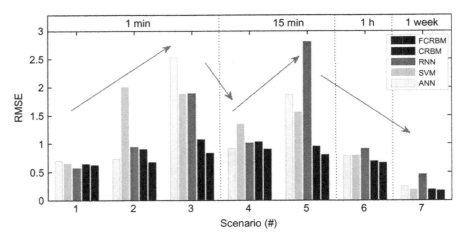

FIG. 7

Comparison of the error obtained in all scenarios using aggregated active power measurements.

this may be reflected by a negative correlation coefficient, while for smoother data such as the aggregated active power in Scenarios 6 and 7 the correlation coefficient is always positive. A short resume of all seven scenarios is depicted in Fig. 7 where the red arrows indicate the variations in multistep prediction. It can be observed that as the number of steps predicted into the future is increasing, the performance of ANN is decreasing in comparison with CRBM and FCRBM.

This observation is well exemplified in the case of Scenarios 3 and 5, where the prediction is made on 1440 and 672 future steps, respectively, and the error for ANN is approximately double than the error of CRBM or FCRBM. In almost all the scenarios, the error given by SVM is a little bigger than CRBM or FCRBM, but smaller than the one of ANN or RNN. Also, we have observed that RNN, even it is very fast in terms of training time, it is not stable and its performance its highly dependent on the number of reservoirs chosen and the type of the data.

The analysis performed showed that FCRBM is a powerful method which outperformed the state-of-the art prediction methods such as ANNs, SVMs, RNNs, and CRBMs. It is worth mentioning that as the prediction horizon is increasing, FCRBMs and CRBMs seem to be more robust and their prediction error is typically half that of the ANN. All methods presented showed comparable prediction time, in the order of few hundred milliseconds, and are therefore suitable for near real-time exploitation in applications such as home and building automation systems. From all the experiments, it can be observed that all methods perform better when predicting the aggregated active power consumption, than predicting the demand of intermittent appliances (e.g., electric water heater) recorded with the three submeterings.

3 Unsupervised Energy Prediction Using Deep Learning

There are many methods for supervised energy prediction. Although they remain at the forefront of academic and applied research, all these methods require labeled data able to faithfully reproduce the energy consumption of buildings. In the remaining of this chapter we refer to the labeled data as to the historical (known) data of the analyzed building. Usually the lack of historical data can be replaced by simulated data. Still, both, historical and simulated data, are employed in these forecasting methods in a nonadaptable way without considering the future events or changes which can occur in the smart grid.

A stronger motivation for this section is given by the not too well exploited fact that sometimes there are not historically data consumption available for a particular building. From the machine learning perspective this is a typical unsupervised learning problem. One of the most used methods of unsupervised learning, RL, was introduced in power system area to solve stochastic optimal control problems [30]. RL methods are used in a wide range of applications, such as system control [31], playing games, or more recently in transfer learning [4, 30]. The advantage of the combination of RL and transfer learning approaches is straightforward. Hence, we want to transfer knowledge from a global to a local perspective to encode the uncertainty of the building energy demand.

Owing to the curse of dimensionality, these methods fail in high dimensions. More recently, there has been a revival of interest in combining deep learning with RL. Therein, RBMs were proven to provide a value function estimation [32] or a policy estimation [33]. More than that, Mnih et al. [3] combined successfully DNNs and Q-learning to create a deep Q-network which successfully learned control policies in a range of different environments. For a more general view, a short bibliometric analysis of the collections of publications related with the Transfer Learning and RL concepts is performed further using specialized queries on Scopus. The overall count obtained for Transfer Learning is 189,524 publications and 113,263 for RL, indexed by Scopus on 26 July 2016. In Fig. 8 three disjoint sets are highlighted by zooming over the last decade: (i) Q-learning with 3174 publications, (ii) SARSA with 332 publications, and (iii) deep reinforcement learning which counts for 207 publications.

In Mocanu et al. [34], we comprehensively explore and extend two RL methods to predict the energy consumption at the building level using unlabeled historical data, namely state-action-reward-state-action (SARSA) [35] and Q-learning [36]. Due to the fact that in the original form both methods cannot handle well continuous state space, this work contributes theoretically to extend them by incorporating a DBN [1] for continuous states estimation and automatically

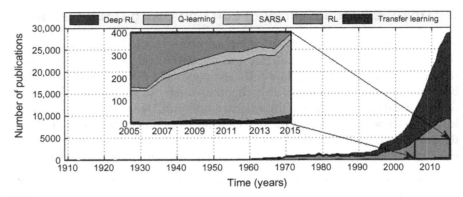

FIG. 8

The number of Scopus-indexed publications with focus on the transfer learning and reinforcement learning concepts between 1909 and 2015.

features extraction in an unified framework. Our proposed RL methods are appropriate when we do not have historical or simulated data, but we want to estimate the impact of changes in smart grid, such as the appearance of a building or several buildings in a certain area, or more commonly, a change in energy consumption due to building renovation. In this section, we have shown the applicability and efficiency of our proposed method in three different situations:

- In the case of a new type of building being connected with the smart grid, thus transferring knowledge from a commercial building to a residential building.
- In the case of a renovated building, thus transferring knowledge from a nonelectric heat building to a building with electric heating.
- Additionally, we propose experiments to highlight the importance of external factors for the estimation of building energy consumption, such as price information. In respect with this the transfer learning is applied, from a building under a static tariff to a building with a time-of-use tariff.

According to our knowledge, this is the first time when the energy prediction is performed without using any information about that building, such as historical data, energy price, physical parameters of the building, meteorological condition, or information about the user behavior.

3.1   Problem Formulation

We propose a method to solve the unsupervised energy prediction problem with cross-building transfer by using machine learning time-series prediction

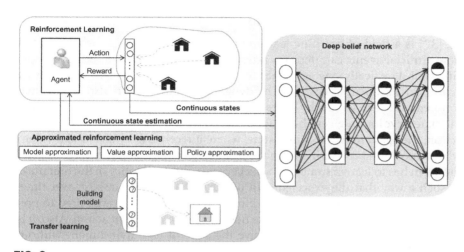

FIG. 9
The unsupervised learning explores and extends reinforcement and transfer learning setup, by including a
deep belief network for continuous states estimation.

techniques. In the most general statement, the proposed Reinforcement and
Transfer Learning setup is depicted in Fig. 9. Given the unevenly distributed
building energy values during time, firstly, a special attention is given to the
question: *How to estimate a continuous state space?* The idea is to find a lower
dimensional representation of the energy consumption data that preserves
the pairwise distances as well as possible.

More formally, the energy prediction using unlabeled data problem presented
in this section is divided into three different subproblems, namely:

1. Continuous state estimation problem:

Given a dataset, $\mathcal{D}: \mathbb{R} \to \mathcal{S}$ **find** a confined space state representation $\mathcal{S}_1$.

2. RL problem:

Given a building model $M_1 = \langle \mathcal{S}_1, \mathcal{A}_1, \mathcal{T}.(\cdot, \cdot), \mathcal{R}_1 \rangle$, **find** an optimal policy, π_1^*.

3. Transfer learning problem:

Given a model, $M_1 = \langle \mathcal{S}_1, \mathcal{A}_1, \mathcal{T}.(\cdot, \cdot), \mathcal{R}_1 \rangle$, a reasonable π_1^* and $M_2 = \langle \mathcal{S}_2, \mathcal{A}_2, \mathcal{T}.(\cdot, \cdot), \mathcal{R}_2 \rangle$, **find** a good π_2^*.

The proposed solution is presented in Section 3.3, where a new method to estimate continuous states in RL using DBN is detailed. Further this state estimation method is integrated in SARSA and Q-learning algorithms in order to
improve the prediction accuracy.

3.2 Reinforcement Learning

RL [37] is a field of machine learning inspired by psychology, which studies how artificial agents can perform actions in an environment to achieve a specific goal. Practically, the agent has to control a dynamic system by choosing actions in a sequential fashion. The dynamic system, known also as the environment, is characterized by states, its dynamics, and a function that describes the state's evolution given the actions chosen by the agent. After it executes an action, the agent moves to a new state, where it receives a reward (scalar value) which informs it how far it is from the goal (the final state). To achieve the goal, the agent has to learn a strategy to select actions, dubbed policy in the literature, in such a way that the expected sum of the rewards is maximized over time. Besides that, a state of the system captures all the information required to predict the evolution of the system in the next state, given an agent action. Also, it is assumed that the agent could perceive the state of the environment without error, and it could make its current decision based on this information. There are two different categories of RL algorithms, (i) Online RL which are *interaction-based* algorithms, such as Q-learning [36], SARSA [35], or Policy Gradient, and (ii) Offline RL, like Least-Square Policy Iteration or fitted Q-iteration. For a more comprehensive discussion of RL algorithms we refer to Busoniu et al. [38]. In the remaining of this chapter we will refer just to online RL.

3.2.1 Markov Decision Process

An RL problem can be formalized using Markov decision process (MDPs). MDPs are defined by a 4-tuple $\langle \mathcal{S}, \mathcal{A}, \mathcal{T}.(\cdot, \cdot), \mathcal{R}.(\cdot, \cdot) \rangle$, where $\mathcal{S}$ is a set of states, $\forall s \in \mathcal{S}$, $\mathcal{A}$ is a set of actions, $\forall a \in \mathcal{A}$, $\mathcal{T} : \mathcal{S} \times \mathcal{A} \times \mathcal{S} \rightarrow [0, 1]$ is the transition function given by the probability that by choosing action a in state s at time t the system will arrive to state s' at time $t + 1$ such that $p_a(s, s') = p(s_{t+1} = s' | s_t = s, a_t = a)$, and $\mathcal{R} : \mathcal{S} \times \mathcal{A} \times \mathcal{S} \rightarrow \mathbb{R}$ is the reward function, where $\mathcal{R}_a(s, s')$ is the immediate reward (or expected immediate reward) received by the agent after it performs the transition to state s' from state s. An important property in MDPs is the Markov property [39] which makes the assumption that the state transitions are dependent just on the last state of the system, and are independent of any previous environment states or agent actions, i.e., $p(s_{t+1} = s', r_{t+1} = r | s_t, a_t)$ for all s', r, s_t, and a_t.

The MDPs theory does not assume that $\mathcal{S}$ or $\mathcal{A}$ are finite, but the traditional algorithms make this assumption. In general, they can be solved by using linear or dynamic programming. The interested reader is referred to Puterman [40] for a more comprehensive discussion about MDPs. Furthermore, in the real world, the state transition probabilities $\mathcal{T}.(\cdot; \cdot)$ and the rewards $R.(\cdot; \cdot)$ are unknown, and the state space $\mathcal{S}$ or the action space $\mathcal{A}$ might be continuous. Thus, RL represents a normal extension and generalization over MDPs for such situations, where the tasks are too large or too ill defined, and cannot be solved using optimal control theory [35].

3.2.2 Q-Learning

First, the Q-learning algorithm [36] is recommended like a standard solution in RL where the rules are often stochastic. This algorithm therefore has a function which calculates the Quality of a state-action combination, defined by $\mathcal{S} \times \mathcal{A} \to \mathbb{R}$. Before learning has started, Q matrix returns an initial value. Then, each time the agent selects an action, and observes a reward and a new state that both may depend on the previous state and the selected action. The action-value function of a fixed policy π with the value function $V^{\pi} : \mathcal{S} \to \mathbb{R}$ is

$$Q^{\pi}(s, a) = r(s, a) + \gamma \sum_{s'} p(s' \mid s, a) V^{\pi}(s'), \forall s \in \mathcal{S}, a \in \mathcal{A} \tag{28}$$

The value of state-action pairs, $Q^{\pi}(s, a)$, represent the expected outcome when one agent is starting from s, executing a and then following the policy π afterward, such that $V^{\pi}(x) = Q^{\pi}(x, \pi(x))$, with their corresponding Bellman equation

$$Q^*(s, a) = r(s, a) + \gamma \sum_{s'} p(s' \mid s, a) \max_b Q^*(s, b) \tag{29}$$

where the discount factor $\gamma \in [0, 1]$ trades off the importance of rewards and $b = \max(a)$. Thus, the optimal value is obtained for $\forall s \in S$, $V^*(s) = \max_a Q^*(s, a)$, and $\pi^*(s) = \arg \max_a Q^*(s, a)$. The value of state-action pairs is given by the same formal expectation value, $\mathbb{E}_{\pi}$, of an expected total return r_t such that $Q(s, a) = \mathbb{E}_{\pi}(r_t \mid s_t = s, a_t = a)$. The off-policy Q-learning algorithm has the update rule defined by

$$Q_{t+1}(s_t, a_t) = Q_t(s_t, a_t) + \alpha_t [r_{t+1} + \gamma max Q_t(s_{t+1}, a) - Q_t(s_t, a_t)] \tag{30}$$

where r_{t+1} is the reward observed after performing a_t in s_t, and where $\alpha_t(s, a)$, with all $\alpha \in [0, 1]$, is the learning rate which may be the same for all pairs.

Q-learning algorithm has problems with big numbers of continuous states and discrete actions. Usually, it needs function approximations, e.g., neural networks, to associate triplets like state, action, and Q-value. Exploration of one MDP can be done under Markov assumption, to take into account just current state and action, but because in the real world we have partially observable MDPs (POMDP), we may have better results if an arbitrary k number of history states and actions $(S_{t-k}, a_{t-k}, \ldots, S_{t-1}, a_{t-1})$ will be considerate [41] to clearly identify a triplet $\langle S_t, A_t, Q_t \rangle$ at time t.

3.2.3 SARSA

An interesting variation for Q-learning is the SARSA algorithm [35], which aims at using Q-learning as part of a Policy Iteration mechanism. The major difference between SARSA and Q-Learning is that the maximum reward for the next state in SARSA is not necessarily used for updating the Q-values. Therefore, the core of the SARSA algorithm is a simple value iteration update. The information required for the update is a tuple $(s_t, a_t, r_{t+1}, s_{t+1}, a_{t+1})$, and the update is defined by

$$Q_{t+1}(s_t, a_t) = Q_t(s_t, a_t) + \alpha_t[r_{t+1} + \gamma Q_t(s_{t+1}, a_{t+1}) - Q_t(s_t, a_t)] \tag{31}$$

where r_{t+1} is the reward and $\alpha_t(s, a)$ is the learning rate. In practice, Q-learning and SARSA are the same if we use a greedy policy (i.e., the agent chooses the best action always), but are different when the ϵ-greedy policy is used, which favors more random exploration.

In traditional RL algorithms, only MDPs with finite states and actions are considered. However, building energy consumption can take nearly arbitrary real value resulting in a very large number of states in MDPs. Due to the fact that building energy consumption can be seen as a time-series problem, a prior discretization of the state space is not very useful. So, we try to find algorithms that work well with large (or continuous) state spaces, as shown next.

3.3 States Estimation via DBNs

Deep Architectures [1] show very good results in different applications, such as to perform nonlinear dimensionality reduction [42], images recognition, video sequences, or motion-capture data [43]. A comprehensive analysis on dimensionality reduction and deep architectures can be referred to van der Maaten et al. [44]. Overall, DBN could be a way to naturally decompose the problem into subproblems associated with different levels of abstraction.

3.3.1 Deep Belief Networks

DBNs are composed of several RBMs stacked on top of each other [42]. An RBM is a stochastic recurrent neural network that consists of a layer of visible units, **v**, and a layer of binary hidden units, **h**. The total energy of the joint configuration of the visible and hidden units (**v**, **h**) is given by

$$E(\mathbf{v}, \mathbf{h}) = -\sum_{i,j} v_i h_j W_{ij} - \sum_i v_i a_i - \sum_j h_j b_j \tag{32}$$

where i represents the indices of the visible layer, j those of the hidden layer, and $w_{i,j}$ denotes the weight connection between the ith visible and jth hidden unit. Further, v_i and h_j denote the state of the ith visible and jth hidden unit, respectively, and a_i and b_j represent the biases of the visible and hidden layers. The first term, $\sum_{i,j} v_i h_j W_{ij}$ represents the energy between the hidden and visible units with their associated weights. The second, $\sum_i v_i a_i$ represents the energy in the visible layer, while the third term represents the energy in the hidden layer. The RBM defines a joint probability over the hidden and visible layer $p(\mathbf{v}, \mathbf{h})$.

$$p(v, h) = \frac{e^{-E(v,\,h)}}{Z} \tag{33}$$

where Z is the partition function, obtained by summing the energy of all possible $(\mathbf{v}, \mathbf{h})$ configurations, $Z = \sum_{\mathbf{v},\mathbf{h}} e^{-E(\mathbf{v},\,\mathbf{h})}$. To determine the probability of a data point represented by a state $\mathbf{v}$, the marginal probability is used, summing out the state of the hidden layer, such that $p(v) = \sum_h p(\mathbf{v}, \mathbf{h})$.

The above equation can be used for any given input to calculate the probability of either the visible or the hidden configuration to be activated. These values are further used to perform inference in order to determine the conditional probabilities in the model. To maximize the likelihood of the model, the gradient of the log-likelihood with respect to the weights must be calculated. The gradient of the first term, after some algebraic manipulations, can be written as

$$\frac{\partial \log \left(\sum_h \exp\left(-E(v, h)\right) \right)}{\partial W_{ij}} = v_i \cdot p\left(h_j = 1 \mid v\right) \tag{34}$$

However, computing the gradient of the second term is intractable. The inference of the hidden and visible layers in RBM can be done accordingly with the next formulas

$$p\left(\mathbf{h}_j = 1 \mid \mathbf{v}\right) = \sigma \left(b_j + \sum_i v_i W_{ji} \right) \tag{35}$$

$$p\left(\mathbf{v}_i = 1 \mid \mathbf{h}\right) = \sigma \left(a_i + \sum_j h_j W_{ji} \right) \tag{36}$$

where $\sigma(\cdot)$ represents the sigmoid function. Moreover, to learn an RBM we can use the following learning rule which performs stochastic steepest ascent in the log probability of the training data [26]:

$$\frac{\partial \log \left(p(\mathbf{v}, \mathbf{h}) \right)}{\partial W_{ij}} = \langle v_i h_j \rangle_0 - \langle v_i h_j \rangle_\infty \tag{37}$$

where $\langle \cdot \rangle_0$ denotes the expectations for the data distribution (p_0) and $\langle \cdot \rangle_\infty$ denotes the expectations under the model distribution.

Overall, a DBN [1] is given by an arbitrary number of RBMs stack on the top of each other. This yields a combination between a partially directed and partially undirected graphical model. Therein, the joint distribution between visible layer $\mathbf{v}$ (input vector) and the l hidden layers h^k is defined as follows:

$$p\left(\mathbf{v}, \mathbf{h}^1, ..., \mathbf{h}^k\right) = \prod_{k=0}^{l-2} P\left(\mathbf{h}^k \mid \mathbf{h}^{k+1}\right) P\left(\mathbf{h}^{l-1}, \mathbf{h}^l\right) \tag{38}$$

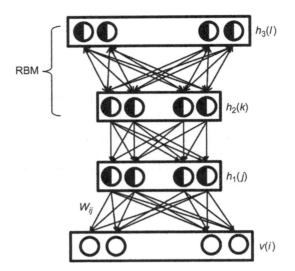

FIG. 10

A general deep belief network structure with three hidden layers. The top two layers have undirected connections and form an associative memory.

where $P(\mathbf{h}^k | \mathbf{h}^{k+1})$ is a conditional distribution for the visible units conditioned on the hidden units of the RBM at level k, and $P(\mathbf{h}^{l-1}, \mathbf{h}^l)$ is the visible-hidden joint distribution in the top-level RBM. An example of a DBN with 3 hidden layers (i.e., $h_1(j)$, $h_2(j)$, and $h_3(j)$) is depicted in Fig. 10. The top-level RBM in a DBN acts as a complementary prior from the bottom level directed sigmoid likelihood function. A DBN can be trained in a greedy unsupervised way, by training separately each RBM from it, in a bottom to top fashion, and using the hidden layer as an input layer for the next RBM [45]. Furthermore, the DBN can be used to project our initial states acquired from the environment to another state space with binary values, by fixing the initial states in the bottom layer of the model, and inferring the top hidden layer from them. In the end, the top hidden layer can be directly incorporated into the SARSA or Q-learning algorithms.

Now that we have considered the problem of state estimation and we incorporated all three subproblems in a unified approach we look into the experimental validation.

3.4   Numerical Results

Dataset characteristics: The proposed solution is experimentally evaluated using a dataset recorded over 7 years, more exactly between 6 January 2007 and 31 January 2014. The load profiles, including different residential and commercial buildings, are made on-line available by Baltimore Gas and Electric Company. For every type of building analyzed the available historical load data in

Table 3 Building Types in Datasets

	R	Residential (Nonelectric Heat)
Residential	R(ToU)	Residential time-of-use (nonelectric heat)
	RH	Residential (electric heat)
	RH(ToU)	Residential time-of-use (electric heat)
Commercial	G	General Service—Commercial, Industrial & Lighting (<60 kW)

kWh represents an average building profile per hour. Overall, there are five different building profiles, as presented in Table 3.

For a more comprehensive view of the datasets used in this chapter we have shown in Fig. 11A the hourly evolution of the electrical energy consumption for a General Service (G) dataset, including Commercial, Industrial & Lighting, and a residential with nonelectric heat (R) building over different time horizons. Moreover, some general characteristics for the entire dataset are graphically depicted in Fig. 11B. In all experiments the data was separated into the training and testing datasets. More precisely, the data collected from 1 June 2007 until 1 January 2013 (2041 days) was used in the learning phase and the remaining data, between January 2013 and 31 January 2014 (396 days) was used to evaluate the performance of the methods. The metrics used to assess the quality of the different buildings energy consumption prediction are described further.

Metrics for prediction assessment: As we mention earlier, the goal is to achieve good generalization by making accurate prediction for new building energy consumption data. Firstly, some quantitative insights into the dependence of the generalization performance of our approach are evaluated using the root-mean-square error defined by $\text{RMSE} = \sqrt{\frac{1}{N}\sum_{i=1}^{N}(v_i - \hat{v}_i)^2}$, where N represents the number of multisteps prediction within a specified time horizon, v_i represents the real values for the time step i, and $\hat{v}_i$ represents the model estimated value at the same time step. Then, by using the Pearson product-moment correlation coefficient (R), insights are given on the degree of linear dependence between the real value and the predicted value. Hence $R(u, v) = \frac{\mathbb{E}[(u - \mu_u)(v - \mu_v)]}{\sigma_u \sigma_v}$, where $\mathbb{E}[\cdot]$ is the expected value operator with standard deviations σ_u and σ_v. The correlation coefficient may take on any value within the range $[-1, 1]$. The sign of the correlation coefficient defines the direction of the relationship, either positive or negative. Finally, we perform the Kolmogorov-Smirnov test [46] in order to gain insights on statistical significance of our results. The Kolmogorov-Smirnov test has the advantage of making no assumption about the distribution of data. This elaborate statistical test is not a typical metric used in the analysis of the prediction accuracy, but is imposed by the fact that the learning and the testing procedure is made using

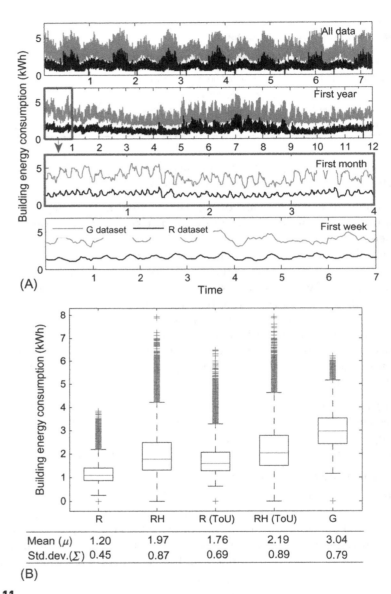

FIG. 11

(A) Electrical energy consumption for a Commercial, Industrial & Lighting (G) dataset and for a residential building with nonelectric heat (R) building over different time horizons and (B) general characteristics of all data sets: A box plot with the exact value for mean and standard deviation encoded in it.

Table 4 Summary of the Experiments

	Notation	Time Horizon	Resolution
Scenario 1	S1	1 h	1 h average
Scenario 2	S2	1 day	1 h average
Scenario 3	S3	1 week	1 h average
Scenario 4	S4	1 month	1 h average
Scenario 5	S5	1 year	1 week average

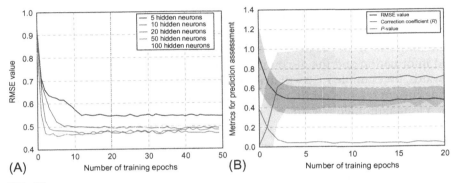

FIG. 12

(A) The RMSE values observed for different RBM configurations in the DBN architecture, with varying number of hidden neurons, as a function of training epochs. (B) Performance metrics for the chosen RBM configuration with 10 hidden neurons.

different building types. Hence, exceeding the statistical significance level, $P < .05$, would be expected and will validate the different probability distribution function from where this data is provided.

Empirical Results: To assess the performance of our extended reinforcement and transfer learning approaches presented in Section 3.3, we have designed five different scenarios. These are selected to cover various multistep prediction at different resolution, and are summarized in Table 4. Further on, before to go into the deep analyses of the numerical results, firstly we present some details of the implementation.

Implementations details: The implementation has been done in two parts. Firstly, a DBN is implemented, and secondly the RL algorithms use the DBN in their implementations for continuous states estimation, as it is shown next.

Continuous states estimations using DBN: We implemented the DBN in MATLAB from scratch using the mathematical details described in Section 3.3. In order to obtain a good prediction we investigate carefully the choice of the optimal number of hidden units in our DBN configuration with respect to the RMSE evolution, see Fig. 12.

Thus, the number of hidden neurons was set to 10 and the learning rate was 10^{-3}. The momentum was set to 0.5 and the weight decay to 0.0002. We trained the model for 20 epochs, but as it can be seen in Fig. 12B, the model converged after approximately 4 epochs. More details about the optimal choice of the parameters can be found in Hinton [27].

SARSA and Q-learning: We implemented the SARSA and Q-learning in MATLAB using the mathematical details described in Section 3.2. In both cases the learning rate was set to 0.4 and the discount factor was set to. Both parameters have a direct influence on the performance of the both algorithms.

The choice of these parameters was made after a thorough examination of the RMSE outcome, as is shown for example in Fig. 13. Overall, the learning rate determines to what extent the newly acquired information will override the old information and the discount factor determines the importance of future rewards, for example $\gamma = 0$ will make the agent "opportunistic" by only considering current rewards, while a discount factor approaching 1 will make it strive for a long-term high reward.

3.4.1 Commercial to Residential Transfer

In this set of experiments, we use Commercial, Industrial & Lighting data to train the DBN model. Furthermore, we use the trained DBN model to predict four different types of unseen residential building consumption, such as residential with electric heat and without electric heat, and residential electric consumption with ToU pricing, as it is shown in Table 5 and Fig. 14. The analysis of the different types of residential buildings advances the insight on the generalization capabilities of our proposed method and studies its robustness by testing the behavior on different probability distributions (see Fig. 11).

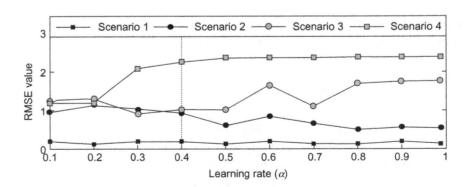

FIG. 13

Analyses of RMSE values obtained from different α values in exploration step, for different scenarios. This involves the prediction of G dataset [Commercial, Industrial & Lighting consumption, General Service (<60 kW)].

Table 5 Using Commercial, General Service (G) (<60 kW) Dataset to Predict Residential Energy Consumption, Such as R, R(ToU), RH and RH(ToU) Values Using SARSA, Q-Learning, SARSA With DBN Extension and Q-Learning With DBN Extension

	Methods	G	R	R(ToU)	RH	RH(ToU)
Scenario 1	SARSA	0.18	0.02	0.10	0.36	0.42
	Q-learning	0.22	0.02	0.04	0.34	0.34
	SARSA and DBN	0.04	0.02	0.06	0.04	0.04
	Q-learning and DBN	0.01	0.03	0.09	0.04	0.02
Scenario 2	SARSA	0.65	0.75	0.47	1.23	1.20
	Q-learning	1.09	0.98	0.40	1.28	1.55
	SARSA and DBN	0.38	0.37	0.37	0.46	0.47
	Q-learning and DBN	0.33	0.37	0.29	0.41	0.66
Scenario 3	SARSA	1.27	1.73	1.36	1.59	1.33
	Q-learning	1.39	1.10	0.83	1.47	1.61
	SARSA and DBN	0.69	1.31	0.55	1.33	1.18
	Q-learning and DBN	0.62	0.98	0.58	1.26	1.30
Scenario 4	SARSA	1.55	3.70	2.39	2.05	1.89
	Q-learning	1.41	1.24	1.14	1.67	1.71
	SARSA and DBN	1.14	1.45	1.17	1.33	1.21
	Q-learning and DBN	0.98	1.40	0.87	1.52	1.55
Scenario 5	SARSA	1.01	2.61	2.04	2.16	1.95
	Q-learning	0.72	2.28	1.81	1.83	1.59
	SARSA and DBN	0.05	0.08	0.10	0.11	0.24
	Q-learning and DBN	0.03	0.02	0.02	0.03	0.03

3.4.2 Residential to Residential Transfer

During these experiments we learn and transfer one type of residential building energy demand profile to another type of residential building with different characteristics. More exactly, we used to train the learning algorithm (i) a residential building profile without electric heat (R), and (ii) a residential building with electric heat (RH). The prediction results of these two building models can be seen in Table 6.

In Tables 5 and 6, the RMSE values show a good agreement between the real values and the model estimated values. In addition, the confidence in our results is formally determined not just by the RMSE values, but also by the correlation coefficient and the number of steps predicted into the future. For example, if there is just one step ahead, such as in Scenario 1, then the Pearson correlation coefficient needs to be very close to 1 or -1 in order to consider it statistically significant. However, in the case of Scenarios 3 and 4, where the prediction is made on 168 and 672 future steps, a coefficient close to 0 can still be considered highly significant. More discussions about the

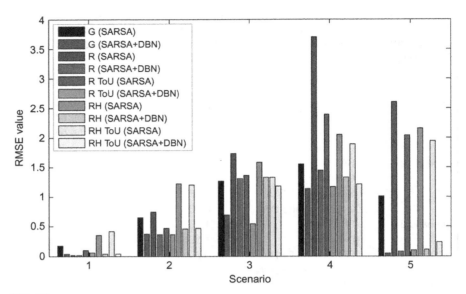

FIG. 14

Overview of errors obtained, where (C) using G dataset we predict R, R(ToU), RH, and RH(ToU) values, (A) using R we predict RH, and (B) using R(ToU) we predict the RH(ToU). Four methods are used: SARSA, Q-learning, SARSA with DBN extension, and Q-learning with DBN extension.

robustness of the correlation coefficient can be found in Devlin et al. [47]. Still, the inaccuracy was reflected in a negative correlation coefficient in 24% of the experiments when we used the simple form of the SARSA and Q-learning methods. By contrast, our two improved approaches, SARSA with DBN extension and Q-learning with DBN extension, show a negative correlation in just 4% of the cases. Overall, the Kolmogorov-Smirnov test in most cases confirms that the data do indeed come from different distributions, which is represented with gray color in Table 6. This is partially due to the unique characteristics of this dataset, given by the presence of a highly nonlinear profile shape and large outlier values, as seen in Fig. 11. All of these observations give a strong argument for employing a more comprehensive examination of the distributions used in the transfer learning. Nevertheless, the results presented in Tables 5 and 6 demonstrate that the energy prediction accuracies in terms of RMSE significantly improve in 91.42% of the cases after using a DBN for automatically computing high-level features from the unlabeled data, as compared to the situation when the counterpart RL methods are used without any DBN extension.

Notably, the proposed approach is also suitable when we have access to historical data. In the scope of this argument, the result obtained in the first column of Table 5 is expected to be equivalent with the results obtained with any supervised learning methods, such as ANN or SVM. Nevertheless, the RMSE accuracy

Table 6 (A) Prediction of Residential Building With Electric Heat Consumption Using Data Collected From a Residential With Nonelectric Heat Building. (B) Prediction of Residential Building Consumption With Electric Heat Using Data Collected From a Residential With Nonelectric Heat Building, Both With ToU Pricing

	Methods	A		B	
		RMSE	R	RMSE	R
S1	SARSA	0.42	0.88	0.50	0.83
	Q-learning	0.44	0.87	0.16	0.99
	SARSA with DBN	0.42	0.88	0.28	0.94
	Q-learning with DBN	0.03	0.99	0.24	0.99
S2	SARSA	2.15	−0.18	1.69	0.33
	Q-learning	1.93	−0.10	0.91	0.83
	SARSA with DBN	1.25	0.61	1.42	0.55
	Q-learning with DBN	0.50	0.64	1.18	0.77
S3	SARSA	2.63	−0.27	2.69	−0.11
	Q-learning	2.57	−0.18	1.65	0.17
	SARSA with DBN	2.67	0.13	1.98	0.27
	Q-learning with DBN	0.69	0.09	1.55	0.21
S4	SARSA	2.23	0.04	2.45	−0.01
	Q-learning	2.14	0.11	1.62	0.17
	SARSA with DBN	1.97	−0.09	2.38	0.24
	Q-learning with DBN	0.71	−0.10	1.60	0.28
S5	SARSA	0.74	0.62	0.67	0.19
	Q-learning	0.57	0.62	0.41	0.47
	SARSA with DBN	0.03	0.43	0.03	0.34
	Q-learning with DBN	0.02	0.51	0.02	0.42

obtained using the Q-learning algorithm with the DBN extension for the long-term forecasting of buildings energy consumption (Scenario 5) is >90% in all the experiments than Q-learning without DBN extension. For example, in Table 6 the RMSE is 0.02 if we use Q-learning with DBN versus 0.57 for Q-learning without DBN, yielding a 96.5% improved RMSE accuracy.

4 CONCLUSIONS

In this chapter, firstly, we present two deep learning methods for supervised energy prediction, namely CRBM and FCRBM. FCRBM has good generalization capabilities and it can be used to accommodate large databases, while its exploitation time in real-world settings is on the order of few milliseconds. On the one hand, the comparative results show that FCRBM outperforms the other methods such as ANN, RNN, SVM, and on the other hand, they suggest that by adding more information to FCRBM, its performance may be improved further, leading to a more full automatic real-time control of electrical energy profiles in the smart grid context.

Secondary, in Section 3, a new paradigm for building energy prediction is introduced, which does not require historical data from the specific building under scrutiny. In a unified approach, we can successfully learn a building model by including a generalization of the state space domain, then we transfer it across other building. This contribution is twofold. First, we present a DBN for automatically feature extraction and second, we extend two standard RL algorithms able to perform knowledge transfer between domains (buildings models), namely SARSA algorithm and Q-learning algorithm by incorporating the states estimated with the DBN. The novel proposed machine learning methods for energy prediction are evaluated over different time horizons with different time resolutions using real data. Notably, it can be observed that as the prediction horizon is increasing, SARSA and Q-learning extensions by including a DBN for states estimation seem to be more robust and their prediction error is approximately 20 times lower than that of their unextended versions. The strength of this method is given by the DBN generalization capabilities over the underlying state space for a new building and the robustness to invariance in the state representation. However, a forthcoming deep investigation can be done at different smart grid levels in order to help the transition to the future energy system.

Nowadays, deep learning methods for power system data analysis are in an incipient phase. Still, the power system transition toward the big data era encourages the use of deep learning, as the most advanced solutions for large-scale applications. For example, the interested reader is referred to Mocanu et al. [19] for a comparison between CRBM and HMMs for energy prediction, and to Mocanu et al. [48] to see FCRBM capabilities in a price-responsive context. In Mocanu et al. [49, 50] deep learning methods are used to perform energy disaggregation and building flexibility detection. Recently, Marino et al. [51] propose the use of Long Short-Term Memory (LSTM) networks for building energy prediction, performing experiments on the same dataset as the one used in Section 2. More details on deep learning for energy prediction can be found in Ryu et al. [52] and in a recent review of Manic et al. [53].

References

[1] Y. Bengio, Learning deep architectures for AI, Foundations and Trends in Machine Learning, 2, Now Publishers, Hanover, MA, 2009, pp. 1–127.

[2] H. Lee, P. Pham, Y. Largman, A.Y. Ng, in: Unsupervised feature learning for audio classification using convolutional deep belief networks, Advances in Neural Information Processing Systems, 2009, pp. 1096–1104.

[3] V. Mnih, K. Kavukcuoglu, D. Silver, A.A. Rusu, J. Veness, M.G. Bellemare, A. Graves, M. - Riedmiller, A.K. Fidjeland, G. Ostrovski, S. Petersen, C. Beattie, A. Sadik, A. Antonoglou,

H. King, D. Kumaran, D. Wierstra, S. Legg, D. Hassabis, Human-level control through deep reinforcement learning, Nature 518 (7540) (2015) 529–533.

[4] H. Bou Ammar, D.C. Mocanu, M.E. Taylor, K. Driessens, K. Tuyls, G. Weiss, in: Automatically mapped transfer between reinforcement learning tasks via three-way restricted Boltzmann machines, Machine Learning and Knowledge Discovery in Databases, 8189, 2013, pp. 449–464.

[5] D.C. Mocanu, H. Bou Ammar, D. Lowet, K. Driessens, A. Liotta, G. Weiss, K. Tuyls, Factored four way conditional restricted Boltzmann machines for activity recognition, Pattern Recogn. Lett. 66 (2015) 100–108.

[6] P. Smolensky, Information processing in dynamical systems: foundations of harmony theory, Parallel Distributed Processing: Volume 1: Foundations, MIT Press, Cambridge, MA, 1987, pp. 194–281.

[7] T. Osogami, M. Otsuka, Restricted Boltzmann machines modeling human choice, Adv. Neural Inf. Proces. Syst. 27 (2014) 73–81.

[8] R. Salakhutdinov, A. Mnih, G. Hinton, in: Restricted Boltzmann machines for collaborative filtering, Proceedings of the 24th International Conference on Machine Learning, ACM, 2007, pp. 791–798.

[9] P.V. Gehler, A.D. Holub, M. Welling, in: The rate adapting poisson model for information retrieval and object recognition, Proceedings of 23rd International Conference on Machine Learning (ICML06), 2006, p. 2006.

[10] H. Larochelle, Y. Bengio, in: Classification using discriminative restricted Boltzmann machines, Proceedings of the 25th International Conference on Machine Learning, Helsinki, Finland, 2008, pp. 536–543.

[11] D.C. Mocanu, E. Mocanu, P.H. Nguyen, M. Gibescu, A. Liotta, A topological insight into restricted Boltzmann machines, Mach. Learn. 104 (2) (2016) 243–270.

[12] M. Krarti, Energy Audit of Building Systems: An Engineering Approach, second ed. Mechanical and Aerospace Engineering Series, CRC Press, 2010. ISBN: 978-1-4398-2871-7.

[13] A.I. Dounis, Artificial intelligence for energy conservation in buildings, Adv. Build. Energy Res. 4 (1) (2010) 267–299.

[14] A. Foucquier, S. Robert, F. Suard, L. Stephan, A. Jay, State of the art in building modelling and energy performances prediction: a review, Renew. Sust. Energ. Rev. 23 (2013) 272–288.

[15] H.X. Zhao, F. Magoulès, A review on the prediction of building energy consumption, Renew. Sust. Energ. Rev. 16 (6) (2012) 3586–3592.

[16] M. Aydinalp-Koksal, V.I. Ugursal, Comparison of neural network, conditional demand analysis, and engineering approaches for modeling end-use energy consumption in the residential sector, Appl. Energy 85 (4) (2008) 271–296.

[17] L.A. Hurtado Munoz, E. Mocanu, P.H. Nguyen, M. Gibescu, W.L. Kling, in: Comfort-constrained demand flexibility management for building aggregations using a decentralized approach, 4th International Conference on Smart Cities and Green ICT Systems, 2015.

[18] L. Xuemei, D. Lixing, L. Jinhu, X. Gang, L. Jibin, in: A novel hybrid approach of kpca and svm for building cooling load prediction, Int. Conf. Knowledge Discovery and Data Mining, 2010.

[19] E. Mocanu, P.H. Nguyen, M. Gibescu, W.L. Kling, in: Comparison of machine learning methods for estimating energy consumption in buildings, Proc. of the 13th Int. Conf. on Probabilistic Methods Applied to Power Systems, 2014.

[20] L.E. Baum, T. Petrie, Statistical inference for probabilistic functions of finite state Markov chains, Ann. Math. Stat. 37 (1966) 1554–1563.

[21] T. Zia, D. Bruckner, A. Zaidi, in: A hidden Markov model based procedure for identifying household electric loads, Annual Conference on IEEE Industrial Electronics Society, 2011, pp. 3218–3223.

[22] V. Mnih, H. Larochelle, G. Hinton, in: Conditional restricted Boltzmann machines for structured output prediction, Proceedings of the International Conference on Uncertainty in Artificial Intelligence, 2011.

[23] E. Mocanu, D.C. Mocanu, H.B. Ammar, Z. Zivkovic, A. Liotta, E. Smirnov, in: Inexpensive user tracking using Boltzmann machines, In IEEE International Conference on Systems, Man and Cybernetics, 2014, pp. 1–6.

[24] J.W. Taylor, Exponentially weighted methods for forecasting intraday time series with multiple seasonal cycles, Int. J. Forecast. 26 (4) (2010) 627–646.

[25] G.W. Taylor, G.E. Hinton, S.T. Roweis, Two distributed-state models for generating high-dimensional time series, J. Mach. Learn. Res. 12 (2011) 1025–1068.

[26] G.E. Hinton, Training products of experts by minimizing contrastive divergence, Neural Comput. 14 (8) (2002) 1771–1800.

[27] G.E. Hinton, A practical guide to training restricted Boltzmann machines, Neural Networks: Tricks of the Trade, second ed., Lecture Notes in Computer Science, 7700, Springer, Berlin, Heidelberg, 2012, pp. 599–619.

[28] K. Bache, M. Lichman, UCI-Machine Learning Repository, University of California, School of Information and Computer Science, Irvine, CA, 2013.

[29] E. Mocanu, P.H. Nguyen, M. Gibescu, W.L. Kling, Deep learning for estimating building energy consumption, Sustain. Energy Grids Netw. 6 (2016) 91–99.

[30] D. Ernst, M. Glavic, F. Capitanescu, L. Wehenkel, Reinforcement learning versus model predictive control: a comparison on a power system problem, IEEE Trans. Syst. Man Cybern. B Cybern. 39 (2) (2009) 517–529.

[31] R. Crites, A. Barto, in: Improving elevator performance using reinforcement learning, Advances in Neural Information Processing Systems 8, 1996, pp. 1017–1023.

[32] B. Sallans, G.E. Hinton, Reinforcement learning with factored states and actions, J. Mach. Learn. Res. 5 (2004) 1063–1088.

[33] N. Heess, D. Silver, Y.W. Teh, in: Actor-critic reinforcement learning with energy-based policies, JMLR Workshop and Conference Proceedings: EWRL, 2012.

[34] E. Mocanu, P.H. Nguyen, W.L. Kling, M. Gibescu, Unsupervised energy prediction under smart grid context using reinforcement cross buildings transfer, Energy Build. 116 (2016) 646–655.

[35] R.S. Sutton, A.G. Barto, Introduction to Reinforcement Learning, first ed., MIT Press, Cambridge, MA, 1998. ISBN: 0262193981.

[36] C.J. Watkins, P. Dayan, Technical note: Q-learning, J. Mach. Learn. Res. 8 (3–4) (1992) 279–292.

[37] M. Wiering, M. van Otterlo, Reinforcement Learning: State-of-the-Art, Springer, Heidelberg, New York, Dordrecht, London, 2012.

[38] L. Busoniu, D. Ernst, B. De Schutter, R. Babuska, in: Approximate reinforcement learning: an overview, IEEE Symposium on Adaptive Dynamic Programming And Reinforcement Learning (ADPRL), 2011, pp. 1–8.

[39] A.A. Markov, The theory of algorithms, in: Collection of Articles. To the Sixtieth Birthday of Academician Ivan Matveevich Vinogradov, Trudy Mat. Inst. Steklov., vol. 38, Acad. Sci. USSR, Moscow, 1951, pp. 176–189.

[40] M.L. Puterman, Markov Decision Processes: Discrete Stochastic Dynamic Programming, first ed., John Wiley & Sons, Hoboken, NJ, 1994.

[41] M. Castronovo, F. Maes, R. Fonteneau, D. Ernst, in: Learning exploration/exploitation strategies for single trajectory reinforcement learning, JMLR Proceedings EWRL, 24, 2012, pp. 1–10.

[42] G.E. Hinton, R.R. Salakhutdinov, Reducing the dimensionality of data with neural networks, Science 313 (5786) (2006) 504–507.

[43] G.E. Hinton, S. Osindero, Y.W. Teh, A fast learning algorithm for deep belief nets, Neural Comput. 18 (2006) 2006.

[44] L.J.P. van der Maaten, E.O. Postma, H.J. van den Herik, Dimensionality reduction: a comparative review, J. Mach. Learn. Res. 10 (1–41) (2009) 66–71.

[45] R. Salakhutdinov, in: Learning deep Boltzmann machines using adaptive MCMC, Proceedings of the 27th International Conference on Machine Learning, 2010, pp. 943–950.

[46] F.J. Massey, The Kolmogorov-Smirnov test for goodness of fit, J. Am. Stat. Assoc. 46 (253) (1951) 68–78.

[47] S.J. Devlin, R. Gnanadesikan, J.R. Kettenring, Robust estimation and outlier detection with correlation coefficients, Biometrika 62 (3) (1975) 531–545.

[48] E. Mocanu, E.M. Larsen, P.H. Nguyen, P. Pinson, M. Gibescu, in: Demand forecasting at low aggregation levels using factored conditional restricted Boltzmann machine, Power Systems Computation Conference, PSCC 2016, 20–24 June, Genoa, Italy, 2016.

[49] E. Mocanu, P.H. Nguyen, M. Gibescu, in: Energy disaggregation for real-time building flexibility detection, IEEE PES General Meeting 2016, 17–21 July Boston, MA, USA, 2016.

[50] D.C. Mocanu, E. Mocanu, H.P. Nguyen, M. Gibescu, A. Liotta, in: Big IoT data mining for real-time energy disaggregation in buildings, Proceedings of the IEEE International Conference on Systems, Man, and Cybernetics, 2016.

[51] D.L. Marino, K. Amarasinghe, M. Manic, in: Building energy load forecasting using deep neural networks, IEEE Industrial Electronics Society, 2016.

[52] S. Ryu, J. Noh, H. Kim, in: Deep neural network based demand side short term load forecasting, IEEE International Conference on Smart Grid. Communications, 2016, pp. 308–313.

[53] M. Manic, K. Amarasinghe, J.J. Rodriguez-Andina, C. Rieger, Intelligent buildings of the future: Cyberaware, deep learning powered, and human interacting, IEEE Ind. Electron. Mag. 10 (4) (2016) 32–49.

Further Reading

[1] L.E. Baum, in: An inequality and associated maximization technique in statistical estimation for probabilistic functions of Markov processes, Proceedings of the Third Symposium on Inequalities, 1972, pp. 1–8.

[2] K. Brügge, A. Fischer, C. Igel, The flip-the-state transition operator for restricted Boltzmann machines, Mach. Learn. 93 (1) (2013) 53–69.

[3] C.M. Bishop, Pattern Recognition and Machine Learning (Information Science and Statistics), first ed., Springer, New York, NY, 2006.

[4] C.C. Chang, C.J. Lin, Libsvm: a library for support vector machines, ACM Trans. Intell. Syst. Technol. 2 (3) (2011) 1–27.

[5] C. Cortes, V. Vapnik, Support-vector networks, Mach. Learn. 20 (3) (1995) 273–297.

[6] G. Desjardins, A. Courville, Y. Bengio, P. Vincent, O. Delalleau, in: Tempered Markov chain Monte Carlo for training of restricted Boltzmann machines, Proceedings of the 13th Int. Conf. on Artificial Intelligence and Statistics, 2010, pp. 145–152.

[7] S. Fan, R.J. Hyndman, Short-term load forecasting based on a semi-parametric additive model, IEEE Trans. Power Syst. 27 (1) (2012) 134–141.

[8] A. Foley, P.G. Leahy, A. Marvuglia, E.J. McKeogh, Current methods and advances in forecasting of wind power generation, Renew. Energy 37 (1) (2012) 1–8.

[9] B.C. Geiger, G. Kubin, in: Signal enhancement as minimization of relevant information loss, Systems, Communication and Coding (SCC), Proceedings of 2013 9th International ITG Conference, 2012.

[10] N. Nicola Jones, Computer science: the learning machines, Nature 505 (7482) (2014) 146–148.

[11] J. Laserson, From neural networks to deep learning: zeroing in on the human brain, ACM Crossroads 18 (1) (2011) 29–34.

[12] A.M. De Livera, R.J. Hyndman, R.D. Snyder, Forecasting time series with complex seasonal patterns using exponential smoothing, J. Am. Stat. Assoc. 106 (496) (2011) 1513–1527.

[13] E.M. Larsen, P. Pinson, G.L. Ray, G. Giannopoulos, in: Demonstration of market-based real-time electricity pricing on a congested feeder, 12th Int. Conf. on the European Energy Market, 2015, pp. 1–5.

[14] E.L. Lehmann, J.P. Romano, Testing Statistical Hypotheses, Springer Texts in Statistics, Springer-Verlag, New York, 2005.

[15] M. Lukoovsevivcius, A practical guide to applying echo state networks, Neural Networks: Tricks of the Trade, Lecture Notes in Computer Science, 7700, Springer, Berlin, Heidelberg, 2012, pp. 659–686.

[16] M. Lukoovsevivcius, H. Jaeger, Reservoir computing approaches to recurrent neural network training, Comput. Sci. Rev. 3 (3) (2009) 127–149.

[17] D.W. Marquardt, An algorithm for least-squares estimation of nonlinear parameters, SIAM J. Appl. Math. 11 (2) (1963) 431–441.

[18] L.R. Rabiner, Readings in speech recognition, A Tutorial on Hidden Markov Models and Selected Applications in Speech Recognition, Morgan Kaufmann Publishers, San Francisco, CA, 1990, pp. 267–296.

[19] T. Tieleman, G. Hinton, in: Using fast weights to improve persistent contrastive divergence, Proceedings of the 26th Annual Int. Conf. on Machine Learning, 2009, pp. 1033–1040.

[20] T. Tieleman, in: Training restricted boltzmann machines using approximations to the likelihood gradient, Proceedings of the 25th Int. Conf. on Machine Learning, 2008, pp. 1064–1071.

[21] M. Wytock, J.Z. Kolter, in: Large-scale probabilistic forecasting in energy systems using sparse gaussian conditional random fields, Proceedings of the 52nd Conference on Decision and Control, 2013, pp. 1019–1024.

[22] M. Welling, M. Rosen-Zvi, G.E. Hinton, in: Exponential family harmoniums with an application to information retrieval, Advances in Neural Information Processing Systems 17 (NIPS 2004), 2004.

[23] L. Yang, H. Yan, J.C. Lam, Thermal comfort and building energy consumption implications—a review, Appl. Energy 115 (2014) 164–173.

Compressive Sensing for Power System Data Analysis

Mohammad Babakmehr*, Mehrdad Majidi†, Marcelo G. Simoes*

*Colorado School of Mines, Golden, CO, United States, †University of Nevada, Reno, NV, United States

CHAPTER OVERVIEW

Within this chapter, we will introduce the applications of a state-of-the-art theorem in signal processing and system identification, named as compressive sensing-sparse recovery (CS-SR), in smart power networks monitoring, data analysis, security, and reliability. The sparse nature of the electrical power grids as well as electrical signals is exploited to introduce alternative mathematical formulations to address some of the most famous system modeling problems in power engineering through a compressive signal processing or a sparse system identification framework. First, a short background on CS-SR theorems and techniques is presented. Next, the state of the art in CS-SR applications in smart grid technology is discussed, and finally, the following three data analyses and power network control problems are specifically addressed in detail. The CS-SR techniques are exploited to propose novel methods for distribution system state estimation (DSSE), single and simultaneous fault location in smart distribution and transmission networks, and partial discharge pattern recognition.

1 INTRODUCTION

During the last two decades, the frontier technologies of sensor networks and data centers deeply affected the field of power engineering. A layer of information flows throughout the interconnected electrical power networks that transfers the measurements of electrical parameters from the system equipment to the decision-making hubs while feedback the monitoring signals from operational rooms to the controllers, within an almost online manner [1]. Besides various inherent advantages in such an online monitoring and recording framework, the huge amount of data arriving at each sample of time (to the operational and control rooms) would result in new technical and analysis issues for system operators (also known as the era of big data). Thus, despite the high resolution of the monitoring picture we extract from this comprehensive dataset a variety of challenges also arise in parallel including data storage, time cost, mathematical complexity, and so on [2].

To deal with the corresponding challenges associated with the big data analysis, various approaches have been developed in the literature. Beyond the state of

Big Data Application in Power Systems. https://doi.org/10.1016/B978-0-12-811968-6.00008-5

the art in the era of big data, the compressive sensing-sparse recovery (CS-SR) techniques have found enormous interest. The fundamental idea of CS-SR emerges from the fact that, if a signal of interest has an alternative representation (in some mathematical domain), with a small number of degrees of freedom, one should be able to capture its overall behavior using a much smaller number of measurements rather than its original time domain representation [3,4]. For example, based on the Shannon sampling theorem, to correctly measure the inherent characteristics of a typical 60 Hz sinusoidal signal (including a possible set of harmonics), the sampling frequency should exceed at least two times of the bandwidth of the signal (resulting in a dense time vector). However, the corresponding Fourier representation of such a signal would at most have a couple of nonzero coefficients within the corresponding frequency spectrum (resulting in a *sparse* frequency vector). During last decade, such a sparse behavioral pattern (which appears in most of the natural and industrial signals) motivated many researchers to investigate new mathematical tools and theorems to develop an alternative sensing approach that captures a small subset of measurements from a signal of interest, *compressive sensing* (roughly in the order of the sparsity level[1] of the signal), and still can recover its original behavior, *sparse recovery*.

Nevertheless, the sparse recovery techniques have been farther studied as an independent mathematical approach to address the analysis and time complexity of any practical problem with an inherent sparsity in its corresponding formulation [5]. Beyond this set of problems, those associated with a kind of graph model-based analysis are notable. Despite outward complexity, most of the corresponding monitoring problems in power networks do have a sparse structure in some sense. For example, as it has been suggested in Ref. [6], usually, the number of simultaneous line outages happening in a power network (PN) is a small subset of total number of transmission lines. Thus, one may assign a *sparse outage vector recovery* problem to identify the location of damaged lines. Moreover, due to local connectivity pattern in electrical networks, within a PN, the average connection degree[2] of a typical electrical bus is a small number; this fact results in a sparse corresponding graph representation for a PN. On the other hand, one may easily observe that most of the power signals such as voltages and currents are lying over a sparse representation in other mathematical domains such as Fourier domain, where most of the Fourier coefficients are usually very small or zero while only a small portion of the coefficients are significantly larger than the others. This inherent sparsity has been exploited as a

---

[1]A sparse signal $x \in \mathbb{R}^N$ is usually defined as a vector with most of its elements equal to zero except a small portion, K, of them, with $K \ll N$. K is also referred to the sparsity level of the signal.

[2]In this work, connection-degree is referred to the total number of neighbor buses directly connected to an electrical bus.

progressive fact in a variety of related signal processing problems, such as partial discharge (PD) pattern recognition [7,8], power quality event analysis, and classification [9].

In the rest of this chapter, we will first give a short introduction to CS-SR definitions, techniques, and theorems (Section 2). Next, we present a general overview on the state of the art in the applications of CS-SR in the field of power engineering (Section 3). Finally, we would investigate, by details, the following CS-SR-based approaches to address three of the monitoring issues in power systems: distribution system state estimation (DSSE), single and simultaneous fault location in smart distribution networks, and PD pattern recognition.

2 MATHEMATICAL MODELING OF A COMPRESSIVE SENSING-SPARSE RECOVERY PROBLEM

CS-SR is a new paradigm in signal processing that tries to introduce new alternative data acquisition protocols to reduce the size of the required storage unit, time cost, and complexity in the subordinate analysis. In the next two following sections, we briefly describe the fundamentals of CS-SR with an emphasis on concepts which are used throughout the rest of this chapter.

2.1 Compressive Sensing

Roughly speaking, there is a simple question that emerged the theory of compressive sensing: "if a signal of interest has small degrees of freedom why should we record too many samples to capture its corresponding behavior?" For example, a sinusoidal signal can be defined as $A\sin(2\pi ft + \varphi)$, with three characteristic parameters including A, f, and φ. However, we would need a sampling frequency as high as at least two times the bandwidth of the signal to avoid the aliasing effect [10]. The number of recorded samples under such a sampling frequency, namely N (also referred to the original domain dimensionality of the signal), is usually much bigger than the number of representative parameters (also interpreted as the real dimensionality of the signal). The theorem of compressive sensing is then trying to develop a superseded measuring framework to capture a much smaller number of measurements, namely M (which changes in the order of K and $M \ll N$), while still be able to reveal the original behavior of the signal or recover its original structure using this partial information.

Definition 1—Sparse signal. A signal $x \in \mathbb{R}^N$ is said to be sparse if most of its elements are equal to zero except a small number, K, where $K \ll N$.

Definition 2—Compressible signal. A signal $x \in \mathbb{R}^N$ is said to be compressible if most of its elements are close to zero except a small number, K, where $K \ll N$.

Theorem 1 [11]. *Consider a signal $x \in \mathbb{R}^N$ is K-sparse or has a K-sparse representation in some domain ($x = D\alpha$, $D \in \mathbb{R}^{N \times N}$, $\alpha \in \mathbb{R}^N$ and K-sparse); one can recover the correct structure of x (alternatively α) from a set of measurements $y = Ax$ (or $y = AD\alpha$), $y \in \mathbb{R}^M$, $A \in \mathbb{R}^{M \times N}$ and $M \ll N$, if the sensing protocol A (or in general AD) satisfies a set of conditions widely known as sparse recovery guarantee conditions and with a number of measurements that scale like $K \log(N/K)$.*[3]

2.2   Sparse Recovery Problem

In this part, we investigate the definition of a sparse recovery problem in addition to the essentials and corresponding theorems (please also refer to Refs. [3,12] and references therein for more details).

Definition 3—Coherence. The coherence of a matrix $A \in \mathbb{R}^{M \times N}$ with normalized columns $a_1, \ldots, a_N$ is defined as follows:

$$\mu_A = \max \langle a_i, a_j \rangle \text{ for } i, j = 1 : N \text{ and } i \neq j \tag{1}$$

Definition 4—Restricted isometry property. The matrix $A \in \mathbb{R}^{M \times N}$ is said to satisfy the restricted isometry property of the order K with isometry constant $\delta_K \epsilon (0, 1)$ if the following inequality holds for all the K-sparse vectors $x \in \mathbb{R}^N$:

$$(1 - \delta_K)\|x\|_2^2 \leq \|Ax\|_2^2 \leq (1 - \delta_K)\|x\|_2^2 \tag{2}$$

Theorem 2 [13]. *Consider $x \in \mathbb{R}^N$ to be a K-sparse signal and a set of measurements $y = Ax \in \mathbb{R}^M$. If A satisfies the RIP condition of the order 2K with some isometry constant $\delta_{2K} < 1$, or alternatively if $\mu_A < \frac{1}{2K-1}$, then the solution.*

$$\hat{x} = \underset{\acute{x}}{\text{argmin}} \|\acute{x}\|_0 \text{ s.t.} y = A\acute{x} \tag{3}$$

is the unique solution to $y = A\hat{x}$ having sparsity level K or less, where $\|x\|_0$ indicates the number of nonzero coefficients in x or alternatively its cardinality. Unfortunately, this problem is known to be mathematically NP-hard [13]. Fortunately, the following relaxed version of such a problem has been introduced to address the corresponding time complexity issue:

$$\hat{x} = \underset{\acute{x}}{\text{argmin}} \|\acute{x}\|_1 \text{ s.t.} y = A\acute{x} \tag{4}$$

The l_1-nrom is also a convex function in x and tends to have a small value for sparse signals.

---

[3]For ease of notation we consider the sensing protocol matrix to be a single matrix A and the K-sparse vector to be x, one may refer to Refs. [3,12] for detail definitions and concepts related to alternative sparse representation in the other domains, $D\alpha$, and the corresponding definition and concepts such as mutual coherence of two matrices A and D.

ALGORITHM 1 ORTHOGONAL MATCHING PURSUIT (OMP)

require: matrix A, measurements y, stopping criterion
initialize: $r^0 = y, x^0 = 0, l = 0, SUP^0 = \varnothing$
  repeat
    1.match: $h^l = A^T r^l$
    2.identify support indicator:
      $sup^l = \{argmax_j \, | h^l(j) | \}$
    3.update the support:
      $SUP^{l+1} = SUP^l \cup sup^l$
    4.update signal estimate:
      $x^{l+1} = argmin_{z:SUP(z) \subseteq SUP^{l+1}} \| y - Az \|_2$
      $r^{l+1} = y - Ax^{l+1}$, $l = l + 1$
  Until stopping criterion met
output: $\hat{x} = x^l$

Theorem 3 [13]. *Consider $x \in \mathbb{R}^N$ to be a K-sparse signal and a set of measurements $y = Ax \in \mathbb{R}^M$. If A satisfies the RIP condition of the order 2K with some isometry constant $\delta_{2K} < 0.4651$, or alternatively $\mu_A < \frac{1}{2K-1}$, then the solution $\hat{x}$ to Eq. (4) correctly returns the K-sparse solution to $y = A\hat{x}$. This problem is widely known as Basis Pursuit.*

Theorem 4 [13]. *Consider x to be any vector $\mathbb{R}^N$ and a set of noisy measurements $y = Ax + \varepsilon \in \mathbb{R}^M$. If A satisfies the RIP condition of the order 2K with some isometry constant $\delta_{2K} < 0.4651$, then the solution $\hat{x}$ to.*

$$\hat{x} = \underset{\hat{x}}{argmin} \| \hat{x} \|_1 \text{ s.t.} \| y - A\hat{x} \|_2 \leq \eta \qquad (5)$$

with $\eta \geq \| \varepsilon \|_2$ obeys:

$$\| x - \hat{x} \|_2 \leq \frac{C_1 \| x - x_K \|_1}{\sqrt{K}} + C_2 \eta$$

where x_K is said to be the nearest K-sparse signal to x and C_1 and C_2 are two constants dependent on δ_{2K}. This problem is widely known as the basic pursuit de-noising (BPDN).

Besides the usual convex optimization-based approaches to solve the BP or BPDN, a set of alternative greedy methods have been developed to find the correct solution x. Rather than searching among possible candidate solutions for Eqs. (4), (5), greedy methods try to directly reconstruct a sparse solution. A famous and well-known greedy method is the orthogonal matching pursuit (OMP) [14], which has been detailed in Algorithm 1. Roughly speaking, this method attempts to find the true support of the signal x (position of nonzero coefficients) within a stepwise correlation minimization procedure.

3 APPLICATIONS OF CS-SR TECHNIQUES IN SMART GRIDS

During last couple of years, CS-SR techniques started penetrating the field of power engineering. Nowadays, a variety of applications have been introduced for CS-SR within the smart power grids technology (SG) [15–28]. Although SGs are usually forming a huge interconnected and complex system there is an inherent sparsity in both the structure and data format within these complex networks. For instance, consider a PN as graph $G(S_N, S_E)$, consisting of a set of N nodes $S_N = \{1, \ldots, N\}$, where each node i represents an arbitrary bus of the SG, and a set of L edges or transmission lines $S_E = \{l_{i,j} : i, j \epsilon S_N\}$, which connect certain nodes and form the general structure or topology of the grid. Considering a variety of standard PN models, it can be observed that the regular maximum connectivity level of an electrical bus in a grid is typically <5%–10%, especially in case of large-scale PNs [29]. As suggested in Ref. [30], the same fact can be observed in case of distribution networks as well. This sparsity in structure has been used in Refs. [5,15], to introduce an alternative compressive sensing-based SG topology identification.

The sparse recovery-based power line outage identification has been first addressed in Ref. [6] and further developed in Ref. [16]. The basic idea comes from the fact that the number of simultaneous outages happening within a power grid would be a small portion of the total number of the transmission lines. Incorporating the linear DC power flow model [17], $p = B\theta \epsilon R^N$, into some linear algebraic matrices properties a failure event has been modeled (with a set of damaged lines ζ_{out}) in terms of a set of linear systems of equations termed as power outage identification-sparse recovery problem (POI-SRP) such that within the POI-SRP formulation we have $y = Ax + n$, where $y \epsilon R^N$ is termed the measurement vector and the matrix $A \epsilon R^{N \times L}$ is termed the sensing matrix. It has been shown that the resulting vector $x \epsilon R^L$ can be considered as a sparse vector with most of its elements equal to zero except on the corresponding L_{out} outage positions. The resulting Sparse Outage Vector (SOV) follows the following mathematical format: $x = \left\{ \begin{array}{ll} x_l & \text{if } l \in \zeta_{out} \\ 0 & \text{otherwise} \end{array} \right\}$.

Clearly, recovering such an SOV, x, one can easily obtain the outage lines index set ζ_{out} by looking at the nonzero positions. In order to address the inherent challenges from the sparse recovery perspective a couple of generalizations have been introduced in Refs. [16,18], such as Binary-POI-SRP and Structured-POI-SRP.

Moreover CS-SR has found application in other SG monitoring and control issues, such as attack recognition [19], communicational protocols [20,21], and power signals processing [7–9]. CS-SR techniques have been also exploited to propose novel methods for single and simultaneous fault location in smart distribution and transmission networks [22–24], DSSE [25,26], and smart grid dynamic behavior modeling [27]. Also, an alternative sparse-based formulation has been developed in Ref. [28] to address the power-flow modeling problem.

Within the rest of this chapter we will take a deeper look at the CS-SR-based frameworks which have been developed to address the following three distinct monitoring issues in the distribution level systems in smart grids:

(1) DSSE in smart grid [25,26].
(2) Fault location in smart grids [22–24].
(3) PD pattern recognition [7,8].

4 SPARSE RECOVERY-BASED DSSE IN SMART GRID

The voltage profiles along the distribution networks show that voltage differences between two buses of each line segment are so small and almost negligible compared with the voltage in the beginning of the distribution feeder. This electrical characteristic in the distribution networks is used to sparsify the voltage profile by a difference transformation. The sparsified voltage profile is recovered by a few micro-phasor measurement units (µPMUs) installed along the network. The proposed DSSE technique does not need system observability and measurement redundancy that are the main requirements for conventional weighted least square state estimators. µPMUs record current passing through adjacent branches to buses and bus voltages synchronously [31]. The captured phasors are related to the system states by:

$$Z_0 = H_0 x_0 + \nu_0 \tag{6}$$

where $Z_0 \in R^{m_0}$, $m_0 = p + q$, is the measurement vector with p three-phase voltage and q three-phase current measurements, $x_0 \in R^{n_0}$ is the three-phase state vector, $\nu_0 \in R^{m_0}$ is the measurement error vector, and $H_0 \in R^{m_0 \times n_0}$ is the constant measurement Jacobian matrix:

$$H_0 = \begin{bmatrix} II \\ \gamma A + \gamma_s \end{bmatrix} \tag{7}$$

where $II \in R^{p \times n_0}$ is the three-phase representation of the voltage measurement-bus incident matrix, $\gamma \in R^{q \times q}$ is the three-phase series admittance matrix, $A \in R^{q \times n_0}$ is the three-phase representation of current measurement-bus incident matrix, and $\gamma_s \in R^{q \times n_0}$ is the three-phase shunt admittance matrix. By writing the Eq. (6) in rectangular coordinates:

$$[Re\{Z_0\} + jIm\{Z_0\}] = [Re\{H_0\} + jIm\{H_0\}] \cdot [Re\{x_0\} + jIm\{x_0\}] + [Re\{\nu_0\} + jIm\{\nu_0\}] \tag{8}$$

Separating the real and imaginary terms, Eq. (8) is rewritten in matrix form:

$$\overbrace{\begin{bmatrix} Re\{Z_0\} \\ Im\{Z_0\} \end{bmatrix}}^{Z} = \overbrace{\begin{bmatrix} Re\{H_0\} & -Im\{H_0\} \\ Im\{H_0\} & Re\{H_0\} \end{bmatrix}}^{H} \overbrace{\begin{bmatrix} Re\{x_0\} \\ Im\{x_0\} \end{bmatrix}}^{x} + \overbrace{\begin{bmatrix} Re\{v_0\} \\ Im\{v_0\} \end{bmatrix}}^{v} \tag{9a}$$

$$Z = H \cdot x + v \tag{9b}$$

where $Z \in R^m$, $H \in R^{m \times n}$, $x \in R^n$, and $v \in R^m$ ($m = 2m_0$, $n = 2n_0$) are representations of Z_0, H_0, x_0, and v_0 in rectangular coordinates, respectively.

In the DSSE problem, the number of measurements is less than the number of states $m < n$. Therefore, Eqs. (9a), (9b) is an underdetermined system of equations with infinite solutions. However, if there is a priori knowledge that x is a sparse vector most of whose entries are zero or negligible, it can be recovered with a few measurements by solving the ℓ_1-regularized least-squares problem [32]

$$\min \| H \cdot x - Z \|_2^2 + \lambda \|x\|_1 \tag{10}$$

where $\|\cdot\|_1$ denotes the ℓ_1 norm and $\lambda > 0$ is the regularization factor.

When (x) is not sparse but is compressible, it can be transformed to a sparse vector by

$$x_s = S \cdot x \tag{11}$$

where $x_s \in R^n$ is the sparse representation of the vector x in the transform domain and $S \in R^{n \times n}$ is a sparsifying transform matrix. When S is nonsingular, the vector (x) is obtained by $x = S^{-1} \cdot x_s$ and substituted in Eq. (9b) to form:

$$Z = \psi \cdot x_s + v \tag{12}$$

where $\psi = H \cdot S^{-1}$. Eq. (12) is solved by

$$\min \| \psi \cdot x_s - Z \|_2^2 + \lambda \|x_s\|_1 \tag{13}$$

The solution of Eq. (13) x_s^* yields $x^* = S^{-1} \cdot x_s^*$ as the solution of Eq. (10).

For the DSSE application, voltages at the two buses of any segment are subtracted from each other by a difference transformation. If bus 1 is the infeed bus and matrix (B) is the three-phase branch-bus incident matrix, the state vector (x_0) in Eq. (6) is sparsified by

$$S = \begin{bmatrix} I_3 & O_{3 \times (n-3)} \\ O_{(n-3) \times 3} & B \end{bmatrix} \tag{14}$$

where $O_{M \times N}$ is an M-by-N matrix of zeros and I_N is an N-by-N identity matrix.

The performance of the proposed DSSE technique is validated through the simulation results of the IEEE 123-bus standard unbalanced network containing three-, double-, and single-phase branches with 256 phase-to-ground voltage variables. The minimum number of µPMUs is optimally placed along the network by solving a binary integer linear programming problem in order to obtain a unique solution for the proposed state estimation formulation. Using the placement algorithm, eight µPMUs are placed at buses 1, 13, 44, 60, 65, 83, 89, and 101 to provide 86 voltage and current phasors.

Actual and estimated voltages in 256 phase-to-ground nodes are illustrated in Fig. 1. The mean absolute error of the estimated states is 8.017×10^{-5} (p.u.). This value implies that the proposed technique accurately estimates $n = 2n_0 = 2 \times 256 = 512$ system states using only $m = 2m_0 = 2 \times 84 = 168$ measurements that are just 32.8% of the system states. In addition, the actual and estimated sparse representations of the voltage profile are shown in Fig. 2 where the sparse representation of the voltage profile includes three

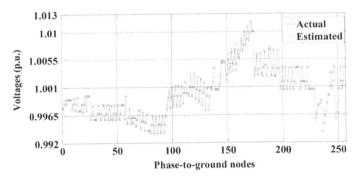

FIG. 1
Actual and estimated voltages in the IEEE 123-bus system.

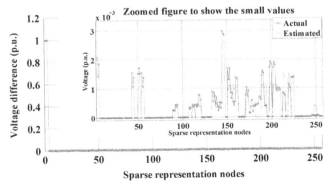

FIG. 2
Actual and estimated sparse representations of the voltage profile.

significant values for the first three nodes and negligible coefficients for other nodes.

As shown, in addition to the dominant nonzero values corresponding to the infeed point voltage, the ℓ_1-regularized least-squares problem will identify certain nonzero voltage drops between the two ends of each segment connected to big loads, capacitor banks, DGs, and main three-phase feeders supplying several laterals.

5   SPARSE RECOVERY-BASED FAULT LOCATION IN SMART DISTRIBUTION AND TRANSMISSION NETWORKS

This section introduces a new paradigm for locating single and simultaneous faults in power distribution and transmission networks. The idea is inspired by the underlying hypothesis that the voltage sag vector, which is the difference between the pre and during-fault voltages at all buses (N buses), is obtained through multiplying the system impedance matrix ($Z_{bus} \in R^{N \times N}$) by the injection fault current vector ($\Delta I \in R^N$) which is a sparse vector whose nonzero value(s) determine the fault location(s). Since measuring the voltage sag at all buses is not plausible and economic specifically in the distribution networks, the voltage sag values are measured by PMUs/smart meters at a few buses ($M \ll N$ buses) and the corresponding rows to those measurement buses are selected from the impedance matrix to form an underdetermined equation system. The compressive sensing and ℓ_1-norm minimization are then exploited to recover the sparse current vector as

$$(\ell^1) : \widetilde{\Delta I}_1 = \text{argmin} \, \|\Delta I\|_1 \text{ s.t.} \Delta V = Z \cdot \Delta I \tag{15}$$

where $\widetilde{\Delta I}_1 \in R^N$ is the estimated current vector, $Z \in R^{M \times N}$ is the modified system impedance matrix, and $\Delta V \in R^M$ is the voltage sag vector obtained by $M \ll N$ measurements in the system.

Due to lack of measurement redundancy, the number of nonzero values in the current vector is not necessarily equal with number of faults.

Therefore, the nonzero values in the recovered current vector are investigated by the fuzzy c-mean clustering algorithm to derive four possible fault locations with this assumption that no more than three simultaneous faults occur. Also, if it is assumed that single faults only occur in the system, a machine learning technique based on the k-nearest neighbor is proposed to analyze the nonzero values and estimate a single fault location. If there are (P) dominant values in the normalized current vector whose row numbers are $Q_i, i = 1, \ldots, P$, the estimated distance of the faulted point to substation is calculated using Eq. (16) based on the k-nearest neighbor technique [33].

$$\tilde{F}_L = \frac{\sum_{i=1}^{P} F_L(i).e^{-(1-\Delta I_n(Q_i))^2}}{\sum_{i=1}^{P} e^{-(1-\Delta I_n(Q_i))^2}} \tag{16}$$

where ΔI_n denotes the normalized current vector and $F_L(i)$ is the distance of the bus associated with the ith dominant value to the substation. The simulation results of a 13.8 kV, 134 bus distribution network are used to validate the effectiveness of the proposed fault location method. The fault location errors are obtained by subtracting the distance of the actual faulted bus to substation from the estimated one and categorized in 100 m steps.

The numbers of buses estimated between 0 and 100 m, 100 and 200 m, 200 and 300 m, 300 and 400 m, and higher than 400 m are counted individually and presented in the result tables. For example, Table 1 presents the results for locating single faults using smart feeder meter data. Method 1 uses Fuzzy-c means to estimate four possible faulted points and Method 2 uses k-nearest neighborhood to find a single faulted point.

In the aforementioned studies, it is assumed that faults occur at buses in distribution networks as the distribution lines are relatively short compared with transmission lines. However, it is necessary to locate the fault along the faulty line(s) in the transmission networks. Therefore, the recovered nonzero values in the current vector reconstructed by solving Eq. (15) are used for faulty line(s) detection. The substitution theorem is then used to pinpoint the fault along the line(s) [34]. If line $i-j$ is identified as the faulty line, the fault current injected from each of its buses to the fault point (I_{ij}, I_{ji}) is modeled with

Table 1 Simulation Results for Locating Single Faults Using Smart Meters

Fault Type	1pH-g				3pH				2pH-g				2pH			
Method No.	1		2		1		2		1		2		1		2	
ohm Error	0.5	10	0.5	10	0.5	10	0.5	10	0.5	10	0.5	10	0.5	10	0.5	10
0–100 m	124	120	124	112	121	118	115	110	126	125	122	116	122	126	110	120
100–200 m	7	10	8	16	8	9	15	14	4	6	10	14	8	4	20	11
200–300 m	2	2	1	4	3	5	2	6	3	2	1	3	3	3	3	2
300–400 m	0	1	0	1	1	1	1	3	0	0	0	0	0	0	0	0
>400 m	0	0	0	0	0	0	0	0	0	0	0	0	0	0	0	0
Total error (km)	4.4	4.6	5	6.7	5.4	5.1	6.8	7.9	4.4	3.3	5.4	6.5	5.1	4.4	6.7	5.7

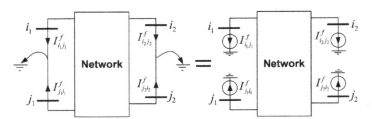

FIG. 3
Schematic of substitution theorem in a double-fault scenario.

an identical current source in each bus after the line removal. For two simultaneous faults along the lines $i_1 - j_1$ and $i_2 - j_2$, the system is modeled as shown in Fig. 3.

In the positive-sequence system, the fault current sources can be related to the voltage sag phasors in the observable buses and the system impedance matrix in which the faulted lines are removed as:

$$\Delta V_{All}^+ = Z_{bus,M}^+ \cdot \Delta I_l^+ \tag{17}$$

where $\Delta I_l^+ = [0\ldots 0\ I_{i_1 j_1}{}^f \ldots 0\ I_{j_1 i_1}{}^f \ldots 0\ I_{i_2 j_2}{}^f \ldots 0\ I_{j_2 i_2}{}^f \ldots 0]^T \in R^N$ is the faulted lines current vector, $\Delta V_{All}^+ = [\Delta V_1^+, \Delta V_2^+ \ldots \Delta V_M^+, \Delta V_{M+1}^+, \ldots \Delta V_{M+K}^+]^T \in R^{M+K}$ is the voltage sag vector in all observable buses including M measurement and K calculated buses, and $Z_{bus,M}^+$ is the modified impedance matrix of the network with lines $i_1 - j_1$ and $i_2 - j_2$ removed. Please note that the recovered nonzero values in the faulty line detection process determine the possible faulty area(s) in the system. Since the current phasors in the lines connected to the PMU buses are also measured by PMUs, the during- and prevoltage phasors in the incident buses to PMU buses placed in the healthy areas are calculated using the measured current phasors. Therefore, K calculated voltage phasors are added to M measured voltage phasors in the fault distance estimation.

Columns i_1, j_1, i_2, and j_2 of $Z_{bus,M}^+$ are used to obtain the nonzero values in ΔI_l^+ by the overdetermined equation system whose number of equations is more than the number of variables

$$\Delta V_{All}^+ = \begin{bmatrix} z_{1,i_1} & z_{1,j_1} & z_{1,i_2} & z_{1,j_2} \\ z_{2,i_1} & z_{2,j_1} & z_{2,i_2} & z_{1,j_2} \\ \vdots & \vdots & \vdots & \vdots \\ z_{M+K,i_1} & z_{M+K,j_1} & z_{M+K,i_2} & z_{M+K,j_2} \end{bmatrix} \cdot \begin{bmatrix} I_{i_1 j_1}^f \\ I_{j_1 i_1}^f \\ I_{i_2 j_2}^f \\ I_{j_2 i_2}^f \end{bmatrix} \tag{18}$$

Eq. (18) is solved by the least-squares method to find the nonzero values in the ΔI_l^+ that are used to calculate the voltage sag phasors at buses i_1, j_1, i_2, and j_2 by Eq. (19) unless they are directly measured by PMUs.

$$
\begin{bmatrix} \Delta V_{i_1}^+ \\ \Delta V_{j_1}^+ \\ \Delta V_{i_2}^+ \\ \Delta V_{j_2}^+ \end{bmatrix} = \begin{bmatrix} z_{i_1,i_1} & z_{i_1,j_1} & z_{i_1,i_2} & z_{i_1,j_2} \\ z_{j_1,i_1} & z_{j_1,j_1} & z_{j_1,i_2} & z_{j_1,j_2} \\ z_{i_2,i_1} & z_{i_2,j_1} & z_{i_2,i_2} & z_{i_2,j_2} \\ z_{j_2,i_1} & z_{j_2,j_1} & z_{j_2,i_2} & z_{j_2,j_2} \end{bmatrix} \cdot \begin{bmatrix} I_{i_1 j_1}^f \\ I_{j_1 i_1}^f \\ I_{i_2 j_2}^f \\ I_{j_2 i_2}^f \end{bmatrix}
\tag{19}
$$

The fault distance ratio is then estimated by the calculated voltage sags and current phasors of the faulted lines.

$$
\delta_{ij} = \frac{1}{\gamma_{ij} l_{ij}}
$$

$$
\tan h^{-1} \left\{ \frac{-\cos h\left(\gamma_{ij} l_{ij}\right) \Delta V_j^+ - Z_{ij}^0 \sin h\left(\gamma_{ij} l_{ij}\right) I_{ji}^f + \Delta V_i^+}{-\sin h\left(\gamma_{ij} l_{ij}\right) \Delta V_j^+ - Z_{ij}^0 \cos h\left(\gamma_{ij} l_{ij}\right) I_{ji}^f - Z_{ij}^0 I_{ij}^f} \right\}
\tag{20}
$$

δ_{ij} is calculated for every faulted line, lines $i_1 - j_1$ and $i_2 - j_2$. Similarly, Eqs. (17)–(20) can be derived for single fault location.

All fault types with 0, 50, and 100 Ω fault resistances are simulated at 90%, 50%, and 10% of every line in the IEEE 39-bus test system to evaluate the effectiveness of the proposed fault location algorithm. The fault location estimation errors are calculated by Eq. (21) and presented in Table 2.

$$
\text{Error}(\%) = \frac{|\text{Actual distance} - \text{Estimated distance}|}{\text{Faulted line length}} \times 100\%
\tag{21}
$$

6 COMPRESSIVE SENSING-BASED PD PATTERN RECOGNITION

Compressive sensing and ℓ_1-norm minimization are also used in the sparse representation classifier (SRC) for the pattern recognition purpose. This section introduces a new application of SRC in PD pattern recognition. PD lead to electrical insulation failures in high voltage (HV) and extra high voltage (EHV) equipment in the long term and consequently electrical equipment outages and cascading failures. Therefore, on-line PD monitoring schemes are essential in academic and industry investigations. Each type of PD source has a particular pattern which can be recognized by some discriminatory features that can be extracted from the raw measured data. A classical PD feature extraction technique is phase-resolved partial discharge (PRPD) by which three main features of PDs are extracted. They are maximum charge, average charge, and number of PD pulses in each degree of a 360-degree cycle [35]. The extracted features are trained by a classifier to recognize different patterns.

Table 2 Performance of Fault Location Method for Single Fault Cases

Fault Type	(ohm)	Line Percentage (%)	Error Ranges (%)					Average	Variance
			0–0.1	0.1–0.2	0.2–0.3	0.3–0.4	0.4–0.5		
Three-phase fault	100	90	11	10	8	1	4	0.1975	0.0179
		50	5	13	8	4	4	0.2291	0.0137
		10	5	13	8	5	3	0.2311	0.0160
	50	99	11	12	6	2	3	0.1844	0.0158
		90	11	7	9	3	4	0.2114	0.0208
		50	5	12	10	3	4	0.2331	0.0148
		10	6	13	6	7	2	0.2182	0.0156
		1	12	11	5	4	2	0.1744	0.0173
	0	90	11	8	11	0	4	0.1998	0.0188
		50	6	10	9	6	3	0.2288	0.0143
		10	6	15	8	4	1	0.2041	0.0116
Single-phase-to-ground fault	100	90	8	12	7	2	5	0.2172	0.0202
		50	7	11	7	6	3	0.2229	0.0157
		10	8	11	7	4	4	0.2286	0.0194
	50	99	2	16	8	5	3	0.2471	0.0125
		90	12	7	3	4	8	0.2323	0.0310
		50	3	10	10	5	6	0.2773	0.0164
		10	4	16	7	0	7	0.2380	0.0182
		1	2	16	6	6	4	0.2448	0.0159
	0	90	10	12	9	1	2	0.1871	0.0145
		50	7	13	6	5	3	0.2133	0.0188
		10	7	13	9	3	2	0.2001	0.0144
Double-phase fault	100	90	10	10	10	0	4	0.1983	0.0181
		50	5	15	7	3	4	0.2225	0.0146
		10	5	15	7	3	4	0.2238	0.0153
	50	99	10	15	6	1	2	0.1621	0.0125
		90	8	13	5	5	3	0.2165	0.0164
		50	5	11	11	4	3	0.2357	0.0141
		10	7	14	5	5	3	0.2210	0.0176
		1	12	11	8	1	2	0.1682	0.0148
	0	90	10	9	10	1	4	0.2030	0.0184
		50	6	11	8	7	2	0.2289	0.0130
		10	6	17	3	5	3	0.2108	0.0154
Double-phase-to-ground fault	100	90	11	11	8	2	2	0.1919	0.0158
		50	6	12	8	5	3	0.2266	0.0151
		10	6	13	8	4	3	0.2255	0.0164
	50	99	8	13	6	4	3	0.1952	0.0159
		90	8	10	7	5	4	0.2301	0.0200
		50	4	12	10	6	2	0.2346	0.0139
		10	5	13	8	5	3	0.2228	0.0137
		1	10	12	6	4	2	0.1861	0.0151
	0	90	10	9	10	1	4	0.2008	0.0197
		50	4	10	11	5	4	0.2393	0.0153
		10	8	14	5	4	3	0.2055	0.0158

In order to assess the SRC performance for PD pattern recognition, 17 samples are created in a high-voltage lab for classifying internal, surface, and corona PDs. The 17 samples include 15 samples associated with internal PDs and the two samples for surface and corona PD measurements. The 15 samples consist of 1–5 air voids with dimensions 1, 1.5, and 2 mm. Using the captured PD signals, the features are extracted by a combination of PRPD and signal norms and fed to SRC for the pattern recognition. As shown in Fig. 4, PDs are measured during 3000 cycles for each sample and 250 cycles, each comprising 3600 measurements, are randomly selected. Note that each cycle does not necessarily include 3600 PD events. However, the recording system can record the possible PD events in each 0.1-degree interval.

The maximum discharge (q_{mx}), the mean discharge (q_{mn}), and the rate of PD repetition (q_n) are calculated for every 10 measurements per degree. Therefore, a 3 × 360 matrix is obtained for each cycle. In Fig. 4, qn_i^j, qmx_i^j, and qmn_i^j are the number of PD pulses, maximum discharge, and mean discharge given by the sum of all discharge magnitudes divided by number of PDs in ith cycle and jth degree, respectively, where $i=1, \ldots, 250$ and $j = 1,\ldots,360$. The figure also shows the qn, qmx, and qmn curves versus cycle numbers and degrees. Next, the 1-norm, 2-norm, and infinity norm are calculated for each row of the matrix to obtain a 9 × 1 vector. The process is repeated for all 250 selected cycles to obtain a 9 × 250 matrix whose columns are the 9 × 1 vectors of each cycle. The 9 × 250 matrix is the representative for each sample from which 200 feature vectors are used for training and 50 vectors are used for testing in the pattern recognition step. As Fig. 5 shows, the training matrices for every sample $T_i \in R^{9\times200}, i=1, \ldots, M=17$ are linked together to form a dictionary matrix (A) and $y \in R^9$ is the test vector which is selected from every 50 test vectors of each sample.

The SRC equation system in Fig. 5 is solved by Eq. (22) to recover the x vector whose nonzero values assign the selected test sample to one specific class.

$$\tilde{x}_1 = \mathrm{argmin}\, \|x\|_1 \text{ subject to } y = Ax \tag{22}$$

Since the nonzero terms in the recovered $\tilde{x}_1$ may correspond to different classes, the following residual function can be formed to minimize the ℓ_2-distance between the actual signal feature vector and the recovered

$$r_i = \|y - A\tilde{x}_1(i)\|_2, i = 1,\ldots,M \tag{23}$$

where $\tilde{x}_1(i)$ is equal to $\tilde{x}_1$ vector but with all entries other than the n_i entries corresponding to the ith class set equal to zero. The residual of Eq. (23) must be calculated for all $M=17$ classes and the test sample y is assigned to the class with the minimum residual. In order to assess the SRC performance for the PD pattern recognition, four different scenarios are defined below:

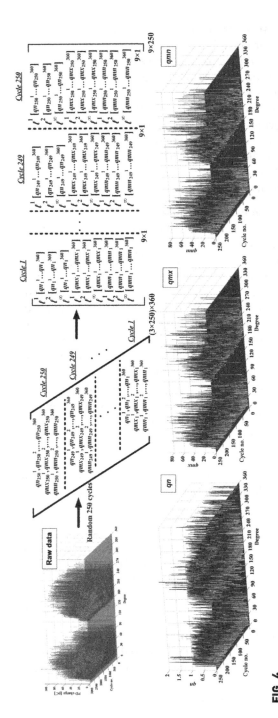

FIG. 4

Schematic of the feature extraction algorithm.

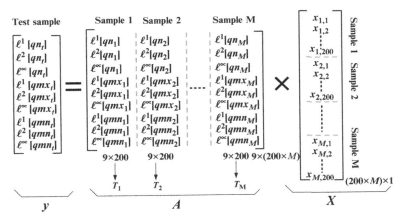

FIG. 5
SRC equation system.

- *Scenario 1 (7 samples)*: samples 1,2,...,5 include PDs of 1,2,...,5 internal voids 1 mm, sample 6 is corona, and sample 7 is surface discharge.
- *Scenario 2 (7 samples)*: samples 1,2,...,5 include PDs of 1,2,...,5 internal voids 1.5 mm, sample 6 is corona, and sample 7 is surface discharge.
- *Scenario 3 (7 samples)*: samples 1,2,...,5 include PDs of 1,2,...,5 internal voids 2 mm, sample 6 is corona, and sample 7 is surface discharge.
- *Scenario 4 (17 samples)*: samples 1,2,...,15 include PDs of 1,2,...,5 internal voids 1 mm, 1,2,...,5 internal voids 1.5 mm, and 1,2,...,5 internal voids 2 mm, sample 16 is corona, and sample 17 is surface discharge.

In order to compare the SRC performance with ANN classifier, a FFBP-NN Multilayer Preceptor (MLP) is also used to identify PD classes in all four scenarios. The identification percentages for each method in all four scenarios are presented in Table 3.

As shown, both methods have comparable performance if training process, tuning options, and other tasks for finding the best result from ANN are not taken into account. Even with this assumption, it is shown that SRC still outperforms ANN in some cases.

7 CONCLUSIONS AND FUTURE OF THE CS-SR IN SMART GRIDS

Compressive sensing is an emerging signal processing technique whose applications in different fields are investigated. This chapter demonstrates some applications of this technique in smart grids. The promising results confirm the practical usefulness of this technique in power systems. The proliferation

Table 3 Identification Percentages in All Scenarios With SRC and ANN Methods

No	Method	Identification Percentage (%)			
		Scenario 1	Scenario 2	Scenario 3	Scenario 4
1	SRC	99.7	92.9	94.0	81.6
4	ANN	90.8	95.9	93.6	82.6

of distribution energy resources (DER), e.g. electric vehicles, distributed generators (DGs), and demand response, necessitates visibility and situational awareness in smart networks. Unlike legacy radial networks with one-way power flow, smart distribution networks with multiple sources contain notable variability and uncertainties that must be continuously observed and actively managed. Distribution systems are not fully telemetered so far due to the huge cost of deploying the required infrastructure in thousands or tens of thousands of feeders. Therefore, pseudo measurements are not widely available in real distribution networks. For this reason, the state estimation and fault location algorithms using a few low-cost and highly accurate μPMUs have more practical benefits. To improve the reliability and resiliency of the power networks, efficient fault location schemes must be provided to facilitate rapid response to abnormal grid conditions. As a result, utilities will be able to better detect and head off potential blackouts, while improving day-to-day grid reliability and enhance the integration of clean renewable sources of energy onto the grid. The reliability requirements necessitate some robust and reliable cyber security modules which can be developed using the compressive sensing concept in future.

References

[1] V.C. Gungor, D. Sahin, T. Kocak, et al., Smart grid technologies: communication technologies and standards, IEEE Trans. Ind. Inf. 7 (4) (2011) 529–539.

[2] X. Fang, S. Misra, G. Xue, D. Yang, Smart grid—the new and improved power grid: a survey, IEEE Commun. Surv. Tutorials 14 (4) (2012) 944–980.

[3] E.J. Candes, M.B. Wakin, An introduction to compressive sampling, IEEE Signal Process. Mag. 24 (2008) 21–30.

[4] E.J. Candes, T. Tao, Decoding by linear programming, IEEE Trans. Inf. Theory 51 (12) (2005) 4203–4215.

[5] M. Babakmehr, M.G. Simoes, M.B. Wakin, F. Harirchi, Compressive sensing-based smart grid topology identification, IEEE Trans. Ind. Inf. 12 (2) (2016) 532–543.

[6] H. Zhu, G.B. Giannakis, Sparse overcomplete representations for efficient identification of power line outages, IEEE Trans. Power Syst. 27 (4) (2012) 2215–2224.

[7] M. Majidi, M.S. Fadali, M. Etezadi-Amoli, M. Oskuoee, Partial discharge pattern recognition via sparse representation and ANN, IEEE Trans. Dielectr. Electr. Insul. 22 (2) (2015) 1061–1070.

[8] M. Majidi, M. Oskuoee, Improving pattern recognition accuracy of partial discharges by new data preprocessing methods, Electr. Power Syst. Res. 119 (2015) 100–110.

[9] M. Sabarimalai Manikandan, S.R. Samantaray, I. Kamwa, Detection and classification of power quality disturbances using sparse signal decomposition on hybrid dictionaries, IEEE Trans. Instrum. Meas. 64 (1) (2015) 27–38.

[10] C.E. Shannon, Communication in the presence of noise, Proc. IRE 37 (1) (1949) 10–21.

[11] E.J. Candes, The restricted isometry property and its implications for compressed sensing, C.R. Math. 346 (9) (2008) 589–592.

[12] M.B. Wakin, Compressive sensing fundamentals, M. Amin (Ed.), Compressive Sensing for Urban Radar, CRC Press, Boca Raton, FL, 2014, pp. 1–47.

[13] D.L. Donoho, Compressed sensing, IEEE Trans. Inf. Theory 52 (4) (2006) 1289–1306.

[14] J. Tropp, Greed is good: algorithmic results for sparse approximation, IEEE Trans. Inf. Theory 50 (10) (2004) 2231–2242.

[15] M. Babakmehr, M.G. Simoes, M.B. Wakin, A. Al Durra, F. Harirchi, Sparse-based smart grid topology identification, IEEE Trans. Ind. Appl. 52 (5) (2016) 4375–4384.

[16] M. Babakmehr, M.G. Simoes, A. Al-Durra, F. Harirchi, Q. Han, Application of compressive sensing for distributed and structured power line outage detection in smart grids, Proc. IEEE American Control Conference ACC 2015, ACC, Chicago, IL, 2015, , pp. 3682–3689. July.

[17] J. Duncan Glover, M. Sarma, Power System Analysis & Design, second ed., PWS Publishing Company, USA, 1994.

[18] M. Babakmehr, M.G. Simoes, A. Al-durra, Compressive sensing for smart grid security and reliability, S. Rahman, S.M. Muyeen (Eds.), Communication, Control and Security for the Smart Grid, IET, London, 2016.

[19] M. Ozay, I. Esnaola, F.T. Vural, S.R. Kulkarni, H.V. Poor, Sparse attack construction and state estimation in the smart grid: centralized and distributed models, IEEE J. Sel. Areas Commun. 31 (7) (2013) 1306–1318.

[20] A.I. Sabbah, A. El-Mougy, M. Ibnkahla, A survey of networking challenges and routing protocols in smart grids, IEEE Trans. Ind. Inf. 10 (1) (2014) 210–221.

[21] W. Li, M. Ferdowsi, M. Stevic, A. Monti, F. Ponci, Cosimulation for smart grid communications, IEEE Trans. Ind. Inf. 10 (4) (2014) 2374–2384.

[22] M. Majidi, M. Etezadi-Amoli, M.S. Fadali, A novel method for single and simultaneous fault location in distribution networks, IEEE Trans. Power Syst. 30 (6) (2015) 3368–3376.

[23] M. Majidi, A. Arabali, M. Etezadi-Amoli, Fault location in distribution networks by compressive sensing, IEEE Trans. Power Deliv. 30 (4) (2015) 1761–1769.

[24] M. Majidi, M. Etezadi-Amoli, M.S. Fadali, A sparse-data-driven approach for fault location in transmission networks, IEEE Trans. Smart Grid 8 (2) (2017) 548–556.

[25] M. Majidi, M. Etezadi-Amoli, H. Livani, Distribution system state estimation using compressive sensing, Int. J. Electr. Power Energy Syst. 88 (2017) 175–186.

[26] M. Majidi, M. Etezadi-Amoli, H. Livani, M.S. Fadali, Distribution systems state estimation using sparsified voltage profile, Electr. Power Syst. Res. 136 (2016) 69–78.

[27] M. Babakmehr, R. Ammerman, M.G. Simoes, in: Modeling and tracking transmission line dynamic behavior in smart grids using structured sparsity, Preprint, to Appear in 54th Allerton Annual Conference on Communication, Control, and Computing, IEEE, Chicago, IL, 2016.

[28] Zhang Z, Nguyen HD, Turitsyn K, Daniel L. "Probabilistic power flow computation via low-rank and sparse tensor recovery". arXiv preprint arXiv:1508.02489.2015 August 11.

[29] R.D. Zimmerman, C.E. Murillo-Sanchez, R.J. Thomas, MATPOWER: steady state operations, planning, and analysis tools for power systems research and education, IEEE Trans. Power Syst. 26 (1) (2011) 12–19.

[30] Z. Wang, A. Scaglione, R. Thomas, Generating statistically correct random topologies for testing smart grid communication and control networks, IEEE Trans. Smart Grid 1 (1) (2010) 28–39.

[31] M. Gol, A. Abur, A fast decoupled state estimator for systems measured by PMUs, IEEE Trans. Power Syst. 30 (5) (2015) 2766–2771.

[32] S.J. Kim, K. Koh, M. Lustig, S. Boyd, D. Gorinevsky, An interior-point method for large-scale l1-regularized least squares, IEEE J. Sel. Top. Sign. Proces. 1 (4) (2007) 606–617.

[33] C.G. Atkeson, A.W. Moore, S. Schaal, Locally weighted learning, Artif. Intell. Rev. 11 (1) (1997) 11–73.

[34] H. Saadat, Power System Analysis, second ed., McGraw-Hill, New York, NY, 2002.

[35] R. Bartnikas, Partial discharges: their mechanism, detection and measurement, IEEE Trans. Dielectr. Electr. Insul. 9 (5) (2002) 763–808.

CHAPTER 9

Time-Series Classification Methods: Review and Applications to Power Systems Data

Gian Antonio Susto, Angelo Cenedese, Matteo Terzi
University of Padova, Padova, Italy

CHAPTER OVERVIEW

The diffusion in power systems of distributed renewable energy resources, electric vehicles, and controllable loads has made advanced monitoring systems fundamental to cope with the consequent disturbances in power flows; advanced monitoring systems can be employed for anomaly detection, root cause analysis, and control purposes. Several machine learning-based approaches have been developed in the past recent years to detect if a power system is running under anomalous conditions and, eventually, to classify such situation with respect to known problems. One of the aspects, which makes Power Systems challenging to be tackled, is that the monitoring has to be performed on streams of data that have a time-series evolution; this issue is generally tackled by performing a features' extraction procedure before the classification phase. The features' extraction phase consists of translating the informative content of time-series data into scalar quantities: such procedure may be a time-consuming step that requires the involvement of process experts to avoid loss of information in the making; moreover, extracted features designed to capture certain behaviors of the system, may not be informative under unseen conditions leading to poor monitoring performances. A different type of data-driven approaches, which will be reviewed in this chapter, allows to perform classification directly on the raw time-series data, avoiding the features' extraction phase: among these approaches, dynamic time warping and symbolic-based methodologies have been widely applied in many application areas. In the following, pros and cons of each approach will be discussed and practical implementation guidelines will be provided.

1 INTRODUCTION

The modern trends in energy generation, transmission, and distribution follow the paradigm of *smart infrastructures* to gain in service flexibility, reliability, and autonomy while not compromising the overall system performance and control. Related for example to the decentralized and/or distributed exploitation of renewable energy resources, the employment of electric vehicle fleets, the management of controllable loads, and these policies have made advanced monitoring (AM) systems fundamental to assess line conditions and utilities' behaviors, in order to grant the requested quality of service to the final user, cope with the presence of disturbances in power flows, and push current generation of power systems toward their limit.

179

Big Data Application in Power Systems. https://doi.org/10.1016/B978-0-12-811968-6.00009-7

Pervasive measurement of such a complex and interconnected system of systems can provide (and do provide) a huge amount of data to be employed for a variety of purposes ranging from failure and anomaly detection [3] to the demand/response analysis and optimization [4–6], from the root cause analysis [7] to the service provider control [8], and from the predictive/preventive maintenance [9] to the physical or cyber-attack prevention [10]. In particular, the development of phasor measurement units (PMUs) [11], frequency disturbance recorder [12], and advanced metering infrastructure (AMI) [13] have allowed the continuous monitoring of the transmission line and the connected power systems, and can be complemented with utility monitoring devices, smart meters, insulation monitoring units, to build a thorough picture of the whole grid structure, health, and dynamic behavior.

Nonetheless, to unleash the full value of these complex data sets, algorithms need to be developed to transform these massive dumb data flows into synoptic smart information and drive the way to manage the energy and power systems [14]. Indeed, these solutions typically constitute the core of energy management systems (EMS), which can be specifically translated toward the final application in factories (FEMS), buildings (BEMS), and home (HEMS) [15]. In this sense, several machine learning (ML)-based approaches [16, 17] have been developed in the past recent years to *characterize* the behavior of smart grid systems and power lines, to *profile* user demand and exploitation of resources and services, to *detect* if a power system is running under anomalous conditions, to *classify* such situations with respect to known problems.

1.1 Contribution

In this chapter, we will try to provide an overview of the main ML techniques that are used in the context of power systems. Without aiming at being exhaustive, the main goal of this contribution is to highlight the differences among the approaches in terms of information they can provide and issues in their usage. In particular, we will use the term *classification* to indicate the subfield of ML in the realm of *supervised learning* where "supervised" indicates that the output is known: given a signal $\mathbf{x}$ belonging to some domain $\mathbb{X}$ as an input and a finite set $\mathbb{Y}$ of different classes (the output), the problem of supervised learning consists on finding a rule that associates $\mathbf{x}$ to one $\mathbf{y} \in \mathbb{Y}$. In general, the set of output classes (also called dictionary) is obtained according to a training procedure where a training input dataset is used to characterize both $\mathbb{Y}$ and the learning rule.

In the context of power systems, we are facing classification problems when dealing with fault detection and isolation (FDI), predictive maintenance, AM, user profiling, and cyber-security applications. Just to provide an example in the case of FDI, one output class can be related to the normal behavior of the power system, while additional classes can be referred to known problems like voltage sags, voltage swells, fault currents, voltage oscillations, and frequency oscillations, while the input signals are time-series generated from multiple continuous data flows such as PMUs data, currents, or voltages: the task of an FDI algorithm is to interpret heterogeneous signals coming from different measurement units in order to discern and understand the state of the overall system [2].

1.2 Notation

Throughout this chapter, we consider a time-series $z_\bullet$ as a (finite-length) sequence of n ordered real values at time instants $t_{\bullet,1}, \ldots, t_{\bullet,n}$. For the sake of simplicity, and without loss of generality, we assume that the time series is obtained through a preprocessing phase that may include sampling and windowing of the continuous data flow coming from a measurement unit. The time series is then characterized by p input descriptors x (whose meaning will be clearer in the following), hence the input space $\mathbb{X}$ is p-dimensional, and a training set composed by N signals $\{x^{(1)}, \ldots, x^{(N)}\}$ allows to define the class set $\mathbb{Y}$. These basic definitions and notation are shown in Fig. 1 and summarized in Table 1.

In Fig. 2 the data flow from sensors (PMUs or other types) to classification is represented with the notations adopted in this chapter to indicate all the related quantities. For the sake of clarity, Table 2 provides the list of acronyms adopted throughout this chapter.

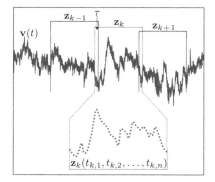

FIG. 1

Windowing procedure. Finite-length windows z_k are extracted from the raw data stream v to obtain time series.

Table 1 Summary Table of the Main Adopted Notation

Symbol	Description
$t \in \mathbb{R}$	Time
$\bar{t}$	Time of interest
R	Number of nodes in the power system cluster
$S \geq R$	Number of data sources generating signals
$\bar{q} \geq R$	Cardinality of raw signal
$\mathbf{v}(t) \in \mathbb{R}^{\bar{q}} \times \mathbb{R}$	Raw signal
$k \in \mathbb{N}$	Window index
$\tau \in [0, 1)$	Window overlap parameter
n	Cardinality of samples per window
$q \leq \bar{q}$	Cardinality of preprocessed signal
$\mathbf{z}_k(t_{k,1}, \ldots, t_{k,n}) \in \mathbb{R}^{n \times q}$	Preprocessed signal
$p \leq n$	Number of signal descriptors
$\mathbb{X} \subseteq \mathbb{R}^p$	Domain of signal descriptors
$\mathbf{x} = x_1, \ldots, x_p \in \mathbb{X}$	Signal descriptors
N	Number of observations available for training
$\{\mathbf{x}^1, \ldots, \mathbf{x}^N\}$	Set of input data for training
$\{y^1, \ldots, y^N\}$	Set of output labels for training
$D \in \mathbb{R}^{N \times (p+1)}$	Design matrix for training
$\tilde{M}$	Number of observations of reduced dataset (e.g., dictionary learning)
M	Number of classes
$\mathbb{Y}$	Class dictionary
$y \in \mathbb{Y}$	Class label
$f(\bullet): \mathbb{X} \rightarrow \mathbb{Y}$	Classifier/association rule

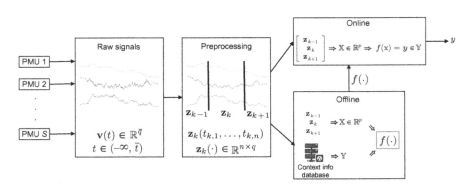

FIG. 2

Data flow. Scheme of the data flow from sensors (PMUs or other types) to classification.

Table 2 Summary Table of the Main Acronyms Used in the Text

Acronyms	Descriptions
1-NN	1 Nearest neighbor
1-NN-DTW	1-NN with DTW distance
AR	Auto regressive
ARMA	Auto regressive moving average
ARIMA	Auto regressive integrated moving average
BEMS	Building energy management system
BoF	Bag of features
BoSS	Bag-of-SFA symbols
BoSS-VS	BoSS-vector space
BoW	Bag of words
DB	Distance based
DBA	DTW Barycenter averaging
DFT	Discrete Fourier transform
DR	Dimensionality reduction
DDTW	Derivative dynamic time warping
DT	Decision tree
DTW	Dynamic time warping
DTWUDC	DTW under dynamic constraints
DWT	Discrete wavelet transform
DNN	Deep neural network
EBC	Ensemble of bundle classifier
EMS	Energy management system
ERP	Edit distance with real penalty
ESN	Echo state network
FB	Feature based
FDI	Fault detection and isolation
FEMS	Factory energy management system
GP	Gaussian process
HEMS	Home energy management system
HMM	Hidden Markov model
ICA	Independent component analysis
IF	Interval feature
k-NN	k-Nearest neighbors
LDA	Linear discriminant analysis
LR	Logistic regression
LSM	Liquid state machine
MCB	Multiple coefficient binning
MDS	Multidimensional scaling
ML	Machine learning
MMCL	Model metric colearning

Continued

Table 2 Summary Table of the Main Acronyms Used in the Text *Continued*

Acronyms	Descriptions
mRmR	Minimum redundancy maximum relevance
NR	Numerosity reduction
PAA	Piecewise aggregate approximation
PCA	Principal component analysis
PDC	Phasor data concentrator
PMU	Phasor measurement unit
RF	Random forest
RVM	Relevance vector machine
SAX	Symbolic aggregate approximation
SFA	Symbolic Fourier approximation
SIFT	Scale invariant feature transform
SMTS	Symbolic multivariate time series
SVM	Support vector machine
TWED	Time warping with edit distance
VAR	Vector autoregressive model
VSM	Vector space model
WDTW	Weighted dynamic time warping

2   THE CLASSIFICATION PROBLEM

The research on classification of time series has been of certain interest for some decades and in various fields, from speech recognition [18] to financial analysis [19], from manufacturing [20] to, of course, power systems [2, 21–23], and it is even more of key importance in this era of big data and pervasive information flow. Specifically, two cornerstone issues need to be addressed:

- How do we compare different time series? In particular, how do we compare time series with different lengths?
- How can we recognize that different time series are *realizations* of a common (unknown) process which represents a certain class?

The last question is particularly relevant in AM applications: if a database of known failures is available, detection of current failures could be performed and exploited in predictive maintenance [24]/fault detection (FD) and FDI solutions. Some works in AM of power systems formalize FD and FDI problems as semisupervised ones, where particular classifiers (like One-Class-SVM) are built on a single group of data: such data are usually associated with normality conditions [25]; the goal of this classifiers is to create a solution that define a "normality space": when a new observation is available, it will be classified as anomaly if it lies outside the boundaries of the normality space. Such problem formulation can also be tackled with some of the methodologies presented in this chapter.

2.1 Classification Methods Taxonomy

For the sake of clarity, we provide a brief introduction to the different methodologies treated in this work that can be employed to solve the classification task with power system data. Time-series classification techniques can be essentially divided into two main branches:

- *Feature-based (FB)*. FB methods perform a *feature extraction* procedure before the classification phase. In general, from the original signal $\mathbf{v}(t)$ a moving window k of fixed length n is considered to obtain a time-series z_k and a set $\mathbf{x}$ of p *features* is calculated over it: to give some examples, commonly chosen features are mean, variance, maximum, minimum, entropy, all related to the time series extracted from the signal.
  The idea underlying these methods is to capture signal statistics that identify a certain class of signals. In theory, if a process is weakly stationary then a second-order statistic is sufficient to characterize that signal; however, signals obtained from real-world scenarios are not stationary due to several nuisance factors and many more features may be necessary to summarize the informative content.
  In this respect, some observations are in order: unfortunately, nonautomatic feature extraction procedures may be a time-consuming step that requires the involvement of process experts to avoid loss of information; moreover, extracted features designed to capture certain behaviors of the system, may not be informative under unseen conditions leading to poor monitoring performances. Finally, the tuning of n is far to be trivial for optimal results: normally, it is estimated through a cross-validation procedure [26]. When dealing with the learning phase in FB methods, the learning rule is based on the definition of a dataset of N observations and of a *design matrix* as

$$D = \begin{bmatrix} \mathbf{x}^{(1)} & y^{(1)} \\ \mathbf{x}^{(2)} & y^{(2)} \\ \vdots & \vdots \\ \mathbf{x}^{(N)} & y^{(N)} \end{bmatrix} \in \mathbb{R}^{N \times (p+1)}. \tag{1}$$

- *Distance-based (DB)*. DB methods avoid the feature extraction phase in favor of the definition of suitable *distances*, among which the most common is dynamic time warping (DTW) [27]. Then, the classification phase is carried out through metric classifiers: one simple and popular choice and, surprisingly, one of the most effective is 1-nearest neighbor classifier (1-NN) [26].

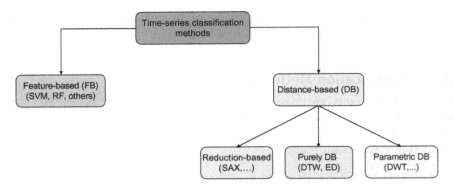

FIG. 3

Classification taxonomy. Time-series classification methodologies' tree highlighting the two families of feature based and distance based.

> This strategy is motivated by the fact that the feature extraction phase could be time consuming and may cause significant loss of information about the original signal [28]. Conversely, though, due to nuisance factors, the DB direct comparison of time series (e.g., by exploiting the Euclidean distance) may lead to ill-posed problems and unsatisfactory performances, thus calling for a careful selection of the distance metrics that trades off between complexity (of the measure) and accuracy (in the classification).

This main categorization is also summarized in Fig. 3.

2.2   Computational Issues

The ML techniques to be employed in big data-related applications strongly depend on the architecture of the EMS infrastructure that delivers the task. In the context of power systems, the architecture can be represented as a main "parent" system unit that monitors and controls the connected "child" nodes. In turn, each node is a smaller unit that processes the measurements derived from multiple PMUs. A typical scheme of an EMS architecture is shown in Fig. 4.

Given this structure, the main unit is provided with powerful hardware in terms of computational and memory resources, while, conversely, the nodes are equipped with resource-limited hardware. The described network architecture must to be taken into account in the classification algorithms design and the algorithmic burden must be distributed over the system; the main system unit will be able to run demanding algorithms in a centralized fashion while nodes will constitute a computational grid with the parallel computation of parsimonious local procedures. In this respect, a further premise is needed to allow a better understanding of the remainder of this chapter. A learning algorithm

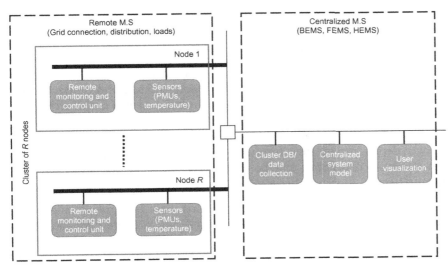

FIG. 4

Energy management system (EMS). Logical block diagram of EMS with the two main parts of remote monitoring and centralized control. These solutions are then specialized into HEMS, BEMS, FEMS, according to their employment within the residential, building, factory environments.

can be characterized in terms of complexity according to two different points of view, namely *space complexity* and *time complexity*: clearly, when dealing with resource-constrained systems, it is crucial to take into account both these aspects. For example, time-complexity is composed by two terms, respectively, related to training complexity and classification complexity: in most applications the training phase is run on systems with high computation and memory capabilities, while the classification complexity can be reduced so that the algorithmic solution can be implemented in the nodes.

There is also another point that forces nodes to be equipped with parsimonious algorithms: generally, in the nodes an *online* monitoring action is required in order to detect anomalies as soon as possible; in these settings, thus, classification must be executed almost in real time. Conversely, this does not necessarily apply for central units, where, generally, *offline* analyses are performed.

An important example of algorithms that are suited only for central units, but not for nodes, is *lazy-learning* approaches. Lazy-learning algorithms are techniques, like nearest neighbors (NN), where all the computational burden is in the evaluation of the classifier and not in its creation: such methods generally exploit comparisons with historical data to perform the classification of a new observations; given these premises, it is apparent that lazy-learning approaches cannot be adopted in nodes since: (i) the evaluation there need to be performed as quick as possible; (ii) nodes do not have access to the whole network data, and therefore the comparisons on that level can be made only on a local, smaller database (not always available) leading to suboptimal classification performances.

Considering this complex scenario, we provide here some general guidelines on which types of algorithms are suited for remote (nodes) and central (main) units. As we will see throughout this chapter, among the two sets of techniques we highlighted before, there are better choices per se and with respect to the specific application; indeed, both FB and DB methods can be simple or cumbersome depending on their formulations. In fact, as FB algorithms with an high number p of features may be prohibitive, the same applies to DB algorithms where an high number N of training examples are considered.

Nonetheless, various *data reduction* procedures may be applied in order to reduce the complexity of learning algorithms. For example, reduction techniques such as symbolic aggregate approximation (SAX) or discrete Fourier transform (DFT) are computationally simple and induce an approximation that may be considered as acceptable for most real applications in power systems. Given the crucial importance of parsimony in power systems, the following section is devoted to discuss data reduction techniques.

3   DATA SOURCES

One of the main practical issues in modern time-series classification is the problem of *time and space complexity* of data. In general, dealing with huge datasets is computationally expensive and, under some conditions, even unfeasible, especially with resource-constrained hardware. On the other hand, in most real-world problems, the informative content of a dataset is generally *sparse* (i.e., the useful information size is much "smaller" than the size of original dataset). For these reasons, research in big data classification has been focused on developing suitable techniques to optimize and reduce the data representation. In the related literature, the procedures related to translating data into simplified and informative representation are called *data reduction*.

In the context of power systems, this problem arises due to huge amount of data coming from PMUs and other sensors. As represented in Fig. 5, PMUs (and other sensors) are connected to local phasor data concentrators (PDCs) and, jointly, to a corporate PDC, receives data from different PDCs. To give insights on how this structure generates an amount of data in the realm of big data, only one PCD collecting data from 100 PMUs of 20 measurements each at 30-Hz sampling rate generates over 50 GB of data 1 day [29]. Hence, data reduction is fundamental to reduce data storage and to best capture the interaction between different PMU locations.

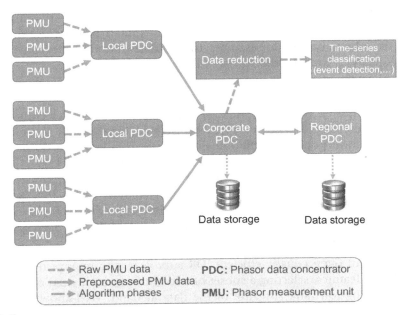

FIG. 5
Typical power system scheme data flow. Detailed structure of a EMS where the data streams and the procedure units are represented (note that other data sources than PMUs may be in place).

For the sake of clarity, and to avoid confusion due to different notations used in related literature, we distinguish two data reduction techniques in which we are interested in: *dimensionality reduction* and *numerosity reduction*. In this section we will present the most important time-series representations for these two types of reduction; for a more exhaustive review, we refer the interested reader to [30, 31].

3.1 Dimensionality Reduction

We refer to dimensionality reduction (DR) when dealing with FB techniques. More in detail, let us consider a design matrix $D \in R^{N \times p}$, with a high-dimensional feature space (e.g., when p is very large; generally $p > 1000$); DR aims at finding a subset of informative features (*feature selection*) or, more generally, informative lower-dimensional structures through linear (e.g., principal/independent component analysis) and nonlinear (e.g., manifold learning) data transformation approaches.

In this context, although the space complexity can be overwhelming, the main issue is the renowned *curse of dimensionality* [26]. It manifest itself in various ways, which all causes high variance and high bias of classifiers resulting in poor classification performance. In fact, when p is high, all feasible training samples sparsely populate the feature space and the concept of locality vanish.

This problem can be easily seen with the k-NN classifier and considering features uniformly distributed in a p-dimensional unit hypercube: when $p = 1000$, in order to capture for example 10% of data to evaluate local average, it is necessary to consider the 99.7% of the range of each feature.

3.1.1   DR Techniques Review

As already stated earlier, the informative content of a signal is often embedded in a low-dimensional space that can be isolated through DR techniques. Among the most popular methodologies, we mention here generalized discriminant analysis [32], independent component analysis (ICA) [33, 34], kernel PCA [35, 36], linear discriminant analysis (LDA) [37], multidimensional scaling (MDS) [38], and principal component analysis (PCA) [39, 40]. In the realm of nonlinear approaches, manifold learning, whose objective is to learn the hidden manifold described by the data [41–44], has been gaining lot of attention in the past recent years.

A simpler, but often equally efficient, DR approach is to remove redundant (correlated) features, selecting a subset of relevant features, instead of finding the underlying low-rank structure of the data at hand. Well-known feature selection techniques are backward feature elimination, forward feature construction, minimum redundancy maximum relevance (mRmR), just to provide some examples. Interestingly, random forests (RFs), besides being among the most effective classifiers, are also powerful instruments for feature selection [45].

For an exhaustive description, we refer the interested readers to [46]. In the context of power systems, recently, Zhou et al. [2, 23] adopted a data-driven feature-based approach combining mRmR reduction technique with an ensemble of bundle classifier (EBC), which combines individual classifiers in order to handle the heterogeneity of the PMU data.

3.2   Numerosity Reduction

We refer as numerosity reduction (NR) when aiming at reducing data volume by choosing alternative, smaller forms of data representation of the signals at hand. It differs from DR in the sense that it aims at finding a different representation of time series and/or reducing the number N of training examples needed for classification without reducing accuracy. For example, consider a collection of N input univariate time-series $\{\mathbf{x} \in \mathbb{R}^n\}_{i=1}^N$. Data reduction techniques aim at reducing n and/or N.

3.2.1   NR Techniques Review

In this framework, to reduce both n and N, parametric and nonparametric approaches may be employed for NR. Parametric approaches model the time series using a parametric model such as discrete wavelet transform (DWT), DFT, and log-linear models to cite the most common examples. Then the complexity space reduces from $O(n)$ to $O(p)$ where $\mathbf{x} \in \mathbb{R}^p$ is p-dimensional vector of parameters of the model. Example of nonparametric approach is histogram or, simply, sampling.

Another family of NR approaches is *symbolic representation*, for which SAX [47, 48] is the most know technique. SAX technique mainly consists of three phases:

- signals standardization in order to obtain a zero mean and unit variance signal;
- piecewise aggregate approximation (PAA) [49] described in the following; and
- symbolic mapping through discretization on amplitude domain.

After normalization, in the PAA phase, a signal $\mathbf{z} = z_1, ..., z_n$ let $\bar{\mathbf{z}} = \bar{z}_1, ..., \bar{z}_p$ of length s is discretized on time in p frames in order to obtain a vector $\bar{\mathbf{z}} = \bar{z}_1, ..., \bar{z}_p \in \mathbb{R}^p$. Formally, the resulting ith element $\bar{z}_i$ is defined by the mean of ith interval:

$$\bar{z}_i = \frac{p}{s} \sum_{j=\frac{s}{p}(i-1)+1}^{\frac{s}{p}i} z_j \qquad (2)$$

Then, the SAX representation procedure (i.e., the discretization on amplitude domain) can be summarized as follows. Let a_i denote the ith element of the alphabet $\mathcal{A}$, with $|\mathcal{A}| = \alpha$. The mapping from the PAA approximation to the correspondent word $\mathbf{x} = x_1, ..., x_p$ of length p is obtained as follow:

$$x_i = a_j \ \text{ iif } \beta_{j-1} \leq \bar{z}_i < \beta_j \qquad (3)$$

where $\left\{ \beta_j \right\}_{j=1}^{\alpha-1}$ are breakpoints tuned to have symbols with equiprobable occurrence. One of the advantages of introducing the *SAX representation* is that a new distance measure—which is a lower bound of Euclidean distance—can be immediately defined. Let $\mathbf{z}^{(1)}$ and $\mathbf{z}^{(2)}$ be two time series of same length n and $\mathbf{x}^{(1)} = x_1^{(1)}, ..., x_p^{(1)}$ and $\mathbf{x}^{(2)} = x_1^{(2)}, ..., x_p^{(2)}$ be their *SAX symbolic representation*; the SAX distance is defined as:

$$D_{\text{SAX}}(\mathbf{z}^{(1)}, \mathbf{z}^{(2)}) = \sqrt{\frac{n}{p} \sum_{i=1}^{p} \text{dist}\left(x_i^{(1)}, x_i^{(2)} \right)^2} \qquad (4)$$

Another popular symbolic approach is the symbolic Fourier approximation (SFA) [50]. The SFA accepts the same parameters p and α as SAX. In this case, p represents the number of Fourier coefficients (real and imaginary) used. Naturally, each sliding window is normalized to have a standard deviation of one to obtain amplitude invariance, before applying SFA. Provided the parameters, the SFA symbolization is carried out in due main steps:

1. preprocessing phase called multiple coefficient binning (MCB) discretization, and
2. SFA transformation.

In the phase (1) the p coefficients (real and imaginary) c_i are extracted for all the training time series and histogram is built for each c_i, where each bin corresponds. Then each histogram is used to infer the breakpoints β_{ij}, $i = 1, \ldots,$ wp, $j = 1, \ldots, \alpha + 1$ in order to make symbols equiprobable. In the phase (2) each coefficient is extracted and it mapped to a symbol according to breakpoints found in the MCB phase. Then the string representing the time series is formed by the sequence of coefficients. Thus, to find a second-order resolution (two coefficients) there are two real plus two imaginary coefficients resulting in a word of length four. SFA presents some important difference from SAX: first of all, the time complexity (to transform a single time-series $\mathbf{z}$) is $O(n \log n)$ while SAX time complexity is $O(n)$. However, provided the same word length, SFA best represents the raw signal as it does not apply any piece-wise discretization, but expresses a linear combination of continuous Fourier functions through the learned coefficients. Moreover, in opposite to SAX, if we wanted a finer resolution increasing p, it would not be necessary to recalculate all DFT coefficients as the symbols of a smaller word length are always a prefix of the larger word lengths.

Regarding the reduction of N, the most popular approaches are clustering and dictionary learning. Time-series clustering aims at finding groups (clusters) in which data can be divided. A way to speed up clustering approaches (that are generally in the realm of *lazy-learning* approaches [26]), "mean" (or centroid) time series for each cluster is usually taken as representative of each group: this generally requires $\tilde{M}$ comparisons instead of N, and, since $N \gg \tilde{M}$, this decreases considerably the time to perform the clustering. Similarly, dictionary learning techniques find a sort of base of $\tilde{M}$ signals, from which a given signal can be represented as a linear combination. Indeed, in the classification context, the *supervised* dictionary learning aims at learning a dictionary containing the elements which best represent the classes and thus they find a *discriminative* representation.

4   CLASSIFICATION METHODS

4.1   Feature-Based Methods

Assuming a dataset of N time-series $\mathbf{z}_k$, $k = 1, \ldots, N$, and M possible target classes $\mathbb{Y} = \{y_1, \ldots, y_M\}$ that jointly describe the classification problem of interest,

FB methods focus on finding a compact description $\mathbf{x} = [x_1, ..., x_p] \in \mathbb{X}$ of the time-series $\mathbf{z}_k$ such that $p \leq n$ (and typically $p \ll n$); all these N observations are collected into a matrix $D \in \mathbb{R}^{N \times (p+1)}$, called *design matrix*, as defined in Eq. (1). In practice, D represents the supervised learning phase of the procedure and it is exploited to define the rule $f(\bullet): \mathbb{X} \to \mathbb{Y}$ according to a chosen classification method [26] as better detailed in the following.

In this category, the most employed classification algorithms are k-nearest neighbor (k-NN) [26], support vector machines (SVMs) [1, 51], relevance vector machine (RVM) [52], decision trees (DT) [53], RF [54], logistic regression (LR) [55], Gaussian processes (GP) [56], and deep neural networks (DNN) [57]. We refer the interested reader to the literature references for further details on the specific methods and the general textbooks [26, 58].

4.1.1 Metrics-Based Approaches
The techniques earlier are not straightforwardly applicable to continuous time series since they need input vectors of fixed length, and this justifies their inclusion in the FB methods.[1] As a matter of fact, even if the time series in input to the classifier were discrete time and of fixed length, the accuracy performance would be poor due to two main reasons: first, it is common to consider long sequences of samples of $n > 100$ or even $n > 1000$; in these cases the space spanned by the time series is too large and sparse incurring in the aforementioned "curse of dimensionality" problem. Second, considering the discrete values as independent features would be not reasonable, since they do not provide any information per se about the characteristics of the signal, since time-series values are strongly highly correlated on time and the feature extraction phase is exactly designed so as to highlight this correlation.

The flowchart of the FB classification procedure is given in Fig. 6: after a preprocessing phase on the raw signal characterized by the presence of a low-pass filtering operation to reduce measurement noise and the windowing procedure, the proper feature extraction task is performed on the time series,

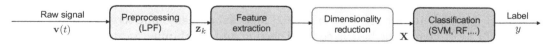

FIG. 6
FB methods. Operation flow of FB classification procedures.

[1]This is partly true for deep learning approaches: recent deep neural networks schemes can avoid the feature extraction phase.

resulting in a compact set of signal descriptors, which undergo the classification phase.

The main advantages of the FB methods clearly reside in the compactness in the representation able to characterize the signal. Typical examples of features are follows.

- Sample features: sample variance/mean/RMS value of the time series.
- Energy/power features: energy value from the DFT coefficients, power spectrum bands.
- Correlation features: number/location/width of (prominent) peaks in the autocorrelation function (repetitive and periodic signals present a peak in the autocorrelation function); correlation parameters among different signal dimensions.

Unfortunately, though, the FB approach presents also several drawbacks:

- features must be defined ad hoc depending to the task[2];
- high dimensionality;
- nonstationarity of time series; and
- time structures are not considered.

It clearly appears from these lists that most of feature-based ML techniques are not adapted to exploit time structures (i.e., patterns), which are an *intrinsic* and *distinctive* characteristic of each time series. In this regard, the first attempts to exploit patterns can be found in Refs. [59–61]. In particular, in Kadous [60] parameterized events are extracted from multivariate time series: these events are clustered in the parameters space and the resulting prototypes are used as basis to build classifiers. Instead, Kudo et al. [61] maps multivariate time series into binary vectors: the space value-time is represented by a grid and each element (of the vector) is associated with one cell of the grid count: if the signal passes through the corresponding cell, the element is true (1) otherwise it is false (0). Then, these converted binary vectors are used as the basis for the classification.

More in general, a limitation to these methods stem from the fact that classification rules are extracted taking into account absolute time values, leading to the inability of handling situations where particular behaviors happen at different time values: Geurts [62] argues that many time-series classification problems can be solved by detecting and combining local properties (patterns) on time series, and proposes a procedure that captures the information of shift-invariant patterns using DTs over piece-wise constant time series.

---

[2]This is one of the main reasons that favors the usage of deep learning in complex problem such as natural language processing and computer vision: in these fields, the definition of informative features has required at least 20 years of research.

More recently, *interval features* (IFs) have been introduced in order to capture temporal information [63–66]. These features are common statistics such as mean, variance, slope but they are calculated over random *intervals* exploiting a boosting procedure. The first idea on IFs was presented in Rodríguez et al. [63], later expanded by Rodríguez and Alonso [64] and Rodríguez et al. [65] using classifiers such as DTs and SVMs applied on the features extracted from binary ensembles. However, as discussed in Deng et al. [66], the number of candidate splits is generally large and thus there can be multiple splits having the same ability of separating the classes. To cope with these issues, Deng et al. [66] introduce an additional measure able to better distinguish among IFs. Another problem in boosting IFs is the size of the relative space that is $O(n^2)$, where n is the length of a time series. In Rodríguez et al. [63] the feature space is reduced to $O(n \log n)$ using only intervals of length equal to powers of two. Deng et al. [66] consider the same approach of random sampling strategy used on RF [54] further reducing the feature space at each node to $O(n)$.

4.1.2 Occurrence Counting Approaches

Another type of approaches to classify time series is the so-called bag-of-words (BoW), also bag-of-features (BoF), nowadays used in image classification and document classification and classically developed in the context of natural language processing. BoW consists in representing data (images in computer vision or documents in natural language processing) using a histogram of word occurrences, where a *word* is a task-dependent element [67], namely a proper textual word in language processing or the image description through intensity local gradients in computer vision. After this encoding, the classification task is reduced to computing a histogram-based similarity (typically using Euclidean distance).

With respect to the two steps of (i) conversion of the time series into a BoW representation (i.e., an histogram of the word occurrences) and (ii) training of a classifier (such as SVM, RF, kNN) upon BoW features, we particularly focus on step (i) in the following, since step (ii) is similarly performed by all methods using RF, SVM, or some other common classifier over the word histogram. Indeed, several BoW-inspired techniques have been recently investigated in order to extract local and global features [68–76]. Hereafter, a brief overview of the main contributions is given.

In the computer vision field, the BoW technique is used for image classification [77] often in combination with the scale invariant feature transform (SIFT) technique. SIFT is a *covariant* detector, which extracts local features (keypoints) that are robust to noise (e.g., changes in illumination) and invariant to affine transformations and scale. As a general note, BoW methods ignore temporal ordering, which may cause that patterns in observed time series or images are not identified. Nonetheless, some BoW-based works try to indirectly remove this lack, although to a limited extent. For example, in Bailly et al. [73, 76] a variant of SIFT for time series is applied and local features (keypoints' descriptors) are extracted with a procedure very similar to SIFT and BoW approach is used over the SIFT descriptors. This choice is motivated by the fact that SIFT captures local structures while BoW allows to describe the global behavior of the time series. Furthermore, in Bailly et al. [76], the same authors adopt dense-SIFT-like descriptor: the main difference with the previous work is that keypoints no longer correspond to extrema but are rather extracted at all scales every time step on Gaussian-filtered time series. This approach in general leads to more robust global descriptors, especially when local extrema can be found (when signal are very smooth).

In Wang et al. [69], DWT is applied on sliding windows of the time series and the resulting DWT coefficients form a word for each window (segment). In the training phase all the DWT segments are clustered through k-means in order to obtain a word dictionary **D**. In the classification phase each DWT window is assigned to the nearest word in **D**, to build a histogram that is used to computed the similarity; the classification is finally carried out using 1-NN.

A BoF framework is proposed also in Bailly et al. [70], which combines IFs and BoW. Here, there are extracted IFs and start/end time points over random subsequences of random length, and a supervised codebook of class probability estimate (CPE) histogram is built in the training phase: for each sequence (of the time series) a CPE is found using an RF classifier. Then, all the CPEs are quantized in order to form a different histogram for each class, which are concatenated into a single histogram of each time series and are used as features in combination with other global features. Finally, the employed classifier is RF.

A symbolic multivariate time series (SMTS) method is discussed in Baydogan et al. [72]: each time series is represented by a feature space, which contains the time instants, the time-series values, and the first difference values, all collected in a design matrix D. Then D is input to a symbolic discretization which is obtained using tree-based classifiers (supervised discretization). Then the classification is performed using a common BoW approach based on histograms of symbols. Its total computational complexity is due to the number of trees, the number of training instances, and the number of time-series subsequences extracted. Then, multivariate time series are mapped into a feature matrix, where features are vectors containing a time index $\bar{t}$, the values, and the gradient of time series at t for all dimensions. The feature space is finally partitioned into regions (i.e., symbols) by an RF classifier. An appreciable property of SMTS is that it does not require tuning parameters, while one main drawback of this representation is the possibly high dimensionality, which limits its application for large datasets.

In Schäfer [74] the bag-of-SFA-symbols (BoSS) is introduced. An univariate time-series z is represented by SFA words and then an histogram is built. However, since this approach is $O(N^2 n^2)$ for training, $O(Nn)$ for classification, in Schäfer [75] it is presented a "scalable version" named BoSS-VS that uses vector space models (VSMs). In this case, once the BoSS histogram is obtained, for each SFA word w, the word (called *term*) frequency *tf* of w in a certain class c_i is computed, together with the ratio *idf* given by the total number of classes divided by the number of classes in which w appears. Then, the *tf-idf* measure is obtained by the product $tf \cdot idf$: this measure is used to weigh the word frequencies in the vector to give a higher weight to representative words of a class. The motivation behind this choice is that an high *tf-idf* for a word w means that w appears with an high frequency in a specific class c_i, while low *tf-idf* values means that w in common in all classes. When a new observation z_{new} arrives, the BoSS histogram and the relative *tf* vector are computed. Then the *cosine similarity*[3] is obtained in order to predict the nearest class. Using this model (named "term frequency inverse document frequency," *tf-idf*, model), the complexities of training and classification reduce to $O(Nn^{3/2})$ and $O(n)$, respectively.

In Lin et al. [68] and Senin and Malinchik [71] time series are mapped into SAX words through a sliding windows partitioning, which are used to build histograms of n-grams: for each time series an histogram counts the frequency of occurrences of each SAX word and thus each time series is represented by its histogram. In particular, in Senin and Malinchik [71], the authors combine SAX and VSM [78] exploiting the *tf-idf* model [79, 80], weighing bags, and cosine similarities as metrics. This technique has a parameter space of dimension $O(n^2)$ and needs to recompute all SAX coefficients for each new choice of parameters p (number of frames in the PAA representation) and α (cardinality of the alphabet); moreover, the training time is $O(Nn^3)$ where N is the number of training instances.

4.1.3 Dynamics-Based Approaches

The last set of FB techniques we present in this review explicitly takes into account the *dynamics* of the signals and comes from *dynamical systems* and *signal processing* theory. In the context of dynamical systems and identification theory, in the past decades much attention has been conveyed on modeling stochastic processes (whose realizations are time series) in order to *predict* their future trends

---

[3] Given two vectors x_1 and x_2, both in , the cosine similarity is defined as $\cos \beta = \frac{\langle x_1, x_2 \rangle}{\|x_1\| \cdot \|x_2\|}$, where $\langle \bullet, \bullet \rangle$ is the inner product.

and values. The most common models are Auto-Regressive (AR), Auto-Regressive-Moving-Average (ARMA), and Auto-Regressive-Integrated-Moving-Average (ARIMA) models [81], just to give some examples. With such approaches, the coefficients themselves of the fitted model are used as features for a suitable classifier [82, 83] or are used to build a more complex generative model [84].

In more detail, Roberts [84] takes a Bayesian point of view and proposes a hierarchical model that consists of a feature extraction stage and a generative classifier, probabilistically linked by a latent feature space. The classifier is implemented as an hidden Markov model (HMM) with Gaussian and multinomial observation distributions defined on a representation of AR models. The HMM is used to model the correlation between adjacent windows (subsequences), that is, this model assumes that time series are consecutively extracted from an unique flow.

In a similar way, signal-processing transformations such as DFT or DWT are applied to the raw signals to obtain coefficients that can be exploited in training suitable classifiers [85–87]. Interestingly, DWT results to be more suitable for nonstationary time series and, conversely w.r.t. DFT, is ideal for identifying highly discriminant local time and scale features [88]. We note that DWT and DFT and dynamical models can be seen as dimensionality reduction procedures: in the next section, we will revise these concepts from the point of view of distance-based methods.

Eruhimov et al. [89] gathered the most known features deriving from the presented methods such as statistical moments, wavelets coefficients, PCA coefficients, Chebyshev coefficients, and the original values of time series and built a classifier from them. However, this method can be accurate at the cost of complexity and a feature selection procedure is needed to reduce the dimensionality.

Although time-pattern (dynamic) information has been considered in literature, most of feature-based models present common limitations due to the nature of time series. In fact, the presence of variability in the time series causes these methods to be ineffective to cope with common issues. The variability arises because of the stochasticity of the process generating the time series, nonstationarity of time series, and nuisance factors. To give a practical example, the most effective FB methods that exploit time-patterns presented in this chapter are not able to deal with variable time-series lengths and the other methods which can handle this issue exploits only global statistics making them ineffective with nonstationarity.

Finally, perhaps the most limiting issue of feature-based methods is that the feature extraction phase can be demanding both in terms of memory and computational burden. These factors could make feature-based methods not

Table 3 Main characteristics of feature-based methods which have been reviewed in this work

Methods	Characteristics
Time-pattern features	First attempt to capture local structures
	Capture temporal information
	Feature space is big ($O(n)$)
Interval features	Feature extraction can be onerous
	Local and global structures are captured
	Histogram extraction is onerous
Bag-of-features	Time-patterns not considered
Dynamic features	Encode information about dynamics
	Capture behavior in the frequency domain
Frequency domain features	FFT/DWT are efficiently implemented
	Good accuracy
Ensemble of features	Feature extraction is very onerous

suitable for resource-constrained devices. In Table 3 the main peculiarities of each FB methods are summarized.

4.2 Distance-Based Methods

DB methods can be clustered into three groups:

- *Purely distance-based*: These methods are based on the direct computation of ad hoc defined distances over raw time series.
- *Reduction distance-based*: Such methods are based on the computation of opportunely defined distances over a reduced representation of raw time series.
- *Parametric distance-based*: With this type of DB approaches, raw signals are represented with a combination (generally linear) of basis signals (e.g., sine functions in the Fourier series representation). The coefficients of different representations (parameters) are used for the computation of ad hoc defined distances.

A general picture of the dataflow for DB methods is given in Fig. 7 and the three groups will be discussed in detail in the following of this section.

4.2.1 Purely Distance-Based Methods

Purely DB methods performs the classification task by adopting a classifier that exploits an opportunely defined distance applied to the raw time-series $\mathbf{z}$. Thus, in this case, we have $p \equiv n$ and we refer to $\mathbf{x}$ as the time-series $\mathbf{z}$ (the map $\mathbf{z} \to \mathbf{x}$ can be seen as the identity map). Here, we consider a set of variable-length

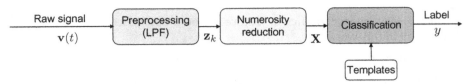

FIG. 7

DB methods. Operation flow of DB classification procedures.

training time series and the corresponding label $D = \{(\mathbf{x}_i, y_i), i = 1, ..., N\}$. As mentioned earlier, purely DB methods are based on the computation of a distance over the raw time series. In choosing a distance, the most straightforward approach is to adopt an *Euclidean distance*; however, this choice has many drawbacks: Euclidean distance to be computed requires time series of equal lengths; moreover, even when comparing two series of equal length, Euclidean distance can be an unfortunate choice since it does not considers common nuisance factors such as warping [90].

For the previous reasons, a more popular approach for distances is DTW [91]. DTW measures the similarity between two time series with, possibly, different lengths by warping the time axis of one (or both) sequences to achieve alignment between the two. DTW provides a *similarity score*, an index on how similar two time series are: in order to define the similarity score, let us consider two time series $\mathbf{x}^{(1)} = \{x_1^{(1)}, ..., x_n^{(1)}\}$ and $\mathbf{x}^{(2)} = \{x_1^{(2)}, ..., x_m^{(2)}\}$ and let us define a grid $\mathcal{G} = [n] \times [m]$. A *warping path wp* in $\mathcal{G}$ is a sequence $wp = (\mathbf{p}_1, ..., \mathbf{p}_l)$ with points $\mathbf{p}_k = (i_k, j_k) \in \mathcal{G}$ s.t.:

$$\mathbf{p}_1 = (1,1) \quad \text{and} \quad \mathbf{p}_l = (n,m) \qquad \text{(boundary conditions)}$$
$$\mathbf{p}_{k+1} - \mathbf{p}_k \in \{(1,0),(0,1),(1,1)\} \quad \text{(warping conditions)}$$

$\forall k | 1 \leq k < l$. The cost of "warping" $\mathbf{x}^{(1)}$ and $\mathbf{x}^{(2)}$ along the warping path s is given by

$$d_s(\mathbf{x}^{(1)}, \mathbf{x}^{(2)}) = \sum_{(i,j) \in s} \left(x_i^{(1)} - x_j^{(2)} \right)^2 \qquad (5)$$

where $(x_i^{(1)} - x_j^{(2)})^2$ is called *local transformation cost*. Then, the DTW similarity score is defined by

$$d(\mathbf{x}^{(1)}, \mathbf{x}^{(2)}) = \min_s d_s(\mathbf{x}^{(1)}, \mathbf{x}^{(2)}) \qquad (6)$$

Regarding the choice of the classifier, the most common choice is 1-NN combined with DTW (NN-DTW). Notably, even if NN is generally considered as one of the simplest approaches to classification, in many papers NN-DTW outperforms more sophisticated approaches when dealing with time-series classification [90, 92, 93]. One of the issues of NN-DTW is its computational cost: the DTW is $O(n^2)$ and it has to be evaluated for each training example in order find the NN. In order to alleviate the aforementioned issue, various approximations of DTW [94–98] have been introduced: the most promising are FastDTW [95] and SDTW [97]. FastDTW adopts a multiscale approach that recursively projects a solution from a coarse resolution and refines the projected solution. FastDTW time and space complexity are $O(n)$; SDTW, instead, extracts keypoint descriptors (similarly to SIFT [67], a popular approach in computer vision) and uses them to reduce complexity.

Beside approximations, several other extensions and modifications of DTW have been proposed: Keogh and Pazzani [99] proposed the derivative dynamic time warping (DDTW), which transforms the original time series into a first-order differences time series, in order is to avoid ill-conditioning; ill-conditioning is a common issue in DTW when dealing with noisy and long signals due to the fact that single points of one of the compared time series could be mapped onto a large subset of the other time series leading to poor alignments. Jeong et al. [100] proposed a penalty-based DTW (WDTW), which adds a multiplicative weight penalty in order to penalize points with higher phase difference between a reference point and a testing point. This has the aim to prevent minimum distance distortion caused by outliers. Other used similarity measures are edit distance with real penalty (ERP) proposed by Chen and Ng [101] and Chen et al. [102] and time warp edit distance (TWED) proposed by Marteau [103]. ERP is a variant of L1-norm, which can support local time shifting. It can also be viewed as a variant of EDR [101] and DTW, but it is a metric distance function. TWE distance is an elastic distance measure (efficiently implemented using dynamic programming) which, unlike DTW, is also a distance. It allows warping in the time axis and combines the edit distance (defined for time series) with L_p-norms. Marteau [103] also provides a lower bound for the TWED measure which allows to operate into down-sampled representation spaces in order to fasten the algorithm.

As we already states, DTW is not a distance measure and this implies that it cannot be employed with kernel methods [104], where kernel must be positive definite. Moreover, time series of different length cannot be compared. In this context, Cuturi et al. [105] propose a new family of kernels between variable-length time series, called alignment kernels, which consider the soft-max of the score of all possible DTW-based alignments to consider the three of the scores of all possible alignments. However, the computation of such kernels can be performed in quadratically, and motivated by this limitation, Cuturi [106] provides an efficient version of it.

Although the usage of kernel methods combined with global alignment allows to consider variable-length time series and disturbances which cause time warping, these do not consider the *dynamics* or patterns. Indeed, albeit the term "dynamic" (deriving form dynamic programming), DTW has nothing which considers the *dynamics* of time series we want to classify. Soatto [107], among other contributes on defining distances for nonstationary time series, also introduces the DTW under dynamic constraints (DTWUDC), which constrains the DTW to follow a dynamical system. More in detail, in Soatto [107] it is assumed that the data (of two time series) are outputs of dynamical models driven by inputs that are warped versions of some common function. Thus, given two univariate time series $\mathbf{x}^{(1)}$ and $\mathbf{x}^{(2)}$ ($i = 1, 2$), he assumes that there exists the dynamical model (linear in the parameters)

$$\begin{cases} \dot{h}_i(t) & = Ah_i(t) + Bu(w_i(t)) \\ \mathbf{x}^{(i)}(t) & = Ch_i(t) + n_i(t) \end{cases}$$

where A, B, C are suitable matrices, h_i are the state functions, $n_i(t)$ are noise processes, $w_i(t)$ are warping functions, u is a common input, and $\mathbf{x}^{(i)}$ is the time series i. Then, the distance can be evaluated in two stages to fit $u_i(t)$ and then $w_i(t)$. In the section dedicated to parametric distance-based methods we will see other methods, mainly kernel methods, on dynamical systems which capture the dynamic essentials of the time series.

4.2.2   Reduction Distance-Based Methods

As we detailed earlier, although the 1-NN-DTW classifier is remarkably difficult to beat, it presents computational issues which prevent its usage in resource-constrained systems. Moreover, shape-based methods typically fail to provide satisfactory results for long time series, as the weight of discriminative "local" structures decreases. In this context, albeit various approximation of DTW was presented, the 1-NN still remains a bottleneck, as it requires the comparison with all the training time series, mostly when the length n and the number N of the time series are large. Moreover, it also requires space to store the entire dataset, which is unfeasible for most resource-constrained devices.

As we have seen in Section 3, data reduction techniques aim at reducing n and/or N. The main idea of reduction distance-based methods is to reduce the time series in a parsimonious in a new representation space and compute a suitable distance in this space. In this part, we will discuss distance-based methods which reduce n or N or both. Before reviewing these techniques, we notice that some of the techniques we have previously exposed may also fall into this category. For example, this is the case of VS-BOSS, in which the time-series $\mathbf{x}$ is mapped into histograms and the histogram-similarity is computed in order to classify $\mathbf{x}$.

As we saw previously, symbolic representations such as SAX are very useful for NR. In this context, the simplest classifier is the 1-NN-SAX classifier, that is the 1-NN on the space of symbolic representation endowed with the metric d_{SAX}.

However, as the Euclidean distance, it is not robust to time distortions or more simply time shifts. Moreover, as we pointed out, when N is large, using SAX approximation may not be enough. Thus, it has been crucial to find some sparse representation of the space of training examples.

The rationale under the reduction of N is to find a subset of canonical $k \ll N$ examples (*templates*) which best describe training set without loss of information (in the sense that they are *sufficient*). Fundamentally there are two directions to find this templates, that is, unsupervised and supervised. In the unsupervised approach, *templates* are found trough clustering techniques regardless the task at hand. On the other hand, supervised approaches aim at finding also the most discriminative templates for the classification tasks, that is, the templates which best represent each class. Naturally, supervised methods are most suited for the classification task. In literature, there are two (at least) different definitions of template, that is *shapelet* and *dictionary*. They rely on the same idea, but is tackled with different approaches.

The concept on *shapelet* has been developed in the recent literature [108–114]. Shapelets are subsequences of time series which are maximally (in some sense) representative of a certain class and thus are useful to classify unlabeled time series; shapelets are maximally representative in the sense of the *information gain* criteria also used to train decision trees and RFs [53].

The main advantages of using shapelets are that 1-NN with all the training instances is avoided in favor of the computation of the distance to the shapelets, which represent each class and it is phase-invariant contrary to simpler techniques such as 1-NN with Euclidean Distance (or SAX distance). On real problems, the speed difference of classification can be greater than three orders of magnitude [108]. However, despite the fast classification, the training of the shapelets is onerous. In the first work where shapelets for classification was introduced, the worst-case scenario for the training time was $O(N^2 n^3)$ where N is the number of time series in the dataset and n is the length of the longest time series in the dataset. In order to reduce the training complexity, various extensions have been proposed. Among all, in [112], SAX is used to find suboptimal shapelets reducing training complexity to $O(Nn^2)$: in this case, time series are mapped to a low-dimensional space of SAX words and shapelets are found directly on this space. Then the distance used for classification is the d_{MAX} defined earlier.

4.2.3 Dictionary Learning

The other approach we find in the literature is called *dictionary learning*, whose aim is to learn a sparse representation of the time series in terms of a basis of signals and express the input signals as a linear combination of basic elements belonging to a set called dictionary. Before continuing we notice that we refer to **x** as the time-series **z**, that is $p = n$. Dictionary learning can be categorized in two approaches: unsupervised and supervised. In order to understand their difference, in the following we briefly present the main concepts of these frameworks which are detailed for example in Refs. [115–117]. Suppose of having N univariate fixed-length training time-series $\{\mathbf{x}^{(i)} \in \mathbb{R}^n\}_{i=1}^N$ associated with binary labels $\{y_i \in -1, +1\}_{i=1}^N$. In order to find an optimal (in the sense of mean square error) and sparse representation, through dictionary $\mathbf{D} = \left[d_1, d_2, ..., d_{\widetilde{M}}\right]$ of signal $\mathbf{x}$ we can solve the convex optimization problem

$$\min_{\alpha, \mathbf{D}} \sum_{i=1}^N \| \mathbf{x}_i - \mathbf{D}\alpha_i \|_2^2 + \lambda \| \alpha_i \|_1, \quad \text{s.t. } \| d_i \|_2 = 1 \tag{7}$$

where ℓ_1-norm is used for α since encourages sparsity [26] inducing noninformative α_i to be zero. Once obtain optimal α^*, $\mathbf{D}^*$, we can solve the classification task solving

$$\min_{\theta} \sum_{i=1}^N L(y_i, f(\mathbf{x}_i, \alpha^*, \mathbf{D}^*, \theta)) + \lambda_2 \| \theta \|_2^2 \tag{8}$$

where $L(\cdot, \cdot)$ and f are an opportune loss function and the predicting function, respectively, which together define a classifier and θ parameterizes the model f. Common choices of f are

1. linear models in α: $f(\mathbf{x}, \alpha, \theta) = w^T \alpha + b$ with $\theta = w \in \mathbb{R}^k, b \in \mathbb{R}$.
2. bilinear models in $\mathbf{x}$ and α: $\mathbf{x}^T W \alpha + b$ where $\theta = W \in \mathbb{R}^{n \times k}, b \in \mathbb{R}$.

This approach is called unsupervised since the dictionary $\mathbf{D}^*$ is obtained to find a sparse representation of N training time series *independently* of the classification task. However, as pointed by Mairal et al. [116], the dictionary found with this procedure is optimal in the sense of *reconstructive* tasks but not for *discriminative* one, that is, classification.

In order to tackle this issues, supervised dictionary learning has been introduced in order to learn a discriminative dictionary exploiting the class label information. In [116], the authors propose an approximation solution of formulation (Eq. 9) which learn jointly $\mathbf{D}$ and θ:

$$\min_{\mathbf{D}, \theta} \sum_{i=1}^N L(S^*(\mathbf{x}_i, \mathbf{D}, \theta, -y_i) - S^*(\mathbf{x}_i, \mathbf{D}, \theta, y_i)) + \lambda_2 \| \theta \|_2^2 \tag{9}$$

where $S^*(\mathbf{x}_i, \mathbf{D}, \theta, y_i) = \min_\alpha L(y_i, f(\mathbf{x}_i, \alpha_i, \mathbf{D}, \theta)) + \lambda_0 \| \mathbf{x}_i - \mathbf{D}\alpha_i \|^2 + \lambda_1 \| \alpha_i \|_1$. Then the classified label $\hat{y}$ of a new time series $\mathbf{x}_{\text{new}}$ is given by

$$\hat{y} = \arg \min_{y \in \{-1; +1\}} S^*(\mathbf{x}_{\text{new}}, \mathbf{D}, \theta, y)$$

Other supervised approaches are discriminative KSVD [118], task-driven dictionary learning [117], Fisher discrimination dictionary learning [119], and label-consistent KSVD (LC-KSVD) [120, 121]. However, all these approaches are not robust to time-shifts or general deformations as it uses the Euclidean distance. In order to overcome this issue, in [122] a family of Gaussian elastic matching kernels was introduced. They use DTW, ERP, and TWED distances to compute the Gaussian kernel

$$K(\mathbf{x}^{(1)}, \mathbf{x}^{(2)}) = \exp - \frac{\|\mathbf{x}^{(1)} - \mathbf{x}^{(2)}\|^2}{2}$$

However, the Gaussian elastic matching kernel cannot be guaranteed to be a positive definite symmetric (PDS) kernel. Thus, proper modifications have to be applied in order to remove the non-PDS part. Although the attempts to embed DTW distance to dictionary learning, several issues such as nonpositive semidefiniteness of "DTW Gaussian kernel" can compromise the robustness of results. Moreover, these approach is computationally expensive.

Recently, in view of these considerations, another sparse representation approach of a dictionary has been considered. It relies on the well-known notion of *centroid* for clustering algorithms. Each centroid is considered as the class representative and thus, if the classification problem involves M classes, M representative time series will be selected. However, DTW does not induce a proper definition of mean and thus the literature attempted to find a definition which is consistent with DTW. The most promising definition was given by Petitjean et al. [123, 124] and it was called DTW Barycenter averaging (DBA). Roughly speaking, it is based on an expectation-maximization scheme and multiple sequence alignment (commonly used in computational biology). This method is very effective as it allows to apply NR reducing N and DTW approximation to speed-up the single comparison. The difference between this approach and dictionary learning is that the "centroid" could not appertain to the dataset. Moreover, this method is supervised in some sense as it exploits the class label information by evaluating the centroid for each class.

4.2.4 *Parametric Distance-Based Methods*

Parametric distance-based methods compute the distance onto a reduced parametric representation of the signal. In this case, each time-series $\mathbf{z}$ is represented by a representation $\mathbf{x} \in \mathbb{R}^p$.

The most common procedure for the training phase is as follows:

- Find a parametric representation of all the training time series. The most used techniques are DWT, DFT stopped at a given coefficient order.
- Find a "centroid" or more generally a representative (template) of each class.

Once obtained the templates for each class, then the classification task is just given by an 1-NN classifier on the representatives $\mathbf{x}^{(\bullet)}$.

These simple methods suffer from various issues as they do not care of the intrinsic "dynamics" information of the signal. This issue has been tackled, mainly, by the computer vision literature [125, 126] and by Cuturi and Doucet [127] and Chen et al. [128] in a general context. Bissacco et al. [126] proposes family of kernels for dynamical systems based on the Binet-Cauchy kernel [129] for recognizing dynamic textures. Bissacco et al. [126] extends the work of Vishwanathan et al. [125] considering phase information, inputs or initial conditions. Essentially, these two works rely on a probabilistic modeling of the time series to define a kernel: in order to compare two time series, first, the dynamic behavior of each time series is learned by learning the parameters of a given state space dynamical systems, and then, the kernel is defined as a kernel between these two sets of parameters. In other words, the distance is computed over the parameters of the fitted dynamical systems. We will see later that a similar approach is followed by other classification methods. The work of Cuturi and Doucet [127] introduced the *Autoregressive Kernels* that are based on the vector autoregressive model (VAR): every multivariate (q-dimensional) time-series $\mathbf{z} \in \mathbb{R}^{q \times n}$ is represented by the feature $L(\theta; \mathbf{z}) = p_\theta(\mathbf{z})$, which is the likelihood function (it is a function of θ for a fixed sample $\mathbf{z}$), modeled by a VAR model. Given a measurable space $\mathcal{X}$ and a model (i.e., a parameterized family of distribution on $\mathcal{X}$ of the form $\{p_\theta, \theta \in \Theta\}$), the kernel K of two time-series $\mathbf{z}^{(1)}$ and $\mathbf{z}^{(2)}$ is defined by

$$K(\mathbf{z}^{(1)}, \mathbf{z}^{(2)}) = \int_{\theta \in \Theta} p_\theta(\mathbf{z}^{(1)}) p_\theta(\mathbf{z}^{(2)}) \omega(d\theta)$$

where, in this case, $\omega(d\theta)$ is the matrix-normal inverse-Wishart prior. Moreover, Cuturi and Doucet [127] have shown that this kernel can be easily computed even when $q \gg n$ due to the fact that it does not resort to the actual estimation of a density. Indeed, all the kernels defined in [125–127] rely on a probabilistic parametric modeling of time series, but the computation of the Autoregressive Kernel avoids the two-step approach presented previously. Finally, Chen et al. [128] presents a model-metric colearning (MMCL) methodology, which differently from the works on [125–127], present a kernel based on nonlinear dynamical systems, named echo state networks (ESN) [130, 131]. For each time series, an ESN-model is trained and the model parameters θ are used to compute an opportune distance, also using kernel methods. For other recent application of ESN and its extension using liquid state machines (LSM), see [132–134]. We can notice that all the methods presented in this section can be seen as feature-based models. Indeed, in [127] the profile likelihood and in [125, 126, 128, 133, 134] the parameters of the dynamical systems can be seen as features.

Although accounting the "dynamics" information using the kernel methods may lead to superior accuracy with respect to the simple parametric distance-based methods, they are more computationally expensive (as they requires the feature extraction phase and the computation of the kernel). This fact would favor the usage of simpler methods which avoid the computation of the kernel. In the context of power systems, simple distance-based methods could be sufficient to capture the difference of signals, to identify, for example, anomalies. Finally, in Table 4 a summary of the most important characteristics of DB methods is presented.

Table 4 Main characteristics of distance-based methods that have been reviewed in this work

	Methods	**Characteristics**
Purely DB	DTW	Invariant to time-warpings Complexity is $O(n^2)$
	DTW approximations	Complexity is $O(n)$
	DDTW	DTW on the first-derivative signal Reduces pathological alignments
	WDTW	Filtering with logistic weight function Favor matching points located in a neighborhood Reduces pathological alignments
	ERP	Supports local time shifting Metric distance function
	TWED	Elastic distance measure Edit distance + L_p norm
	Alignment kernels	DTW-based alignment kernel
	DTWUDC	DTW + dynamic constraints
Red-DB	SAX	Reduce into symbolic space
	1-NN-SAX	SAX representation + SAX distance Suitable for simple signals α and p must be tuned Prediction is $O(N)$
	1-NN-SAX with k templates	Prediction is $O(\tilde{M})$
	1-NN-SFA	Fourier representation More adherence to the shape w.r.t. SAX Finer resolution with total recomputation
	Shapelets	Shift-invariant templates Training is $O(N^2 n^3)$
	Dictionary learning	Learn a sparse representation In general use Euclidean distance Chen et al. [122] add kernel representation

Continued

Table 4 Main characteristics of distance-based methods that have been reviewed in this work *Continued*

	Methods	Characteristics
Par-DB	DBA	DTW centroids are defined and used as templates
	1-NN-DWT	Better than DFT to handle nonstationarity Good frequency and temporal resolution
	Binet-Cauchy kernels	Kernel to embed dynamic behavior Rely on a probabilistic parametric modeling of time series
	Autoregressive kernels	Easily computed even when $q \gg n$ Rely on a probabilistic parametric modeling of time series AR model
	MMCL	Kernel based on nonlinear dynamical systems (ESN)
	LSM	Distance on LSM parameters

4.3 Methods Comparison

In Table 5 we summarize the general characteristics of FB and DB methods.

Moreover, in Table 6 we present a general overview of the complexity of the methods reviewed in this work and we provide an indication of whether they

Table 5 Main Characteristics of Feature and Distance-based Time-Series Classification Approaches

Class of Methods	Characteristics
FB	Feature extraction can be onerous Difficult to define a "distance" between classes Easy interpretation of results
Purely DB	Complexity grows at least linearly with n Kernel-based methods can be onerous Distance has to be chosen tailored to the problem Comparisons with all N training examples (best case: $O(Nn)$)
Red-DB	Distance computed in the reduce space Reduce numerosity of time series
Par-DB	Distance is computed over the space of parameters Kernel-based methods can be onerous

Table 6 Categorization of time-series methods

	Methods	Complexity	Type
FB	Time-pattern features	$O(n)$, low if p and n low	Remote
	Interval features	$O(n) - O(n^2)$, low if p and n low	Remote
	Bag of features	High ($O(Nn) - O(n)$)	Centralized
	Dynamic features	High, low with simple AR models	Remote
	Frequency domain features	Low	Remote
Purely DB	DTW	High ($O(Nn^2)$)	Centralized
	DTW approximations	High (lower bound $O(Nn)$)	Centralized
	DDTW	High ($O(Nn^2)$)	Centralized
	WDTW	High ($O(Nn^2)$)	Centralized
	ERP	High ($O(Nn^2)$)	Centralized
	TWED	High ($O(Nn^2)$)	Centralized
	Alignment kernels	High	Centralized
Red-DB	DTWUDC	High (DTW + dynamic model)	Centralized
	1-NN-SAX	High ($O(Nn)$)	Centralized
	1-NN-SAX with k templates	Low ($O(kn)$)	Remote
	1-NN-SFA	High $O(Nn)$	Centralized (RWT)
	Shapelets	Low (reduction methods)	Remote
	Dictionary learning	$O(k)$ (cardinality of dictionary)	Remote
	DBA	Low ($O(n^2) - O(n)$), DTW approx	Remote
Par-DB	1-NN-DWT	High, low *with* templates	Centralized (RWT)
	Binet-Cauchy kernels	High	Centralized
	Autoregressive kernels	High ($O(N(n^2 p^3))$), p: AR order	Centralized
	MMCL	High	Centralized
	LSM	High	Centralized

Notes: *Methods as DTW which admits variable-length timeseries has complexity which depends on the lengths (*n *and* m *of two timeseries). However, for the sake of simplicity with consider only* n *assuming that* m *is very similar to* n *(very common in real scenarios). RWT, remote with template.*

are suitable for *remote* or *centralized* tasks. With "remote" we refer to classification tasks than can be delivered without any kind of information coming from other nodes or locations, while with "centralized" we refer to algorithms or framework that can be executed in a central system that is aware of all nodes data. Notice that in doing the distinction remote/centralized, we assume that the training phase is carried out offline and thus in Table 6 only classification complexity is concerned.

Table 7 Categorization of time-series classification methods

	Methods	References
FB	Time-pattern features	[60–62]
	Interval features	[63–66]
	Bag of features	[68–76]
	Dynamic features	[82–84]
	Frequency domain features	[85–87]
	Ensemble of features	[89]
Purely DB	Miscellaneous	[90, 92, 93]
	DTW	[91]
	DTW approximations	[94–98]
	DDTW	[99]
	WDTW	[100]
	ERP	[101, 102]
	TWED	[103]
	Alignment kernels	[105, 106]
	DTWUDC	[107]
Red-DB	1-NN-SAX	[47, 48]
	1-NN-SAX with templates	[135]
	1-NN-SFA	[50]
	Shapelets	[108–114]
	Dictionary learning	[115–117]
	DBA	[123, 124]
Par-DB	1-NN-DWT	[136]
	Binet-Cauchy kernels	[125, 126, 129]
	Autoregressive kernels	[105, 127]
	MMCL	[128]
	LSM	[132–134]

Finally in Table 7, we summarize the references that correspond to each group of methods.

5   APPLICATIONS

In Table 8 we provide a list of some power-system applications whose issues have been addressed adopting some of the methodologies presented in this chapter. As can be seen, most of the applications concern event, anomaly and FD problems, exploiting FB techniques, which appear of an immediate application.

Table 8 Power-system applications

Year	Refs.	Method	Methodology	Data	Application
2016	[2]	Kernel PCA Partial SVM	FB	PMU	FDI
2016	[23]	mRmR Ensemble of bundle classifier SVM	FB	PMU	FDI
2016	[137]	DFT, DWT, FDST, PCA, Shapelet 1-NN, SVM	FB	PMU	FDI
2016	[13]	DTW, MDTW	Purely DB	AMI AM	FD
2016	[138]	Decision tree, SVM	FB	AMI	TD
2016	[139]	DWT (for preprocessing) Gaussian mixture models Parzen density estimator k-means clustering, k-NN Standard SVDD SVDD with negative examples	FB	PMU	ND
2016	[140]	Semisupervised SVM Adaboost, Multiple kernel learning	FB	PMU	AD
2015	[141]	Wavelet	Par-DB	PMU	AD/ED
2015	[142]	DWT, neural networks	FB	PMU	Event/FDI
2015	[143]	Adaptive neuro-fuzzy inference system Neural networks, SVM	FB	PMU	FD
2015	[144]	One class classifier	DB	Smart sensors	FDI
2014	[12]	Hidden Markov Models (HMM)	FB	PMU	FD
2014	[145]	Clustering, outlier detection Recursive feature elimination Multiple linear regression, ARIMA Support vector regression, k-NN RF, Boosting tree MARS Ensemble of methods	FB	Meteo	Prediction
2014	[146]	Naive Bayes Rule induction (OneR, NNge, JRipper) Decision tree learning (RFs) Binary classification (SVM) Boosting (Adaboost)	FB	PMU	FD AD
2013	[147]	Iterative Hilbert Huang Transform SAX	Red-DB	PMU	Monitoring

Continued

Table 8 Power-system applications *Continued*

Year	Refs.	Method	Methodology	Data	Application
2012	[148]	SVM One class SVM (semisupervised)	FB	PMU	FDI
2009	[17]	LS-SVM, DWT	FB	PMU	FDI
2008	[149]	DWT (for feature extraction)	Par-DB	PMU	FD
2006	[16]	SVM, neural networks	FB	PMU	FDI
2006	[22]	LR, neural networks	FB	Fault logs Meteo	FDI
2001	[21]	DWT	Par-DB	PMU	FDI

Notes: *A list of applications in the field of power systems is given, together with the main adopted time-series classification techniques. In "Application" column, the acronyms AD, TD, ND stand for anomaly, theft, and novelty detections.*

6 CONCLUDING REMARKS

Surfing through these methodologies applied to the power systems time series highlights the synergy between big data analysis and cyber-physical systems within the *smart paradigm*. These methodologies have paved a golden way toward new frontiers in scientific innovation and quality of service, leveraging continuous technological advances. Nonetheless, still some issues remains to be solved and to gain a main role in industrial and academic research.

A first fundamental aspect regard the control and architecture at all scales, meaning that through the pervasive monitoring of the systems the aim is to reach a full knowledge of the whole power system pipeline (from the first energy transformation to the final user) and at all levels (from the wide area grid network to the residential installations): on the one side this means even higher volumes and complexity of the data, while on the other it calls for the interoperability of the systems, the buzzword in the world of Internet of Things.

Also, it appears how the information and communication technologies, the cyber part, have gained a predominant role with respect to the physical counterpart, leveraging big data so as to offer new services and an increased performance of the systems: conversely, this ICT-mediated technology opens new scenarios in the field of faults, malicious behaviors, and more in general *cybersecurity* maintenance [150]. Concurrently, the governance of data and the privacy concerns need to be taken into account in this context, so as to guarantee the accurate and continuous knowledge of plants and behaviors without being invasive through the definition of policies of information management [151].

References

[1] B.E. Boser, I.M. Guyon, V.N. Vapnik, A training algorithm for optimal margin classifiers, in: Proceedings of the Fifth Annual Workshop on Computational Learning Theory, ACM, 1992, pp. 144–152.

[2] Y. Zhou, R. Arghandeh, I. Konstantakopoulos, S. Abdullah, A. von Meier, C.J. Spanos, Abnormal event detection with high resolution micro-PMU data, in: Power Systems Computation Conference (PSCC), 2016, pp. 1–7.

[3] J. Valenzuela, J. Wang, N. Bissinger, Real-time intrusion detection in power system operations, IEEE Trans. Power Syst. 28 (2) (2013) 1052–1062.

[4] A. Guerini, G. De Nicolao, Long-term electric load forecasting: a torus-based approach, in: 2015 European Control Conference (ECC), IEEE, 2015, pp. 2768–2773.

[5] F. Javed, N. Arshad, F. Wallin, I. Vassileva, E. Dahlquist, Forecasting for demand response in smart grids: an analysis on use of anthropologic and structural data and short term multiple loads forecasting, Appl. Energy 96 (2012) 150–160.

[6] P. Siano, Demand response and smart grids—a survey, Renew. Sust. Energ. Rev. 30 (2014) 461–478.

[7] K. Fischer, T. Stalin, H. Ramberg, J. Wenske, G. Wetter, R. Karlsson, T. Thiringer, Field-experience based root-cause analysis of power-converter failure in wind turbines, IEEE Trans. Power Electron. 30 (5) (2015) 2481–2492.

[8] D. Karlsson, M. Hemmingsson, S. Lindahl, Wide area system monitoring and control—terminology, phenomena, and solution implementation strategies, IEEE Power Energy Mag. 2 (5) (2004) 68–76.

[9] S. Yang, D. Xiang, A. Bryant, P. Mawby, L. Ran, P. Tavner, Condition monitoring for device reliability in power electronic converters: a review, IEEE Trans. Power Electron. 25 (11) (2010) 2734–2752.

[10] A.G. Phadke, P. Wall, L. Ding, V. Terzija, Improving the performance of power system protection using wide area monitoring systems, J. Mod. Power Syst. Clean Energy 4 (3) (2016) 319–331.

[11] A.G. Phadke, J.S. Thorp, Synchronized Phasor Measurements and Their Applications, Springer US, 2008. ISBN: 9780387765372.

[12] H. Jiang, J.J. Zhang, W. Gao, Z. Wu, Fault detection, identification, and location in smart grid based on data-driven computational methods, IEEE Trans. Smart Grid 5 (6) (2014) 2947–2956.

[13] N. Zhou, J. Wang, Q. Wang, A novel estimation method of metering errors of electric energy based on membership cloud and dynamic time warping, IEEE Trans. Smart Grid 8 (3) (2016) 1318–1329.

[14] N. Yu, S. Shah, R. Johnson, R. Sherick, M. Hong, K. Loparo, Big data analytics in power distribution systems, in: 2015 IEEE Power Energy Society Innovative Smart Grid Technologies Conference (ISGT), 2015, pp. 1–5, https://doi.org/10.1109/ISGT.2015.7131868.

[15] M. Manic, D. Wijayasekara, K. Amarasinghe, J.J. Rodriguez-Andina, Building energy management systems: the age of intelligent and adaptive buildings, IEEE Ind. Electron. Mag. 10 (1) (2016) 25–39.

[16] P. Janik, T. Lobos, Automated classification of power-quality disturbances using SVM and RBF networks, IEEE Trans. Power Delivery 21 (3) (2006) 1663–1669.

[17] Q.-M. Zhang, H.-J. Liu, Application of LS-SVM in classification of power quality disturbances, Proc. Chinese Soc. Electr. Eng. 28 (1) (2008) 106.

[18] L. Rabiner, B.-H. Juang, Fundamentals of Speech Recognition, Prentice Hall, Upper Saddle River, NJ, 1993.

[19] R.S. Tsay, Analysis of Financial Time Series, vol. 543, John Wiley & Sons, London, 2005.

[20] G.A. Susto, A. Beghi, Dealing with time-series data in predictive maintenance problems, in: 2016 IEEE 21st International Conference on Emerging Technologies and Factory Automation (ETFA), IEEE, 2016, pp. 1–4.

[21] O.A.S. Youssef, Fault classification based on wavelet transforms, in: 2001 IEEE/PES Transmission and Distribution Conference and Exposition, vol. 1, IEEE, 2001, pp. 531–536.

[22] L. Xu, M.-Y. Chow, A classification approach for power distribution systems fault cause identification, IEEE Trans. Power Syst. 21 (1) (2006) 53–60.

[23] Y. Zhou, R. Arghandeh, I.C. Konstantakopoulos, S. Abdullah, A. von Meier, C. J. Spanos, Distribution Network Event Detection with Ensembles of Bundle Classifiers, in: IEEE PES General Meeting 2016, 2016.

[24] G.A. Susto, A. Schirru, S. Pampuri, D. Pagano, S. McLoone, A. Beghi, A predictive maintenance system for integral type faults based on support vector machines: an application to ion implantation, in: 2013 IEEE International Conference on Automation Science and Engineering (CASE), IEEE, 2013, pp. 195–200.

[25] A. Beghi, L. Cecchinato, C. Corazzol, M. Rampazzo, F. Simmini, G.A. Susto, A one-class SVM based tool for machine learning novelty detection in HVAC chiller systems, IFAC Proc. 47 (3) (2014) 1953–1958.

[26] J. Friedman, T. Hastie, R. Tibshirani, The Elements of Statistical Learning, Springer Series in Statistics, vol. 1, Springer, Berlin, 2009.

[27] M. Müller, Dynamic time warping, Inf. Retr. Music Motion 2007, pp. 69–84.

[28] A. Schirru, G.A. Susto, S. Pampuri, S. McLoone, Learning from time series: supervised aggregative feature extraction, in: 2012 IEEE 51st Annual Conference on Decision and Control (CDC), IEEE, 2012, pp. 5254–5259.

[29] L. Xie, Y. Chen, P.R. Kumar, Dimensionality reduction of synchrophasor data for early event detection: linearized analysis, IEEE Trans. Power Syst. 29 (6) (2014) 2784–2794.

[30] L. Van Der Maaten, E. Postma, J. Van den Herik, Dimensionality reduction: a comparative, J. Mach. Learn. Res. 10 (2009) 66–71.

[31] T.-C. Fu, A review on time series data mining, Eng. Appl. Artif. Intel. 24 (1) (2011) 164–181.

[32] B. Scholkopft, K.-R. Mullert, Fisher discriminant analysis with kernels, in: Neural Networks for Signal Processing IX, vol. 1, 1999, p. 1.

[33] A. Hyvärinen, J. Karhunen, E. Oja, Independent Component Analysis, vol. 46, John Wiley & Sons, New York, 2004.

[34] A. Subasi, M.I. Gursoy, EEG signal classification using PCA, ICA, LDA and support vector machines, Expert Syst. Appl. 37 (12) (2010) 8659–8666.

[35] B. Schölkopf, A. Smola, K.-R. Müller, Kernel principal component analysis, in: International Conference on Artificial Neural Networks, Springer, 1997, pp. 583–588.

[36] S. Mika, B. Schölkopf, A.J. Smola, K.-R. Müller, M. Scholz, G. Rätsch, Kernel PCA and De-Noising in Feature Spaces, in: NIPS, vol. 11, 1998, pp. 536–542.

[37] G. McLachlan, Discriminant analysis and statistical pattern recognition, 544, John Wiley & Sons, New York, 2004.

[38] J.B. Kruskal, M. Wish, Multidimensional Scaling, vol. 11, SAGE, Thousand Oaks, CA, 1978.

[39] H. Hotelling, Analysis of a complex of statistical variables into principal components, J. Educ. Psychol. 24 (6) (1933) 417.

[40] I. Jolliffe, Principal Component Analysis, Wiley Online Library, New York, 2002.

[41] J.B. Tenenbaum, V. De Silva, J.C. Langford, A global geometric framework for nonlinear dimensionality reduction, Science 290 (5500) (2000) 2319–2323.

[42] S.T. Roweis, L.K. Saul, Nonlinear dimensionality reduction by locally linear embedding, Science 290 (5500) (2000) 2323–2326.

[43] L. Cayton, Algorithms for manifold learning, Univ. of California at San Diego Tech. Rep. (2005) 1–17.

[44] H. Narayanan, S. Mitter, Sample complexity of testing the manifold hypothesis, in: Adv. Neural Inf. Process. Syst., 2010, pp. 1786–1794.

[45] G. Biau, Analysis of a random forests model, J. Mach. Learn. Res. 13 (Apr) (2012) 1063–1095.

[46] J. Tang, S. Alelyani, H. Liu, Feature selection for classification: a review, in: Data Classification: Algorithms and Applications, CRC Press, Boca Raton, FL, 2014, p. 37.

[47] J. Lin, E. Keogh, S. Lonardi, B. Chiu, A symbolic representation of time series, with implications for streaming algorithms, in: Proceedings of the 8th ACM SIGMOD Workshop on Research Issues in Data Mining and Knowledge Discovery, ACM, 2003, pp. 2–11.

[48] J. Lin, E. Keogh, L. Wei, S. Lonardi, Experiencing SAX: a novel symbolic representation of time series, Data Min. Knowl. Disc. 15 (2) (2007) 107–144.

[49] E. Keogh, K. Chakrabarti, M. Pazzani, S. Mehrotra, Dimensionality reduction for fast similarity search in large time series databases, Knowl. Inf. Syst. 3 (3) (2001) 263–286.

[50] P. Schäfer, M. Högqvist, SFA: a symbolic Fourier approximation and index for similarity search in high dimensional datasets, in: Proceedings of the 15th International Conference on Extending Database Technology, ACM, 2012, pp. 516–527.

[51] C. Cortes, V. Vapnik, Support-vector networks, Mach. Learn. 20 (3) (1995) 273–297.

[52] M.E. Tipping, Sparse Bayesian learning and the relevance vector machine, J. Mach. Learn. Res. 1 (Jun) (2001) 211–244.

[53] L. Breiman, J. Friedman, C.J. Stone, R.A. Olshen, Classification and Regression Trees, CRC Press, Boca Raton, FL, 1984.

[54] L. Breiman, Random forests, Mach. Learn. 45 (1) (2001) 5–32.

[55] D.R. Cox, The regression analysis of binary sequences, J. R. Stat. Soc. Ser. B Methodol. 20 (1958) 215–242.

[56] C.E. Rasmussen, C.K.I. Williams, Gaussian Processes for Machine Learning, vol. 1, MIT Press, Cambridge, 2006.

[57] I. Goodfellow, Y. Bengio, A. Courville, Deep Learning, MIT Press, 2016.

[58] C.M. Bishop, Pattern Recognition, Mach. Learn. 128 (2006) 1–58.

[59] S. Manganaris, Supervised Classification With Temporal Data, Vanderbilt University, Nashville, TN, 1997.

[60] M.W. Kadous, Learning comprehensible descriptions of multivariate time series, in: ICML, 1999, pp. 454–463.

[61] M. Kudo, J. Toyama, M. Shimbo, Multidimensional curve classification using passing-through regions, Pattern Recogn. Lett. 20 (11) (1999) 1103–1111.

[62] P. Geurts, Pattern extraction for time series classification, in: European Conference on Principles of Data Mining and Knowledge Discovery, Springer, 2001, pp. 115–127.

[63] J.J. Rodríguez, C.J. Alonso, H. Boström, Boosting interval based literals, Intell. Data Anal. 5 (3) (2001) 245–262.

[64] J.J. Rodríguez, C.J. Alonso, Interval and dynamic time warping-based decision trees, in: Proceedings of the 2004 ACM Symposium on Applied Computing, ACM, 2004, pp. 548–552.

[65] J.J. Rodríguez, C.J. Alonso, J.A. Maestro, Support vector machines of interval-based features for time series classification, Knowl.-Based Syst. 18 (4) (2005) 171–178.

[66] H. Deng, G. Runger, E. Tuv, M. Vladimir, A time series forest for classification and feature extraction, Inf. Sci. 239 (2013) 142–153.

[67] D.G. Lowe, Object recognition from local scale-invariant features, in: The proceedings of the Seventh IEEE International Conference on Computer vision, vol. 2, IEEE, 1999, pp. 1150–1157.

[68] J. Lin, R. Khade, Y. Li, Rotation-invariant similarity in time series using bag-of-patterns representation, J. Intell. Inf. Syst. 39 (2) (2012) 287–315.

[69] J. Wang, P. Liu, M.F.H. She, S. Nahavandi, A. Kouzani, Bag-of-words representation for biomedical time series classification, Biomed. Signal Process. Control 8 (6) (2013) 634–644.

[70] M.G. Baydogan, G. Runger, E. Tuv, A bag-of-features framework to classify time series, IEEE Trans. Pattern Anal. Mach. Intell. 35 (11) (2013) 2796–2802.

[71] P. Senin, S. Malinchik, SAX-VSM: interpretable time series classification using SAX and vector space model, in: 2013 IEEE 13th International Conference on Data Mining, IEEE, 2013, pp. 1175–1180.

[72] M.G. Baydogan, G. Runger, Learning a symbolic representation for multivariate time series classification, Data Min. Knowl. Disc. 29 (2) (2015) 400–422.

[73] A. Bailly, S. Malinowski, R. Tavenard, T. Guyet, L. Chapel, Bag-of-temporal-SIFT-Words for time series classification, in: ECML/PKDD Workshop on Advanced Analytics and Learning on Temporal Data, 2015.

[74] P. Schäfer, The BOSS is concerned with time series classification in the presence of noise, Data Min. Knowl. Disc. 29 (6) (2015) 1505–1530.

[75] P. Schäfer, Scalable time series classification, Data Min. Knowl. Disc. volume 30 (2016) 1273–1298.

[76] A. Bailly, S. Malinowski, R. Tavenard, L. Chapel, T. Guyet, Dense bag-of-temporal-SIFT-words for time series classification, in: International Workshop on Advanced Analytics and Learning on Temporal Data, Springer International Publishing, September, 2015, pp. 17–30.

[77] G. Csurka, C. Dance, L. Fan, J. Willamowski, C. Bray, Visual categorization with bags of keypoints, in: ECCV Workshop on Statistical Learning in Computer Vision, vol. 1, Prague, 2004, pp. 1–2.

[78] G. Salton, A. Wong, C.-S. Yang, A vector space model for automatic indexing, Commun. ACM 18 (11) (1975) 613–620.

[79] H.P. Luhn, A statistical approach to mechanized encoding and searching of literary information, IBM J. Res. Dev. 1 (4) (1957) 309–317.

[80] K.S. Jones, A statistical interpretation of term specificity and its application in retrieval, J. Doc. 28 (1) (1972) 11–21.

[81] L. Ljung, System identification, in: Signal Analysis and Prediction, Springer, New York, 1998, pp. 163–173.

[82] D. Garrett, D.A. Peterson, C.W. Anderson, M.H. Thaut, Comparison of linear, nonlinear, and feature selection methods for EEG signal classification, IEEE Trans. Neural Syst. Rehabil. Eng. 11 (2) (2003) 141–144.

[83] Z.-Y. He, L.-W. Jin, Activity recognition from acceleration data using AR model representation and SVM, in: 2008 International Conference on Machine Learning and Cybernetics, vol. 4, IEEE, 2008, pp. 2245–2250.

[84] P.S.S. Roberts, Bayesian time series classification, Adv. Neural Inf. Process. Syst. 14 (2002) 937.

[85] P. Jahankhani, V. Kodogiannis, K. Revett, EEG signal classification using wavelet feature extraction and neural networks, in: IEEE John Vincent Atanasoff 2006 International Symposium on Modern Computing (JVA'06), IEEE, 2006, pp. 120–124.

[86] A. Subasi, EEG signal classification using wavelet feature extraction and a mixture of expert model, Expert Syst. Appl. 32 (4) (2007) 1084–1093.

[87] E.D. Übeyli, Combined neural network model employing wavelet coefficients for EEG signals classification, Digital Signal Process. 19 (2) (2009) 297–308.

[88] N.E. Huang, Z. Shen, S.R. Long, M.C. Wu, H.H. Shih, Q. Zheng, N.-C. Yen, C.C. Tung, H. H. Liu, The empirical mode decomposition and the Hilbert spectrum for nonlinear and non-stationary time series analysis, in: Proceedings of the Royal Society of London A: Mathematical, Physical and Engineering Sciences, vol. 454, The Royal Society, 1998, pp. 903–995.

[89] V. Eruhimov, V. Martyanov, E. Tuv, Constructing high dimensional feature space for time series classification, in: European Conference on Principles of Data Mining and Knowledge Discovery, Springer, 2007, pp. 414–421.

[90] G.E. Batista, X. Wang, E.J. Keogh, A complexity-invariant distance measure for time series, in: SDM, vol. 11, SIAM, 2011, pp. 699–710.

[91] H. Sakoe, S. Chiba, Dynamic programming algorithm optimization for spoken word recognition, IEEE Trans. Acoust. Speech Signal Process. 26 (1) (1978) 43–49.

[92] X. Xi, E. Keogh, C. Shelton, L. Wei, C.A. Ratanamahatana, Fast time series classification using numerosity reduction, in: Proceedings of the 23rd International Conference on Machine Learning, ACM, 2006, pp. 1033–1040.

[93] J. Lines, A. Bagnall, Time series classification with ensembles of elastic distance measures, Data Min. Knowl. Disc. 29 (3) (2015) 565–592.

[94] E. Keogh, C.A. Ratanamahatana, Exact indexing of dynamic time warping, Knowl. Inf. Syst. 7 (3) (2005) 358–386.

[95] S. Salvador, P. Chan, Toward accurate dynamic time warping in linear time and space, Intell. Data Anal. 11 (5) (2007) 561–580.

[96] G. Al-Naymat, S. Chawla, J. Taheri, SparseDTW: a novel approach to speed up dynamic time warping, in: Proceedings of the Eighth Australasian Data Mining Conference, vol. 101, Australian Computer Society, Inc., 2009, pp. 117–127.

[97] K.S. Candan, R. Rossini, X. Wang, M.L. Sapino, sDTW: computing DTW distances using locally relevant constraints based on salient feature alignments, Proc. VLDB Endowment 5 (11) (2012) 1519–1530.

[98] D.F. Silva, G.E. Batista, Speeding up all-pairwise dynamic time warping matrix calculation, in: Proceedings of the 2016 SIAM International Conference on Data Mining, SIAM, 2016, pp. 837–845.

[99] E.J. Keogh, M.J. Pazzani, Derivative dynamic time warping, in: SDM, vol. 1, SIAM, 2001, pp. 5–7.

[100] Y.-S. Jeong, M.K. Jeong, O.A. Omitaomu, Weighted dynamic time warping for time series classification, Pattern Recogn. 44 (9) (2011) 2231–2240.

[101] L. Chen, R. Ng, On the marriage of LP-norms and edit distance, in: Proceedings of the Thirtieth International Conference on Very Large Data Bases, vol. 30, VLDB Endowment, 2004, pp. 792–803.

[102] L. Chen, M.T. Özsu, V. Oria, Robust and fast similarity search for moving object trajectories, in: Proceedings of the 2005 ACM SIGMOD International Conference on Management of Data, ACM, 2005, pp. 491–502.

[103] P.-F. Marteau, Time warp edit distance with stiffness adjustment for time series matching, IEEE Trans. Pattern Anal. Mach. Intell. 31 (2) (2009) 306–318.

[104] B. Scholkopf, A.J. Smola, Learning With Kernels: Support Vector Machines, Regularization, Optimization, and Beyond, MIT Press, Cambridge, MA, 2001.

[105] M. Cuturi, J.-P. Vert, O. Birkenes, T. Matsui, A kernel for time series based on global alignments, in: 2007 IEEE International Conference on Acoustics, Speech and Signal Processing—ICASSP'07, vol. 2, IEEE, 2007, pp. 413.

[106] M. Cuturi, Fast global alignment kernels, in: Proceedings of the 28th International Conference on Machine Learning (ICML-11), 2011, pp. 929–936.

[107] S. Soatto, On the distance between non-stationary time series, in: Modeling, Estimation and Control, Springer, 2007, pp. 285–299.

[108] L. Ye, E. Keogh, Time series shapelets: a new primitive for data mining, in: Proceedings of the 15th ACM SIGKDD International Conference on Knowledge Discovery and Data Mining, ACM, 2009, pp. 947–956.

[109] L. Ye, E. Keogh, Time series shapelets: a novel technique that allows accurate, interpretable and fast classification, Data Min. Knowl. Disc. 22 (1–2) (2011) 149–182.

[110] A. Mueen, E. Keogh, N. Young, Logical-shapelets: an expressive primitive for time series classification, in: Proceedings of the 17th ACM SIGKDD International Conference on Knowledge Discovery and Data Mining, ACM, 2011, pp. 1154–1162.

[111] J. Lines, L.M. Davis, J. Hills, A. Bagnall, A shapelet transform for time series classification, in: Proceedings of the 18th ACM SIGKDD International Conference on Knowledge Discovery and Data Mining, ACM, 2012, pp. 289–297.

[112] T. Rakthanmanon, E. Keogh, Fast shapelets: a scalable algorithm for discovering time series shapelets, in: Proceedings of the 13th SIAM International Conference on Data Mining, SIAM, 2013, pp. 668–676.

[113] J. Hills, J. Lines, E. Baranauskas, J. Mapp, A. Bagnall, Classification of time series by shapelet transformation, Data Min. Knowl. Disc. 28 (4) (2014) 851–881.

[114] J. Grabocka, N. Schilling, M. Wistuba, L. Schmidt-Thieme, Learning time-series shapelets, in: Proceedings of the 20th ACM SIGKDD International Conference on Knowledge Discovery and Data Mining, ACM, 2014, pp. 392–401.

[115] J. Mairal, F. Bach, J. Ponce, G. Sapiro, A. Zisserman, Discriminative learned dictionaries for local image analysis, in: IEEE Conference on Computer Vision and Pattern Recognition, 2008. CVPR 2008, IEEE, 2008, pp. 1–8.

[116] J. Mairal, J. Ponce, G. Sapiro, A. Zisserman, F.R. Bach, Supervised dictionary learning, in: Adv. Neural Inf. Process. Syst., 2009, pp. 1033–1040.

[117] J. Mairal, F. Bach, J. Ponce, Task-driven dictionary learning, IEEE Trans. Pattern Anal. Mach. Intell. 34 (4) (2012) 791–804.

[118] Q. Zhang, B. Li, Discriminative K-SVD for dictionary learning in face recognition, in: 2010 IEEE Conference on Computer Vision and Pattern Recognition (CVPR), IEEE, 2010, pp. 2691–2698.

[119] M. Yang, L. Zhang, X. Feng, D. Zhang, Fisher discrimination dictionary learning for sparse representation, in: 2011 IEEE International Conference on Computer Vision (ICCV), IEEE, 2011, pp. 543–550.

[120] Z. Jiang, Z. Lin, L.S. Davis, Learning a discriminative dictionary for sparse coding via label consistent K-SVD, in: 2011 IEEE Conference on Computer Vision and Pattern Recognition (CVPR), IEEE, 2011, pp. 1697–1704.

[121] Z. Jiang, Z. Lin, L.S. Davis, Label consistent K-SVD: learning a discriminative dictionary for recognition, IEEE Trans. Pattern Anal. Mach. Intell. 35 (11) (2013) 2651–2664.

[122] Z. Chen, W. Zuo, Q. Hu, L. Lin, Kernel sparse representation for time series classification, Inf. Sci. 292 (2015) 15–26.

[123] F. Petitjean, A. Ketterlin, P. Gançarski, A global averaging method for dynamic time warping, with applications to clustering, Pattern Recogn. 44 (3) (2011) 678–693.

[124] F. Petitjean, G. Forestier, G.I. Webb, A.E. Nicholson, Y. Chen, E. Keogh, Faster and more accurate classification of time series by exploiting a novel dynamic time warping averaging algorithm, Knowl. Inf. Syst. 47 (1) (2016) 1–26.

[125] S.V.N. Vishwanathan, A.J. Smola, R. Vidal, Binet-Cauchy kernels on dynamical systems and its application to the analysis of dynamic scenes, Int. J. Comput. Vis. 73 (1) (2007) 95–119.

[126] A. Bissacco, A. Chiuso, S. Soatto, Classification and recognition of dynamical models: the role of phase, independent components, kernels and optimal transport, IEEE Trans. Pattern Anal. Mach. Intell. 29 (11) (2007) 1958–1972.

[127] M. Cuturi, A. Doucet, Autoregressive kernels for time series, 2011 (arXiv preprint arXiv:1101.0673).

[128] H. Chen, F. Tang, P. Tino, A.G. Cohn, X. Yao, Model metric co-learning for time series classification, in: Proceedings of the Twenty-Fourth International Joint Conference on Artificial Intelligence, AAAI Press, 2015, pp. 3387–3394.

[129] S.V.N. Vishwanathan, A.J. Smola, et al., Binet-Cauchy kernels, in: NIPS, 2004, pp. 1441–1448.

[130] H. Jaeger, The "echo state" approach to analysing and training recurrent neural networks-with an erratum note, vol. 148, German National Research Center for Information Technology GMD Technical Report, Bonn, Germany, 2001, p.34.

[131] H. Jaeger, Adaptive nonlinear system identification with echo state networks, in: Adv. Neural Inf. Process. Syst., 2002, pp. 593–600.

[132] W. Aswolinskiy, R.F. Reinhart, J. Steil, Time series classification in reservoir-and model-space: a comparison, in: IAPR Workshop on Artificial Neural Networks in Pattern Recognition, Springer, 2016, pp. 197–208.

[133] Q. Ma, L. Shen, W. Chen, J. Wang, J. Wei, Z. Yu, Functional echo state network for time series classification, Inf. Sci. 373 (2016) 1–20.

[134] Y. Li, J. Hong, H. Chen, Sequential data classification in the space of liquid state machines, in: Joint European Conference on Machine Learning and Knowledge Discovery in Databases, Springer, 2016, pp. 313–328.

[135] P. Siirtola, H. Koskimäki, V. Huikari, P. Laurinen, J. Röning, Improving the classification accuracy of streaming data using SAX similarity features, Pattern Recogn. Lett. 32 (13) (2011) 1659–1668.

[136] P. Fryzlewicz, H. Ombao, Consistent classification of nonstationary time series using stochastic wavelet representations, J. Am. Stat. Assoc. 104 (2012) 299–312.

[137] S. Brahma, R. Kavasseri, H. Cao, N.R. Chaudhuri, T. Alexopoulos, Y. Cui, Real time identification of dynamic events in power systems using PMU data, and potential applications—models, promises, and challenges, IEEE Trans. Power Delivery 32 (2017) 294–301.

[138] A. Jindal, A. Dua, K. Kaur, M. Singh, N. Kumar, S. Mishra, Decision tree and SVM-based data analytics for theft detection in smart grid, IEEE Trans. Ind. Inf. 12 (3) (2016) 1005–1016.

[139] A.E. Lazzaretti, D.M.J. Tax, H.V. Neto, V.H. Ferreira, Novelty detection and multi-class classification in power distribution voltage waveforms, Expert Syst. Appl. 45 (2016) 322–330.

[140] M. Ozay, I. Esnaola, F.T.Y. Vural, S.R. Kulkarni, H.V. Poor, Machine learning methods for attack detection in the smart grid, IEEE Trans. Neural Networks Learn. Syst. 27 (8) (2016) 1773–1786.

[141] D.-I. Kim, T.Y. Chun, S.-H. Yoon, G. Lee, Y.-J. Shin, Wavelet-based event detection method using PMU data, IEEE Trans. Smart Grid 8 (2017) 1154–1162.

[142] S. Alshahrani, M. Abbod, B. Alamri, Detection and classification of power quality events based on wavelet transform and artificial neural networks for smart grids, in: Smart Grid (SASG), 2015 Saudi Arabia, IEEE, 2015, pp. 1–6.

[143] P. Gopakumar, J.B. Reddy, D.K. Mohanta, Adaptive fault identification and classification methodology for smart power grids using synchronous phasor angle measurements, IET Gener. Transm. Distrib. 9 (2) (2015) 133–145.

[144] E. De Santis, L. Livi, A. Sadeghian, A. Rizzi, Modeling and recognition of smart grid faults by a combined approach of dissimilarity learning and one-class classification, Neurocomputing 170 (2015) 368–383.

[145] C. Fan, F. Xiao, S. Wang, Development of prediction models for next-day building energy consumption and peak power demand using data mining techniques, Appl. Energy 127 (2014) 1–10.

[146] R.C.B. Hink, J.M. Beaver, M.A. Buckner, T. Morris, U. Adhikari, S. Pan, Machine learning for power system disturbance and cyber-attack discrimination, in: 2014 7th International Symposium on Resilient Control Systems (ISRCS), IEEE, 2014, pp. 1–8.

[147] M.J. Afroni, D. Sutanto, D. Stirling, Analysis of nonstationary power-quality waveforms using iterative Hilbert Huang transform and SAX algorithm, IEEE Trans. Power Delivery 28 (4) (2013) 2134–2144.

[148] N. Shahid, S.A. Aleem, I.H. Naqvi, N. Zaffar, Support vector machine based fault detection & classification in smart grids, in: 2012 IEEE Globecom Workshops (GC Wkshps), IEEE, 2012, pp. 1526–1531.

[149] N.I. Elkalashy, M. Lehtonen, H.A. Darwish, A.-M.I. Taalab, M.A. Izzularab, DWT-based detection and transient power direction-based location of high-impedance faults due to leaning trees in unearthed MV networks, IEEE Trans. Power Delivery 23 (1) (2008) 94–101.

[150] L. Langer, F. Skopik, P. Smith, M. Kammerstetter, From old to new: assessing cybersecurity risks for an evolving smart grid, Comput. Secur. 62 (2016) 165–176.

[151] M. Buchmann, Governance of data and information management in smart distribution grids: increase efficiency by balancing coordination and competition, Util. Policy (2017), https://doi.org/10.1016/j.jup.2017.01.003.

Put the Power of Big Data into Power Systems

CHAPTER 10

Future Trends for Big Data Application in Power Systems

Ricardo J. Bessa
INESC Technology and Science—INESC TEC, Porto, Portugal

CHAPTER OVERVIEW

The technological revolution in the electric power system sector is producing large volumes of data with pertinent impact in the business and functional processes of system operators, generation companies, and grid users. Big data techniques can be applied to state estimation, forecasting, and control problems, as well as to support the participation of market agents in the electricity market. This chapter presents a revision of the application of data mining techniques to these problems. Trends like feature extraction/reduction and distributed learning are identified and discussed. The knowledge extracted from power system and market data has a significant impact in key performance indicators, like operational efficiency (e.g., operating expenses), investment deferral, and quality of supply. Furthermore, business models related to big data processing and mining are emerging and boosting new energy services.

1   INTRODUCTION

The advent of Smart Grids with advances in information and communication technologies (ICT) and installation of new measurement devices, such as phasor measurement unit (PMU) and remote terminal unit (RTU) in secondary substations (MV/LV), allied to additional information collected by SCADA, will generate a large volume of data streams.

Equipment installed in MV/LV substations collects imported/exported active power, voltage magnitude, and reactive power in four quadrants, and a distribution system operator (DSO) can easily operate more than 10,000 secondary substations. In HV/MV substations, which can be more than 1000 in one DSO, additional data is collected through the SCADA, such as current, active and reactive power flow in the network feeders, switcher and capacitor banks status, as well as variables related to electric transformers (e.g., input/output voltage temperature, tap changer position, transformer oil level, insulation level of transformer oil, load). This high volume of grid data has different constraints in terms of communications' latency and availability. For instance, significant technical and economic constraints are expected in the real-time communication between smart meters and secondary substation, which requires new

223

approaches for the real-time monitoring of low voltage (LV) networks. More-over, the time resolution collected by different equipment differs, PMU collects high-frequency data, while RTU, in general, collects low-frequency data (e.g., 15-min average).

PMU can provide high-update rate data to a transmission system operator (TSO). For instance, the Texas Synchrophasor Network collects 30 measurements per second from each PMU (e.g., voltage/current magnitude and phase, frequency), which means 108,000 lines of comma-separated data per hour and 2.6 million lines for a 24 h' period; for 15 PMUs, file storage is about 1 GB per day [1].

This data, collected at different voltage levels, is essential to revisit classical TSO and DSO grid management functions, such as forecasting, state estimation, operational planning, and develop new tools to increase real-time awareness of operators and design predictive maintenance strategies for network components.

The renewable energy sources (RES) industry is also installing and operating monitoring sensors at the wind turbine and photovoltaic panel level, which generates a large volume of data that needs to be preprocessed and analyzed in realtime and transferred to upstream decision centers. For instance, a 2.5 MW wind turbine has more than 120 sensors inside the rotor, the generator, and on the blades, which gather 10,000 of data points every second. They feed the information to a remote database, which stores 4 TB from 25,000 turbines around the world.[1] The same is valid for gas turbine engine that generates 520 GB per day, in contrast to Twitter where a day of real-time feeds represents around 80 GB.[2] This data can be used for reliability and performance monitoring, predictive maintenance, and asset management of conventional and RES power plants. Eventually, the outcome of the data analysis at the power plant level can feed power system reliability assessment tools [2], by providing, for instance, data-driven time-varying failure rates.

In addition to all these electrical and mechanical variables, there are also exogenous variables with significant impact on the power system and power plants operation and planning, such as measured and predicted weather variables (e.g., wind speed, temperature, and solar irradiance) that can form a grid of spatial-temporal weather information in a region and/or country.

Electricity markets are already generating large volumes of data like offers curves (per unit) in different sessions, energy and ancillary services prices, as

---

[1] Source: http://www.gereports.com/post/118712460090/move-over-slow-food-slow-wind-might-be-the-latest/ (accessed on October 2016).

[2] Source: http://www.computerweekly.com/news/2240176248/GE-uses-big-data-to-power-machine-services-business (accessed on October 2016).

well as locational marginal prices (LMP) for each node of the transmission network. The foreseen creation of flexibility markets at the distribution level will increase the volume of data and its spatial scale. The planned investment in interconnection capacity between different control areas, and the increase integration of RES in power systems with LMP, makes spatial-temporal modeling of large-scale time series vital for operational and planning purposes. Therefore, knowledge extraction from big data can create additional value for both market players and system operators.

All these problems require different layers of data handling: (i) data acquisition and transmission; (ii) data management (e.g., frameworks like Hadoop or Spark); (iii) data analytics, which can comprise knowledge extraction from data, optimization, and decision-aid methods. The first two layers already achieved a high-technology readiness level, with different solutions available in the market [3,4]. However, standardization of the data model, ICT for real-time data transmission, and cybersecurity issues remain areas of significant improvement.

The scope of this chapter is the big data analytics layers and the overall objective is to discuss the main challenges related to knowledge extraction in different power system-related problems and cover new (and evolving) problems, such as distributed learning and optimization, spatial-temporal modeling of time series, data reduction, assimilation, and visualization methods. The entire electric power system is covered, going from Extra HV to LV, without overlooking the wholesale and retailing electricity market.

This chapter is organized as follows: Section 2 describes the data-driven techniques for dynamic and steady-state analysis of transmission systems, as well as the interaction between transmission and distribution system operators; in Section 3, the additional monitoring and control capabilities provided by advanced data mining techniques are discussed in a Smart Grids context; Section 4 discusses the knowledge extraction from failure data to support asset management strategies of system operators and generation companies; the added value of big data techniques for electricity market bidding and simulation is discussed in Section 5, while Section 6 discusses its application to boost demand-side flexibility. The conclusions are presented in Section 7.

2 TRANSMISSION SYSTEM

At the transmission system level, the increasing penetration of RES is demanding for new monitoring and management tools for both interconnected and isolated systems. A new generation of decision-aid tools will supply the operator with valuable information to check the security level of the economic dispatch and/or electricity market-clearing, considering RES variability and uncertainty, as well as to increase the real-time awareness and derive recommendations to support preventive decisions.

2.1 Dynamic Behavior Analysis

The installation of PMU in different voltage levels generates important information to warn operators and system level controllers about impending transient stability issues, support their preventive decisions, and perform *postmortem* analysis. The California independent system operator (CAISO) defined use cases that describe the inclusion of PMU data for grid operations, control and modeling tasks [5]. The use cases identified seven scenarios to demonstrate the value of PMU data:

1. The PMU network triggers an alarm (e.g., rate of frequency change, modes of oscillation, rate of damping) for a recommendation system that generates a set of control actions for the operator.
2. Measure the frequency difference between main and isolated grids for system restoration after a disturbance and determine how much generation must be changed to reconnect the separated grids.
3. *Postmortem* analysis of system events to understand the causes of disturbance, which is used to validate offline dynamic models and contingency simulation tools.
4. Validation of gridcode and market models for new types of resources, such as RES and storage.
5. Detect transient instability and derive preventive control actions that can respond to specific or wind-area grid problems, e.g., angular and voltage stability, low-frequency oscillations.
6. Identify poorly damped interarea oscillations and design smart control actions to mitigate the oscillations, e.g., use PMU to tune power system stabilizers.
7. Increase the line rating of transmission lines in realtime. The PMU data can detect postcontingency technical problems and activate the preventive control actions from scenario (5) to mitigate in realtime the violations by reconfiguring the system (e.g., increase generation or decrease load).

The electric power research institute (EPRI) identifies the following applications for PMU data [6]: (i) improvement of state estimation; (ii) oscillation detection and control; (iii) voltage stability monitoring and control; (iv) load model validation; (v) system restoration and event analysis.

It should be stressed that the use of PMU data demands for a portfolio of different tools at the control center level, which corresponds to the enhancement of classical functions and to the development of new functions. Examples of related tools are the state estimator, voltage stability analysis, volt/Var control, and RES dispatch. A PMU network combined with decision trees can be used to match the generator trips signature with the overall system dynamic, aiming at finding the most likely location of an event in realtime [7]. The data processing and machine learning fitting were performed offline and in a controlled environment since the training consisted of 53 events that match known generator trips. An industrialization of this solution would require machine learning algorithms for classification problems able to cope with high-speed data streams and detect concept drift [8].

Other potential applications are: line trip detection that requires postprocessing methods, such as a low-pass filter to remove high-frequency noise and a second one to get the trend of frequency data [9]; online prediction of transient stability (i.e., three phase faults at different buses) with decision tree algorithm in order to derive corrective control rules [10].

The seemly integration of PMU in power system operational tools will require a data analytics platform that integrates batch, real-time, and iterative data processing. Apache Spark is emerging as the cluster computing platform for future power systems [11]. The trend is toward distributed computing for data collection and analytics. However, there is the need to develop algorithms that are parallelizable to distribute the computational load across multiple nodes [12].

Furthermore, this efficient computational framework does not waive the application of data reduction and compression techniques, which should be flexible to the different operating conditions, e.g., compress less data under disturbance conditions [13]. Classical techniques, such as principal component analysis and discrete wavelet transform, can be extended to this problem to have time-varying (potentially combined with change detection) and situational-dependent characteristics. Clustering algorithms can be also used to group the dynamic response of generators (i.e., transient responses of generator rotor angles) and use a classification algorithm to forecast the dynamic signature of a system using a dataset of postdisturbance responses [14].

Failure in communication creates missing values in the power system dynamic response. The state of the art consists in using the linear auto-regressive with exogenous input model to estimate system dynamics, together with an input location selection methodology based on a coherency function [15]. The spatial and temporal dependencies between the system variables can be further exploited with the different families of covariance functions associated to Gaussian processes theory and improve the missing values estimation tasks [16].

Machine learning algorithms can be also used to give a real-time quantitative security evaluation of the current operating state system (i.e., expected frequency deviation) based on historical states and observations of the power system variables [17]. This research line was further explored in microgrids and isolated systems [18].

2.2 Steady-State Analysis

The tools for steady-state analysis of power systems, such as power flow and state estimation algorithms, reached a high-technological readiness level and several commercial solutions are already available. The current challenge is to integrate new and diverse types of information in these classical algorithms, capture the spatial-temporal structure of variables dependency, while guaranteeing a high scalability.

Past development in state estimation algorithms already included information from load forecasts to predict the future states of the power systems. For instance, modeling the dependency between nodal injections forecast errors with a covariance matrix [19,20]. The load forecast and state estimation theories can be merged to forecast the future values of the power system state variable (bus voltage magnitude and phase) and then calculate the load values as a function of the state parameters [21]. This new load forecast paradigm enables the use of additional data, such as voltage phase from PMU or electrical variables collected from multiarea networks, and the construction of local forecast models for different subnetworks.

However, the modeling of spatial-temporal dependencies is indispensable and requires a method suitable for a large-scale implementation. Gaussian copulas can be employed to model the spatial-temporal dependency structure between random variables [22], but have two limitations: (i) lack of flexibility in modeling different types of tail's dependency; (ii) low scalability when the number of random variables increases.

The effect of RES and load uncertainty (and variability) in state estimation, together with frequent topological changes, leads to significant state shift in power system operation. This problem can be mitigated by developing data-driven solutions, instead of using single data point (last state estimation). Kernel ridge regression with a Bayesian framework that uses historical data collected by the energy management system can tackle this problem [23].

Another relevant trend is the use of distributed learning approaches for robust state estimation that results in minimum data exchanges between neighboring areas [24], mitigates privacy issues, and can run locally in grid equipment. This distributed learning paradigm relies in the alternating direction method of multipliers (ADMM) that combines the decomposability offered by the dual ascent method with the superior convergence properties of the method of multipliers, which means that problems with nondifferentiable objective functions can be easily addressed and it is possible to perform parallel optimization [25]. It is also possible to apply other variants, such as the Douglas-Rachford and block coordinate descent methods [26,27]. It is important to stress the nonlinear nature of the AC power system, which results in a nonconvex problem for the state estimator.

The same paradigm can be applied to RES forecast to explore geographically distributed time-series information [28]. The vector autoregression (VAR) framework can be applied to forecast thousands of time series in a distributed fashion by combining ADMM with LASSO framework to explore the sparsity in the model's coefficients.

The practical implementation of the distributed learning paradigm requires an adequate choice of the distributed processing platform, which can be divided into two types [29]: (i) horizontal scaling: distribute the workload by several servers—decentralized and distributed cluster (cloud) computing framework; (ii) vertical scaling: involves installing more processors, memory, and faster hardware inside a single machine.

For horizontal scaling, message passing interface (MPI) was the first communication protocol to distribute and exchange the data between peers, Apache Hadoop with MapReduce as the data processing scheme emerged later, and Apache Spark is the prevalent solution. For iterative algorithms like ADMM, MapReduce is not adequate due to disk I/O limitations, while Spark performs in-memory computations that overcome these limitations for iterative processes [29]. The most popular vertical scale up technologies are high-performance computing clusters, multicore processors, and graphics processing unit (GPU). The ADDM algorithm and variants can be implemented in these platforms.

2.3 TSO-DSO Cooperation

The data exchange between TSO and DSO will contribute to increase the security of both systems in different time-scales, ranging from real-time to long-term planning. The European project evolvDSO developed a usecase for the TSO-DSO cooperation, which firstly means bidirectional exchange of information, both historical and real-time data, regarding the operating conditions of the transmission and distribution systems [30]. Secondly, it can also mean the DSO supporting the TSO operational and planning tasks, for instance, by controlling the active and reactive power in the primary substation or elaborating a joint expansion plan of both systems. Cooperation is needed since presently the distribution system is a blackbox to the TSO and viceversa. Moreover, considering the increasing integration of distributed energy resources in the distribution system, the operation of both networks becomes challenging and cannot be decoupled. The new flexible resources (e.g., demand response—DR) are also at the distribution system level, which requires new TSO-DSO technical protocols for its activation and management.

This increasing cooperation will mean additional data to be integrated and explored in the managing tasks of both TSO and DSO. One trend is the development of tools capable of estimating the flexibility range of active and reactive power in the TSO-DSO boundary and separating this flexibility by total cost [31]. The same exercise can be conducted for lower voltage levels of the power system [32].

For dynamic analysis, the trend is to estimate the dynamic response of load aggregated at the network node level for a time domain between one and several seconds. One example is probabilistic methodologies based on processing and classifying large amounts of historical load data at each bus and standard dynamic signatures of individual load categories obtained from laboratory/ fieldtests [33]. Another is dynamic equivalent models constructed for the distribution networks that are able to reflect the aggregated behavior of different resources with respect to system requirements such as frequency containment reserve. Machine learning algorithms, such as artificial neural networks, can be used as surrogate models for the dynamic equivalents [34].

3 DISTRIBUTION SYSTEM

The big data trends in the distribution system are mainly driven by two objectives. Firstly, increase the monitoring capability of MV and LV networks and develop fast decision-aid methods for operators. Secondly, implemented predictive active management strategies that take advantage of flexibility from distributed energy resources to mitigate the impact of RES uncertainty and variability.

3.1 Monitoring and Situational Awareness

The smart grid paradigm increases the monitoring capability of the distribution system. However, it might be unmanageable to have real-time monitoring of all the devices in the distribution system, particularly at the LV level. Machine learning algorithms installed in intelligent electronic devices can support power system monitoring by providing several functionalities, such as reconstruction of missing signals, state estimation, asset monitoring and diagnosing, and fault location. These functions should have low computational requirements (e.g., no need to store data, capacity of running in low cost processors) and the possibility to adjust under evolving conditions.

For LV grids, the trend is to explore data collected from smart meters and RTU installed in MV/LV substations for close to real-time situational awareness of operators and with low communication costs. Smart meter data can be used to increase the knowledge about the LV network topology and characteristics. For instance, it can be used to reduce geographical information system errors (e.g., connectivity errors in the network topology) and for phase detection [35].

Data-driven methods, such as autoencoder extreme learning machines (AE-ELM), can be employed to estimate, close to real-time, voltage magnitude and active power for all nodes of the LV network by using only a subset of meters with real-time communication capability [36,37]. This new smart grid function can generate under/overvoltage alarms to operators and trigger control management functions to solve the technical problems. These techniques provide accurate information about voltage magnitude. Only with 30% of the total meters with real-time communication, the AE-ELM state estimator estimates [38]: (i) voltage magnitude values with a mean absolute error (MAE) of 0.49 V; (ii) active power quantities with an MAE of 0.35 kW. The largest MAE was 0.79 V.

The challenge is on how to monitor the operating conditions of multiple LV networks at the same time and derive control strategies to solve detected technical problems. This problem requires new techniques for data streaming visualization and dimension reduction that summarize the operating conditions of each network and present the information to the operator in a readable way. An example of a different data visualization method for power system is the "sparklines" that can display time-varying power system data placed in a geographical map of the system by using methods of graph drawing [39].

The visualization of the electrical network needs to be also revised to better display and identify the branches and nodes with technical problems. One possibility is to project the electrical distance metrics into a 2-dimensional plane using dimensional scaling and graph theory [40], which offers new insights on the electrical network structure and voltage performance.

Social media data (spatial-temporal real-time tweets) can be exploited to detect and locate electricity power outages with a supervised topic model that uses a heterogeneous information network [41]. The next generation of tools should be able to combine data collected from the distribution network, state estimation tools, and social media to improve the quality of supply indices of the DSO.

When alarms are trigged, the operators have a very limited time to take decisions regarding mitigation actions. In this case, techniques that search for analogs (i.e., similar operating conditions in the historical data), and conduct spatial-temporal analysis of events, are needed to generate a set of simple control rules that the operator can implement in a few seconds, e.g., change the secondary substation on-load tap changer position, reduce in x% the PV generation in node Y, etc.

A new generation of grid support and operators' training tools can be developed to exploit large databases of network measurements and corresponding events. One example is a tool that performs a pro-active analysis of grid control actions by analyzing past events and actual real data and resimulate them for improving grid management rules [42]. The fast access to historical data and the capability to extract (and match) relevant events and patterns in the low running time are key functional requirements.

Similar to transmission networks, distributed state estimation algorithms running locally at each smart grid equipment, such as RTU or distributed transformer controller, will pave the way toward peer-to-peer data exchange for improved monitoring at MV level. The uncertainty regarding the distribution system model parameters (e.g., topology, impedances, connected loads/generators) must be reduced for a proper application of state estimation and control tools. This uncertainty can be reduced by applying a parameter estimation method that takes advantage of historical AMI and other sensors data. For instance, the transformer and line series impedance parameters in 3-phase and 1-phase circuits can be estimated with active, reactive power and voltage measurements collected by an advanced metering infrastructure [43].

3.2 Predictive Control and Management

The smart grid paradigm brings new challenges in terms of load forecast. In the past, load forecast was classified as a "solved problem'" with highly accurate predictions (e.g., mean absolute percentage error between 2% and 4%). However, the emerging role of "prosumers'" is making the consumption pattern more volatile and therefore more difficult to predict. In a near future, the load profiles in distribution systems will vary with the weather conditions due to self-consumption from photovoltaic panels, dynamic electricity tariffs, consumer preferences, and behavior (e.g., demand response, storage devices). In the case of self-consumption from PV generation, the DSO does not have information about the PV generation profile at the residential level and its impact in the net-load profile. Data-driven techniques based on weather data and PV time series from neighbored sites, and that combine fuzzy theory and clustering algorithms, can be used to estimate the generation of "invisible" PV sites [44]. Moreover, change detection algorithms can be used to detect unauthorized PV installations and verify the existence of such system with permutation tests [45].

In distribution grids, modeling the spatial-temporal dependency structure of nodal net-load forecast errors is essential to design predictive management strategies with information about uncertainty. To meet this requirement, recent research developed techniques to: (i) explore dependencies between geographically distributed time series to improve the RES forecasting skill [46,47]; (ii) model the spatial-temporal dependency structure of forecast errors (uncertainty) and derive random vectors or joint density functions [22,48]; (iii) apply feature engineering techniques to spatial-temporal numerical weather predictions [49]. The goal is to generate highly accurate probabilistic forecasts that can feed DSO management tools. Moreover, at the LV level, the number of time series to forecast will be very high, which requires distributed computing solutions.

A classical power system optimization problem is the optimal power flow (OPF), which, presently, is being extended to a multiperiod version that includes storage and demand response control actions. Stochastic versions for this problem are being proposed and developed by several authors [50], but the big data-related trend is the distributed OPF supported by techniques that are also used in distributed statistical learning, such as ADMM and auxiliary problem principle [51,52]. This optimization problem can also benefit from deep learning frameworks like TensorFlow that waive the use of linear algebra libraries, which are the source of high-computational times in the OPF [53].

4 ASSET MANAGEMENT

Presently, TSO and DSO are improving their asset data management systems to include multiple time series related to their health and operating conditions, taking advantage of new information collected by the advanced metering infrastructure. This data is valuable for the following tasks: (i) decide about life extension measures or asset renewal; (ii) design predictive management strategies and evaluate their impact in the power system reliability. Information about the failure rates and the impact of different maintenance strategies are essential for reliability assessment tools based on Monte Carlo simulation methods [54].

The identification of the critical assets, i.e., assets that require a more appropriate maintenance strategy, can be made with statistical analysis of failure data selected with Laplace test and correlation coefficient techniques [55]. Moreover, techniques to analyze outage indices are required to evaluate the effect of different processes in the outage rate, evaluate human mistakes effects and planning strategies [56]. The following needs were identified for this problem: (i) estimate distinctive failure rates for each asset, conditioned by weather variables, age, and maintenance strategies; (ii) assess the benefit of maintenance on the failure rates; (iii) evaluate the consequences of a failure of contingency [57].

Predictive management strategies/conditional-based monitoring aim at predict/ infer if *component X of machine Y is about to fail in N days with probability M%*, or in other words, *probability of failure* and *time to failure*. These tasks combine feature engineering techniques with domain knowledge, feature reduction, and selection and a base learner for classification and/or regression. Some examples are: fuzzy-logic techniques applied to evaluate the health index of a transformer using real observations (water content, acidity, etc.) and expert rules based on linguistic expressions [58]; combination of auto-encoders and information theoretic learning (mean shift algorithm) for condition diagnosis in power transformers with online monitoring of dissolved gases in oil [59]; nonparametric regression that combines age and lifecycle data (e.g., health index, manufacture, location) that gives asset-oriented failure rate, health condition, and risk factors analysis [60]. Data mining techniques can be also applied to other assets, like distribution circuit breakers [61].

This research domain has the following challenges for further work: (i) combine features generated by domain knowledge with "automatic" feature extraction algorithms such as deep learning techniques, which might result in a large number of input variables; (ii) propose evaluation metrics that include cost-sensitive actions, such as cost of false alarm, cost of missing failure, and reward of detecting/predicting a failure; (iii) communicate results and uncertainty to the decision maker using appropriate data visualization techniques; (iv) handle a low number of instances with failure mode that results in imbalanced datasets for classification problems; (v) apply nonparametric techniques, in contrast to parametric techniques such as the Weibull model.

Renewable and conventional generation companies, with the advent of Internet of Things (IoT) technology, are hiring data scientists to develop new predictive maintenance and conditioning monitoring data-driven methods for their assets. The high cost of offshore wind power plant's maintenance demands for new monitoring and maintenance planning methods [62]. Machine learning algorithms, such as neural networks and Gaussian processes, can be used to construct power curves (wind speed vs power) from SCADA measurements and control charts for the individual turbine monitoring using standard x chart plots and extreme value statistics that generate alarm thresholds [63]. Short-term probabilistic forecast of weather variables can be integrated in decision-making problems under risk (cost-loss mode) for finding access windows for offshore wind power plants maintenance [64].

Finally, new grid assets like battery storage and variable speed pump power storage will require similar methodologies to infer and predict the current and future operating conditions based on several data sources.

5   ELECTRICITY MARKETS

The electricity market can be divided into two different types: the spot market, where the electrical energy is traded for immediate physical delivery, and the futures market, where the delivery is at a later date and normally does not involve physical delivery. The futures market is normally used for risk hedging.

Presently, the research work is concentrated in designing new market rules and frameworks to boost the Smart Grid paradigm and integrate the active demand-side participation. For instance, studying the effect on the economic efficiency of the day-ahead market from load shifting behavior of consumers with price-responsive bids [65] and simulating a demand response electricity market with price-responsive commercial buildings by using agent-based modeling [66].

The trend at the European level is the construction of pan-European electricity markets by increasing the interconnection capacity between countries. This will make the analysis of the stochastic dependency of spot prices at the European level very important [67]. Therefore, tools for a spatial-temporal analysis of distributed price time series will be needed. Moreover, in countries with locational marginal prices (such as United States) the spatial-temporal pattern of market prices encompasses useful information, such as the network topology [68], and can be explored by machine learning algorithms to unveil operating conditions and improve price forecast.

The power system also has ancillary services separated from the electrical energy market, which are used to support reliability and power quality of the power system. Related to this market, the main research goals consist in modeling and forecasting the direction and magnitude of activated reserve and corresponding price, which, in general, are irregular time series [69].

The optimal participation of market agents (from demand and supply side) in these markets requires price forecast information for different time horizons, ranging from hours-ahead to year-ahead. The literature is rich in point forecasting algorithms, but a lack of works related to the characterization of the uncertainty associated to the price forecast was identified [70]. The increasing penetration of variable renewable energy in the electricity market is impacting the market price level, e.g., zero and negative prices are starting to occur frequently. Therefore, exogenous variables and their intrinsic uncertainty, such as renewable power forecasts and weather predictions [71], should be modeled in price-forecasting algorithms. The dependency between prices of the spot and futures markets is essential to design mid and long-term risk hedging strategies and requires complex dependency structures from copula theory [72].

An important variable for understanding and forecasting the electricity market dynamics is the market agents' strategic bidding behavior. This information is useful for two different types of users, market competitors and energy regulators. Market agents aim to derive the market bid that maximizes individual profit taking into consideration the behavior of the competitors, such as using an analytical approach based on a state space model representation [73] or metaheuristic optimization techniques [74]. Moreover, supply functions can represent an optimal response to the offers of the other market participants [75] and its estimation can benefit from recent research in functional data analysis [76]. Another example of forecasting the competitors' behavior is to forecast the residual demand curve by combining feature reduction and machine learning algorithms or by using functional data time-series theory [77].

Energy regulators are mainly interested in assessing market efficiency and testing new market rules and regulatory frameworks. Reinforcement learning algorithms are used to construct price and quantity bids of market agents (i.e., their bidding strategies) that simulate a clearinghouse auction and assess market efficiency [78]. Reinforcement learning strategies can also be employed to design data-driven bidding strategies that select the optimal quantile (i.e., minimizes the expected value of imbalance costs) for a wind power plant participating in the electricity market and that fully considers the forecast uncertainty [79]. This is a first approach toward data-driven market bidding strategies and autonomous decision-making aiming to explore historical data from the electricity markets [80].

6   DEMAND-SIDE FLEXIBILITY

The smart grid paradigm combined with consumers' engagement in demand response (DR) programs will create new challenges in price forecasting. In this context, forecasting extreme prices will be very useful to identify hours where demand reduction or shift is economically attractive and provides this information to consumers and market agents [81]. In fact, with the growing number of DR programs, the electricity consumption and price forecast problems cannot be detached, e.g., consumption reduction when electricity price is high. Therefore, it will be necessary to forecast both variables by capturing their dependency structure [82].

Estimating the elasticity (or response) of each consumer to dynamic retailing prices is essential to control electricity consumption using a one-way price signal and requires new load forecasting algorithms [83,84]. The big data challenge is not in the number of explanatory variables, it is in the number of consumers to handle at the same time and in time-varying nature of consumption profiles. The trend is to design online learning and optimization approaches that learn (in realtime) the price elasticity at the aggregated and individual level [85,86]. For this problem, recent advances in online data processing can be applied, such as stochastic approximation and online convex optimization [26,27].

Electricity retailers and aggregators, prior to engaging consumers in DR products, need to analyze large volumes of active power observations collected from the smart meters. Yet, in some situations, the active power measurement might be only available for each LV feeder or secondary substation. The goal is to estimate the degree of flexibility of each consumer to shift a share of its consumption from one period to another. The current trend consists in applying data-driven algorithms to relate DR potential with exogenous variables (e.g., ambient temperature) or to uncover the parameters of thermal models per the aggregated consumption. One example is to fit a physically based model of the HVAC consumption with a linear regression framework and by using active power per smart meter, aiming to estimate the DR potential of each potential client [87]. Another example is the application of regression models with a set of explanatory variables (hour of day, set point change, and outside air temperature) to estimate the DR potential [88]. Another area of research is to analyze large datasets of building energy data to evaluate building performance and estimate energy savings due to retrofit actions [89].

Real-time control strategies of appliances can be data-driven and batch reinforcement learning (Q-learning) is a technology suitable for practical implementation. However, the aggregation of multiple controllable loads can result in a high-dimensional state space that leads to the curse of dimensionality. Deep learning techniques developed to handle big data (e.g., large number of input variables), such as convolutional neural networks [90], can be used as regression algorithm to approximate the state-action value function (or Q-function) and extract state-time features [91], making it adequate for a real-world application. This development is called deep reinforcement learning and can be applied to different power system-related problems, such as coordinated voltage control and virtual power plants.

7   CONCLUSIONS AND FUTURE CHALLENGES

Technological solutions for metering and control are now available in the market and installed in several large-scale pilots. The Internet of Things and smart grids concepts can be realized from the technological point of view and the future challenges consist of data intelligent functions and new business models on the top of the component, information and communication layers.

The future generation of big data functions will combine spatial-temporal information and distributed learning techniques that exploit recent advances in high performance and distributed computing. The output should be probabilistic information and with high value for integration in decision-aid methods under risk. Deep learning techniques represent an added value for automatic feature extraction and reduction, but manual feature creation with domain knowledge cannot be abandoned. Data-driven techniques are not exclusively used for estimation and forecasting. Machine learning algorithms can be used to control grid assets, for instance embedded in reinforcement learning techniques or to create surrogate models for complex physical systems.

The creation of new business models for knowledge extraction from data is also expected in a near future. Some examples are analysis of the demand response potential of grid users, big data preprocessing from grid sensors, large-scale simulation of electricity markets, and predictive maintenance of electrical equipment. Forecasting of price, load, and renewable generation time series is presently a business with high-technology readiness levels.

The potential for big data techniques in the power system industry is very high, but several threats can be identified. In general, the quality of data is very low (e.g., high percentage of missing values and gross errors) and, in some cases, is not available close to realtime (e.g., data transfer delays above 24 h). Moreover, in general, power systems engineers have a lack of advanced knowledge in statistics, which undermines the value of the new big data functions. Finally, large-scale dissemination of "ready-to-use" machine learning libraries eases the learning curve but, at the same time, may defer the interest in disruptive solutions as well as its industrialization.

Acknowledgments

This work is financed by the ERDF—European Regional Development Fund through the Operational Programme for Competitiveness and Internationalisation—COMPETE 2020 Programme, and by National Funds through the Portuguese funding agency, FCT—*Fundação para a Ciência e a Tecnologia*, within project *ESGRIDS—Desenvolvimento Sustentável da Rede Elétrica Inteligente/* SAICTPAC/0004/2015-POCI-01-0145-FEDER-016434.

References

[1] M. Grady, in: Texas synchrophasor network, IEEE-PES Fort Worth Chapter Meeting, February, 2016. http://web.ecs.baylor.edu/faculty/grady/_2016_Texas_Synchrophasor_Network_Reports_ Updated_160716.pdf.

[2] M. Matos, J.P. Lopes, M. Rosa, R. Ferreira, A.L. da Silva, W. Sales, et al., Probabilistic evaluation of reserve requirements of generating systems with renewable power sources: the Portuguese and Spanish cases, Int. J. Electr. Power Energy Syst. 31 (9) (2009) 562–569.

[3] D. Singh, C.K. Reddy, A survey on platforms for big data analytics, J. Big Data 2 (1) (2014) 1–20.

[4] Y. Yan, Y. Qian, H. Sharif, D. Tipper, A survey on smart grid communication infrastructures: motivations, requirements and challenges, IEEE Commun. Surv. Tutor. 15 (1) (2013) 5–20.

[5] D. Hawkins, M. Varghese, D. Dieser, Y. Osoba, H. Sanders, IP-1 ISO Uses Synchrophasor Data for Grid Operations, Control, Analysis and Modelling, (2010), California ISO Document Version 3.1.

[6] P. Zhang, J. Chen, M. Shao, Phasor Measurement Unit (PMU) Implementation and Applications, EPRI, Palo Alto, CA, 2007. 1015511.

[7] J.N. Bank, R.M. Gardner, J.K. Wang, A.J. Arana, Y. Liu, in: Generator trip identification using wide-area measurements and historical data analysis, 2006 IEEE PES Power Systems Conference and Exposition, 2006.

[8] J. Gama, P. Medas, P. Rodrigues, in: Learning decision trees from dynamic data streams, Proceedings of the 2005 ACM Symposium on Applied Computing, 2005.

[9] D. Zhou, Y. Liu, J. Dong, in: Frequency-based real-time line trip detection and alarm trigger development, 2014 IEEE PES General Meeting Conference & Exposition, 2014.

[10] T. Guo, J.V. Milanović, Probabilistic framework for assessing the accuracy of data mining tool for online prediction of transient stability, IEEE Trans. Power Syst. 29 (1) (2014) 377–385.

[11] R. Shyam, S. Kumar, P. Poornachandran, K.P. Soman, Apache spark a big data analytics platform for smart grid, Procedia Technol. 21 (2015) 171–178.

[12] D. Zhou, J. Guo, Y. Zhang, J. Chai, H. Liu, Y. Liu, Y. Liu, Distributed data analytics platform for wide-area synchrophasor measurement systems, IEEE Trans. Smart Grid 7 (5) (2016) 2397–2405.

[13] P.H. Gadde, M. Biswal, S. Brahma, H. Cao, Efficient compression of PMU data in WAMS, IEEE Trans. Smart Grid 7 (5) (2016) 2406–2413.

[14] T. Guo, J.V. Milanović, in: Identification of power system dynamic signature using hierarchical clustering, 2014 IEEE PES General Meeting, 2014.

[15] F. Bai, Y. Liu, Y. Liu, K. Sun, N. Bhatt, A. Del Rosso, X. Wang, Measurement-based correlation approach for power system dynamic response estimation, IET Gener. Transm. Distrib. 9 (12) (2015) 1474–1484.

[16] C.E. Rasmussen, C.K. Williams, Gaussian Processes for Machine Learning, The MIT Press, Cambridge, 2006.

[17] H. Vasconcelos, J.P. Lopes, in: ANN design for fast security evaluation of interconnected systems with large wind power production, International Conference on Probabilistic Methods Applied to Power Systems, PMAPS 2006, 2006.

[18] H. Vasconcelos, C. Moreira, A. Madureira, J.P. Lopes, V. Miranda, Advanced control solutions for operating isolated power systems: examining the Portuguese islands, IEEE Electrification Mag. 3 (1) (2015) 25–35.

[19] M.B. Do Coutto Filho, J.C. de Souza, Forecasting-aided state estimation—Part I: Panorama, IEEE Trans. Power Syst. 24 (4) (2009) 1667–1677.

[20] A.K. Sinha, J.K. Mondal, Dynamic state estimator using ANN based bus load prediction, IEEE Trans. Power Syst. 14 (4) (1999) 1219–1225.

[21] A. Tajer, Load forecasting via diversified state prediction in multi-area power networks, IEEE Trans. Smart Grid (2017) (In Press).

[22] J. Tastu, P. Pinson, H. Madsen, Space-time trajectories of wind power generation: parameterized precision matrices under a Gaussian copula approach, in: Lecture Notes in Statistics: Modeling and Stochastic Learning for Forecasting in High Dimension, Springer, Cham, 2015, pp. 267–296.

[23] Y. Weng, R. Negi, C. Faloutsos, M.D. Ilić, Robust data-driven state estimation for smart grid, IEEE Trans. Smart Grid 8 (4) (2017) 1956–1967.

[24] V. Kekatos, G.B. Giannakis, Distributed robust power system state estimation, IEEE Trans. Power Syst. 28 (2) (2013) 1617–1626.

[25] S. Boyd, N. Parikh, E. Chu, B. Peleato, J. Eckstein, Distributed optimization and statistical learning via the alternating direction method of multipliers, Found. Trends Mach. Learn. 3 (1) (2011) 1–122.

[26] K. Slavakis, G.B. Giannakis, G. Mateos, Modeling and optimization for big data analytics: (statistical) learning tools for our era of data deluge, IEEE Signal Process. Mag. 31 (5) (2014) 18–31.

[27] K. Slavakis, S.J. Kim, G. Mateos, G.B. Giannakis, Stochastic approximation vis-a-vis online learning for big data analytics, IEEE Signal Process. Mag. 31 (6) (2014) 124–129.

[28] L. Cavalcante, R.J. Bessa, M. Reis, J. Dowell, LASSO vector autoregression structures for very short-term wind power forecasting, Wind Energy 20 (4) (2017) 657–675.

[29] X. Liu, X. Wang, S. Matwin, N. Japkowicz, Meta-map reduce for scalable data mining, J. Big Data 2 (1) (2015) 1.

[30] A. Ulian, M. Sebastian, Business use cases definition and requirements, (2014). Deliverable D2.1 evolvDSO project.

[31] M. Heleno, R. Soares, J. Sumaili, R.J. Bessa, L. Seca, M.A. Matos, in: Estimation of the flexibility range in the transmission-distribution boundary, IEEE PowerTech 2015, 2015.

[32] E. Polymeneas, S. Meliopoulos, in: Aggregate modeling of distribution systems for multi-period OPF, Power Systems Computation Conference (PSCC 2016), 2016.

[33] J.V. Milanović, Y. Xu, Methodology for estimation of dynamic response of demand using limited data, IEEE Trans. Power Syst. 30 (3) (2015) 1288–1297.

[34] A.M. Azmy, I. Erlich, P. Sowa, Artificial neural network-based dynamic equivalents for distribution systems containing active sources, IEE Proc. Gener. Transm. Distrib. 151 (6) (2004) 681–688.

[35] W. Luan, J. Peng, M. Maras, J. Lo, B. Harapnuk, Smart meter data analytics for distribution network connectivity verification, IEEE Trans. Smart Grid 6 (4) (2015) 1964–1971.

[36] P. Barbeiro, H. Teixeira, J. Pereira, R.J. Bessa, in: An ELM-AE state estimator for real-time monitoring in poorly characterized distribution networks, Proc. of the IEEE PowerTech 2015, Eindhoven, 29 June–2 July, 2015.

[37] H. Teixeira, P. Barbeiro, J. Pereira, R.J. Bessa, P. Matos, D. Lemos, C. Morais, M. Caujolle, M. Sebastian-Viana, in: A state estimator for LV networks: results from the evolvDSO project, Proc. of the CIRED 2016 Workshop, Helsinki, 14–15 June, 2016.

[38] J. Pereira, J. Sumaili, R.J. Bessa, L. Seca, A. Madureira, J. Silva, et al., Business Use Cases Definition and Requirements, (2015) Deliverable D3.4 evolvDSO project.

[39] Dutta, S., Data Mining and Graph Theory Focused Solutions to Smart Grid Challenges (Ph.D. thesis), University of Illinois, Urbana-Champaign, 2013.

[40] P. Cuffe, A. Keane, Visualizing the electrical structure of power systems, IEEE Syst. J. (2017) 1–12.

[41] H. Sun, Z. Wang, J. Wang, Z. Huang, N. Carrington, J. Liao, Data-driven power outage detection by social sensors, IEEE Trans. Smart Grid 7 (5) (2016) 2516–2524.

[42] D. Clerici, G. Viganò, R. Zuelli, B. Swaminathan, V. Debusschere, R. D'Hulst, et al., Advanced Tools and Methodologies for Forecasting, Operational Scheduling and Grid Optimisation, (2015) evolvDSO project deliverable D3.2.

[43] J. Peppanen, M.J. Reno, R.J. Broderick, S. Grijalva, Distribution system model calibration with big data from AMI and PV inverters, IEEE Trans. Smart Grid 7 (5) (2016) 2497–2506.

[44] H. Shaker, H. Zareipour, D. Wood, Estimating power generation of invisible solar sites using publicly available data, IEEE Trans. Smart Grid 7 (5) (2016) 2456–2465.

[45] X. Zhang, S. Grijalva, A data driven approach for detection and estimation of residential PV installations, IEEE Trans. Smart Grid 7 (5) (2016) 2477–2485.

[46] R.J. Bessa, A. Trindade, V. Miranda, Spatial-temporal solar power forecasting for smart grids, IEEE Trans. Ind. Inf. 11 (1) (2015) 232–241.

[47] J. Dowell, P. Pinson, Very-short-term probabilistic wind power forecasts by sparse vector auto-regression, IEEE Trans. Smart Grid 7 (2) (2016) 763–770.

[48] J.B. Iversen, P. Pinson, in: RESGen: renewable energy scenario generation platform, 2016 IEEE PES General Meeting, 2016.

[49] J.R. Andrade, R.J. Bessa, Improving renewable energy forecasting with a grid of numerical weather predictions, IEEE Trans. Sustain. Energy 8 (4) (2017) 1571–1580.

[50] A. Alqurashi, A.H. Etemadi, A. Khodaei, Treatment of uncertainty for next generation power systems: state-of-the-art in stochastic optimization, Electr. Power Syst. Res. 141 (2016) 233–245.

[51] B.H. Kim, R. Baldick, A comparison of distributed optimal power flow algorithms, IEEE Trans. Power Syst. 15 (2) (2000) 599–604.

[52] Q. Peng, S.H. Low, Distributed Optimal Power Flow Algorithm for Balanced Radial Distribution Networks, (2014) arXiv preprint arXiv:1404.0700.

[53] M. Wytock, S. Diamond, F. Heide, S. Boyd, A New Architecture for Optimization Modeling Frameworks, (2016) arXiv preprint arXiv:1609.03488.

[54] Silva, J., Definition of Maintenance Policies in Power Systems (M.Sc. thesis), Universidade do Porto, Porto, 2014.

[55] A.U. Adoghe, C.O.A. Awosope, J.C. Ekeh, Asset maintenance planning in electric power distribution network using statistical analysis of outage data, Int. J. Electr. Power Energy Syst. 47 (2013) 424–435.

[56] R. Dashti, S. Yousefi, Reliability based asset assessment in electrical distribution systems, Reliab. Eng. Syst. Saf. 112 (2013) 129–136.

[57] R. Clement, P. Tournebise, A. Weynants, S. Perkin, K. Johansen, S. Khuntia, et al., Functional Analysis of Asset Management Processes, (2015) Deliverable D5.1, EU Project GARPUR.

[58] A.E. Abu-Elanien, M.M.A. Salama, M. Ibrahim, Calculation of a health index for oil-immersed transformers rated under 69 kV using fuzzy logic, IEEE Trans. Power Deliv. 27 (4) (2012) 2029–2036.

[59] V. Miranda, A.R.G. Castro, S. Lima, Diagnosing faults in power transformers with autoassociative neural networks and mean shift, IEEE Trans. Power Deliv. 27 (3) (2012) 1350–1357.

[60] J. Qiu, H. Wang, D. Lin, B. He, W. Zhao, W. Xu, Nonparametric regression-based failure rate model for electric power equipment using lifecycle data, IEEE Trans. Smart Grid 6 (2) (2015) 955–964.

[61] S.M. Strachan, S.D. McArthur, B. Stephen, J.R. McDonald, A. Campbell, Providing decision support for the condition-based maintenance of circuit breakers through data mining of trip coil current signatures, IEEE Trans. Power Deliv. 22 (1) (2007) 178–186.

[62] I. Antoniadou, N. Dervilis, E. Papatheou, A.E. Maguire, K. Worden, Aspects of structural health and condition monitoring of offshore wind turbines, Philos. Trans. A Math. Phys. Eng. Sci. 373 (2035) (2015).

[63] E. Papatheou, N. Dervilis, A.E. Maguire, I. Antoniadou, K. Worden, A performance monitoring approach for the novel Lillgrund offshore wind farm, IEEE Trans. Ind. Electron. 62 (10) (2015) 6636–6644.

[64] J. Dowell, I. Dinwoodie, D. McMillan, in: Forecasting for offshore maintenance scheduling under uncertainty, European Safety and Reliability Conference, Glasgow, 2016.

[65] C.L. Su, D. Kirschen, Quantifying the effect of demand response on electricity markets, IEEE Trans. Power Syst. 24 (3) (2009) 1199–1207.

[66] Z. Zhou, F. Zhao, J. Wang, Agent-based electricity market simulation with demand response from commercial buildings, IEEE Trans. Smart Grid 2 (4) (2011) 580–588.

[67] Mihaylova, I., Stochastic Dependencies of Spot Prices in the European Electricity Markets (M.Sc. thesis), Universidad de St. Gallen, St. Gallen, 2009.

[68] V. Kekatos, G.B. Giannakis, R. Baldick, in: Grid topology identification using electricity prices, 2014 IEEE PES General Meeting, 2014.

[69] R.J. Bessa, M.A. Matos, Forecasting issues for managing a portfolio of electric vehicles under a smart grid paradigm, 3rd IEEE PES Innovative Smart Grid Technologies Europe (ISGT Europe 2012), Berlin, Germany, 2012.

[70] R. Weron, Electricity price forecasting: a review of the state-of-the-art with a look into the future, Int. J. Forecast. 30 (4) (2014) 1030–1081.

[71] T. Jónsson, P. Pinson, H. Madsen, On the market impact of wind energy forecasts, Energy Econ. 32 (2) (2010) 313–320.

[72] Fischbach, P., Copula-Models in the Electric Power Industry (M.Sc. Thesis), Universidad de St. Gallen, St. Gallen, 2010.

[73] P. Giabardo, M. Zugno, P. Pinson, H. Madsen, Feedback, competition and stochasticity in a day ahead electricity market, Energy Econ. 32 (2) (2010) 292–301.

[74] T.C. Price, Using co-evolutionary programming to simulate strategic behaviour in markets, J. Evol. Econ. 7 (3) (1997) 219–254.

[75] E.J. Anderson, A.B. Philpott, Using supply functions for offering generation into an electricity market, Oper. Res. 50 (3) (2002) 477–489.

[76] J.O. Ramsay, B.W. Silverman, Functional Data Analysis, Springer, New York, 2005.

[77] J. Portela, A. Muñoz, E. Alonso, in: Day-ahead residual demand curve forecasting in electricity markets, The 32nd Annual International Symposium on Forecasting (ISF 2012), Boston, MA, 2012.

[78] J. Nicolaisen, V. Petrov, L. Tesfatsion, Market power and efficiency in a computational electricity market with discriminatory double-auction pricing, IEEE Trans. Evol. Comput. 5 (5) (2001) 504–523.

[79] N. Mazzi, P. Pinson, in: Purely data-driven approaches to trading of renewable energy generation, Proc. of the 13th International Conference on the European Energy Market (EEM), Porto, 6–9 June, 2016.

[80] Peters, M., Machine Learning Algorithms for Smart Electricity Markets (Ph.D. thesis), Erasmus University Rotterdam, Rotterdam, 2015.

[81] J.H. Zhao, Z.Y. Dong, X. Li, K.P. Wong, A framework for electricity price spike analysis with advanced data mining methods, IEEE Trans. Power Syst. 22 (1) (2007) 376–385.

[82] L. Wu, M. Shahidehpour, A hybrid model for integrated day-ahead electricity price and load forecasting in smart grid, IET Gener. Transm. Distrib. 8 (12) (2014) 1937–1950.

[83] O. Corradi, H. Ochsenfeld, H. Madsen, P. Pinson, Controlling electricity consumption by forecasting its response to varying prices, IEEE Trans. Power Syst. 28 (1) (2013) 421–429.

[84] A. Garulli, S. Paoletti, A. Vicino, Models and techniques for electric load forecasting in the presence of demand response, IEEE Trans. Control Syst. Technol. 23 (3) (2015) 1087–1097.

[85] L. Jia, L. Tong, Q. Zhao, An Online Learning Approach to Dynamic Pricing for Demand Response, (2014) arXiv preprint arXiv:1404.1325.

[86] S.J. Kim, G. Giannakis, An online convex optimization approach to real-time energy pricing for demand response, IEEE Trans. Smart Grid (2017) (In Press).

[87] J. Kwac, R. Rajagopal, Data-driven targeting of customers for demand response, IEEE Trans. Smart Grid 7 (5) (2016) 2199–2207.

[88] R. Yin, E.C. Kara, Y. Li, N. DeForest, K. Wang, T. Yong, M. Stadler, Quantifying flexibility of commercial and residential loads for demand response using setpoint changes, Appl. Energy 177 (2016) 149–164.

[89] P.A. Mathew, L.N. Dunn, M.D. Sohn, A. Mercado, C. Custudio, T. Walter, Big-data for building energy performance: lessons from assembling a very large national database of building energy use, Appl. Energy 140 (2015) 85–93.

[90] Y. LeCun, Y. Bengio, G. Hinton, Deep learning, Nature 521 (7553) (2015) 436–444.

[91] B.J. Claessens, P. Vrancx, F. Ruelens, Convolutional Neural Networks for Automatic State-Time Feature Extraction in Reinforcement Learning Applied to Residential Load Control, (2016) arXiv preprint arXiv:1604.08382.

CHAPTER 11

On Data-Driven Approaches for Demand Response

Akin Tascikaraoglu
Mugla Sitki Kocman University, Mugla, Turkey

CHAPTER OVERVIEW

This chapter presents a detailed investigation on the resources, use, and benefits of big data analytics in smart grid activities that enable the participation of demand side in energy management. It starts by elucidating these activities called demand-side management and demand response (DR) and their role in providing higher saving potential for both system operators and end users. It then explains the use of big data management techniques in order to handle the huge amount of data required for efficient DR applications. Afterward, the benefit of various clustering methods and classification methods in DR applications is evaluated by classifying them into four main groups according to their objectives. First, the role of big data analytics on the energy consumption behavior of end users and on the electric load classification is examined. Then, the support of DR programs relying on big data analytics is evaluated for demand and renewable energy generation forecasting as well as dynamic pricing.

1 INTRODUCTION

Electric power demand has been steadily increasing across the world, with an annual rate of nearly 4% during the last decade [1]. The yearly growth in total demand poses various challenges to system operators such as optimizing operational efficiency, and maintaining system stability and reliability. In order to compensate the increasing power demand taking also the energy costs and environmental concerns into account, higher penetration of renewable energy sources has been considered as one of the initial steps [2]. The integration of intermittent and stochastic renewable energy into power systems, however, has caused an additional uncertainty, which is already present in the systems due to daily variations of power demand, in providing the balance between supply and demand. In order to manage this balance in all time periods, the available generation facilities have been adapted to varying load demand since demand has been traditionally considered to be inelastic, i.e., cannot be interrupted and deferred.

Smart grid environment, which has been recently emerged as a modernized power grid using digital information and communication technologies, has led to power

Big Data Application in Power Systems. https://doi.org/10.1016/B978-0-12-811968-6.00011-5

systems to be more adaptive to the increase of renewable energy penetration, and enabled the participation of demand-side resources in energy savings and also in balancing services such as peak load leveling, valley filling, and frequency regulation. These methods, referred to as demand response (DR) or in a broader sense demand-side management (DSM) strategies, allow making use of the flexible demand from end users' appliances for critical load conditions. In other words, DR strategies aim to change the electric usage of end users when needed by adjusting the price of electricity or providing incentives to them.

At the earliest stage, only large commercial and industrial consumers have been considered for DR programs due to their high power capacities and already existing infrastructure for remote measurement and control. In the last years, the application of these programs at household level and even a single appliance level has been also enabled by the widespread use of smart grid technologies. Together with the increase in DR implementations at high-resolution level, load serving entities (LSEs) have faced new challenges such as collecting, storing, and processing of such a huge amount of data, which generally includes power consumption of various appliances and a high number of residential end users, in addition to the data from industrial and commercial consumers. LSEs should also consider the expected (forecasted) consumption values of these consumers and expected generation values of available generation units, as well as the varying electric energy prices, in the decision-making process of scheduling and resulting load reductions, which further complicates the data management problem. In order to handle such an enormous data so that the savings potential for DR programs is maximized, LSEs have recently started to employ "big data" management techniques.

Considering the increasing use of big data and its significant benefits within smart grid environment, particularly in DR applications, this chapter of the book deals with the implementation of DR solutions and the state-of-the-art methods already used or can possibly be used by LSEs in this regard. Particularly, the benefits of various clustering methods and classification methods in determining optimal segmentation of consumers are examined. Besides, the contributions of big data management techniques on improving the forecasts of renewable energy generation and load demand are investigated referring the results presented in the literature. Furthermore, the support of DR programs relying on big data mining and analysis is evaluated for dynamic pricing applications.

The rest of the chapter is organized as follows: Section 2 presents the resources of big data in smart grid context and Section 3 elucidates the implementations of big data analytics in DR applications and summarizes the challenges in the application of big data-driven DR. The studies presented in the literature on using big data in the management of demand side are presented and the real-world implementations are reported in Section 4. Section 5 summarizes the most important remarks and provides the directions for future studies on this topic.

2 SOURCES OF BIG DATA IN DR

The main source of big data in smart grids is the advanced metering infrastructure (AMI) which collects data from measuring units at the consumer side such as smart meters, smart thermostats, and sensors with different sampling and reporting rates varying generally from seconds to 1 h. Given a grid with thousands of data sources and that these devices collect information (e.g., power consumption, voltage, phase angles. etc.) from a great number of consumers, mostly at a time-scale of 15 min, there will be a vast amount of data measured per day or even per hour. In addition to the power and energy data, weather data might also improve the forecasting accuracies and therefore has an important role in supporting the energy management. For instance wind speed and direction data is used for the data-driven wind speed/power forecasting approaches, and solar irradiance and temperature data is used for solar power forecasting approaches. Moreover, the advanced energy demand forecasting approaches use different type of information, such as dwelling type, occupancy level, household income, and education level, for more accurate forecasts. Besides, the geographic features of an area, which are taken from geographic information system, might be used for more effective decisions in forecasting applications and also in power systems.

The characteristic of the big data used in smart grid environment can be given using four different criteria that are abbreviated as 4V characteristics. The first characteristic is the volume of data, which has been significantly increased together with the widely used advanced metering devices in power systems. Handling a vast amount of data is a challenging task; however, it provides many opportunities at the same time. The second characteristic, namely velocity, refers to the speed requirement for data collection and processing. The complexity of the data in energy systems is characterized by variety and the value, which is the last characteristic, provides insights about the potential benefits of data. Regarding the value of any data, it might be possible to determine its contribution for different objectives, such as increasing end-user participation into DR programs, improving forecasting performance, and foreseeing end-user energy consumption.

3 BIG DATA APPLICATIONS IN DR

Large-scale DR implementations have improved the conventional power systems in terms of various aspects such as integration of more renewable power, improving the efficiency of power transmission and distribution, enabling the incorporation of large-scale storage systems, reducing the possibility of failure due to overloading and congestion, and providing the infrastructure required for the connection of electric vehicles (EVs) to the distribution systems. According to their structure, DR programs are generally divided into two groups; those based on different electricity pricing approaches and those based on providing incentives to the end users accepted to participate in programs.

In order to fully make use of the technologies given above, a huge amount of data from both generation and consumption sides is collected in these environments. In addition to its significant benefits, such a big data brings about also new challenges and additional computational burden, which are shown in Fig. 1. First of all, processing big data for real-time DR implementations requires advanced measurement systems, high communication resources, and high storage capabilities since traditional measurement and collection units are not scalable for such a finer granularity data, the bandwidth of available communication tools is insufficient for acquisition of such a large data, and available storage units are deficient for higher amounts of data. The widely used technologies for these commitments, such as centralized SCADA and client-server data network models, however, are not scalable for such data-intensive applications.

As stated above, the data is collected by an AMI network in smart grids, which is capable of measuring energy usage through smart meters as well as storing and analyzing the measured data. This network consists of different applications such as home area networks (HANs) that allow the communication between smart meters and electrical devices/appliances, and also enable to control the certain electrical devices via smart meters, neighborhood area networks (NANs) that transfer the smart meter data collected from multiple HANs in a neighborhood to utility (or a data concentrator unit (DCU)) for various analyses, and wide-area networks (WANs) that transmit the data measured to central control units and enable real-time wide-area monitoring, protection, and control for effective power system operations. When it comes to communication, the communication tools with a relatively low-frequency data transfer rate (up to a few hundred kbps) and short coverage distance (up to hundreds of meters) are generally sufficient for the HAN applications as these applications only consist of data transfers between appliances and a smart meter (or a controller) within end-user premises [3,4]. Instead, HAN applications require

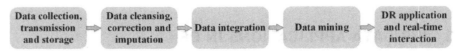

FIG. 1

General process of big data driven DR application.

secure communication with low-cost and low-power consumption. Hence the technologies satisfying all the specifications mentioned above, such as IEEE802.15.4 (known as ZigBee), power line carrier (PLC), IEEE802.11 (known as Wi-Fi) and Z-wave, are mostly employed for HANs. On the contrary, NAN applications require higher data transfer rate (up to tens of Mbps) and larger coverage distance (up to tens of km) since the data is transmitted between a high number end users and a utility (or a data concentrator) in these applications [3,4]. Therefore, ZigBee and Wi-Fi mesh networks, PLC, IEEE802.16 m (known as WiMAX) and Cellular wireless networks such as GPRS and LTE are mostly used for such long distances. Due to their much higher data transfer rate (up to 1 Gbps) and much wider coverage distance (up to hundreds of km) requirements for dynamic power system stability control, Cellular and WiMAX are generally used for WAN applications [3,4].

Various technological and economic constraints, however, have limited to employ these advanced technologies in the case of a great amount of data. Therefore, various techniques are applied to the data sets obtained with already available systems in order to exploit the data. With this objective, after the collection and transmission processes, all the available data is stored using a multistage storage architecture where frequently accessed data is stored on the fastest storage system the LSE has and rarely accessed data is stored on the slowest storage system. Afterward, the quality of the stored data is checked using different methods such as bad data detection, the erroneous data are then corrected, and the missing data are imputed. Subsequently, the data coming from different sources in different data types and specifications are integrated, which is one of the most challenging tasks in processing of big data. It should be noted that energy big data are collected by various data acquisition devices and applications, and hence the specifications of the data collected such as time and space resolutions, size, structure, and format are different from each other. The integration of these heterogeneous and disparate data is, therefore, of great importance for exploiting all the available data in big data analytics. Then, data mining approaches are used for different purposes such as reducing the amount of data without losing information by the means of distributed data mining and dimensionality reduction techniques. Lastly, using these large data sets in real time for DR applications by extracting useful information from the data is another challenge, for which machine learning (ML) algorithms such as artificial neural networks (ANN) and Bayesian networks are generally used.

In addition to the methods mentioned above, advanced computing techniques, such as virtualization and in-memory computing, are generally used in big data management for the purpose of reducing the computational burden in all these stages. Alternatively, high computing resources can be used for handling fast data processing in real-time applications. All these methods and tools enable the big data obtained to effectively use for various objectives in smart grid environment. For instance, these data can provide effective decision support in optimizing power generation and operation regarding the end users' energy consumption behavior, can enable to choose the optimal consumers for DR applications, can improve the forecasting performances of renewable energy and load demand, and can also allow developing efficient dynamic pricing mechanisms. Each of these benefits is elucidated in the following subsections.

3.1 Assessment of Energy Consumption Behavior

One of the main factors affecting the efficiency of DR programs is the energy consumption behavior of consumers. Knowing this behavior in time and space domain can help system operators better manage the energy supply-demand balance in a power system. This information provides additional flexibility to system operators in decision-making process of balancing services. Furthermore, being aware of their consumption behavior through real-time communication with the system operator, the end users can also change their consumption behavior and use energy wisely so that their energy cost is reduced.

The reason underlying the end users' energy consumption behavior depends on various objective and subjective factors. Objective factors can be gathered into two groups: the internal factors, which are related to the house and household such as income and education levels, housing structure and occupancy, and the external factors such as energy policies, energy prices and weather conditions. The subjective factors are the factors that are based on the intention and awareness of end users.

A great deal of valuable information can be discovered from the energy measurements, particularly using the temporal and spatial correlations among the households dispersed in a large area. Analyzing these data can help reveal the energy use patterns of different consumers and the corresponding energy usage behavior can be determined. However, the large volume of data measured by smart meters causes various problems in terms of data transmission, storage, and processing. Therefore, various approaches are used particularly in the real-world applications in order to decrease the data size providing that it conveys almost the same information. These methods, called dimensionality reduction methods, might facilitate the processing of data with an acceptable error, reduce the storage capacity required, decrease the related computational times, and improve the performance by removing the redundant and repetitive features. Various methods can be used to perform dimension reduction. For instance, any variable with missing values can be neglected if data imputation is not possible for its data set, the variables having low variance or with high correlations can be dropped, decision trees and random forest can be used for handling multiple challenges as given above, and the contribution of each variable can be examined by adding them in an order.

3.2 Electric Load Classification

In DR applications, the information that can be used for improving the efficiency of the programs, such as energy consumption values of end users, generation values for conventional and renewable energy sources, energy prices and weather conditions can be filtered, analyzed, and classified for different purposes. In terms of energy consumption, each consumer has a different reaction to varying energy prices and other factors such as weather conditions and time of day. A huge number of energy consumption patterns might, therefore, be gathered into predefined groups for developing effective marketing strategies and providing personal energy services to the consumers. This process of data partitioning is called as electric load classification.

In order to carry out an effective classification of loads with different profiles, data mining, which is defined as the process of analyzing a data set, extracting the information required for special purposes, and converting the data into a certain structure in order to use more effectively, is widely considered both in the literature and real-world applications. Among data mining techniques, the approaches based on ML techniques have been widely used in load classification due to the effectiveness of these approaches in case there is no exact mathematical model for describing the given data. Also, K-means, Fuzzy c-means, hierarchical clustering, and self-organizing map (SOM) methods have especially shown good performances in load classification.

As a nonhierarchical clustering algorithm, K-means clustering method basically partitions data sets into a certain number of clusters by taking the distance between each observation and cluster center with the nearest mean into account. This method has a wide application area in the literature of load classification due to its simple and efficient operation; however, the selection of the number of clusters and initial cluster centers might substantially affect the model performance. Contrary to this method, one data point can belong to every cluster in Fuzzy c-means with some degree of belonging or membership. These membership degrees are then updated iteratively in the minimizing process of objective function. Similar to K-means method, the selection of initial cluster centers and cluster numbers is considered as a challenging task. Due to its relatively simple implementation, hierarchical clustering method has recently gained increasing interest in load classification, in which each observation starts in a cluster by itself and then different clusters are merged with respect to a distance (e.g., Euclidean distance of the furthest neighbors) among clusters until a termination criteria (known as agglomerative strategy), or on the contrary, all observations start in one cluster and this cluster is then split into clusters recursively (known as divisive strategy). An unsupervised neural network method, called SOM network, in which the weights are trained using a competitive-learning algorithm differently from conventional ANN models, has also various applications in load classification and provides favorable results for optimum factors such as network weights and neighborhood functions.

New methods, such as support vector clustering, iterative refinement clustering, honey bee mating optimization and follow the leader have been also recently introduced for load classification. It can be also noted that the distributed data analysis methods have started to be used widely as centralized frameworks require exchange of a large amount of data between measurement devices at the end user side and main processor, which is economically not feasible. Instead, decentralized data mining algorithms require smaller computational and communication resources.

3.3 Demand and Renewable Energy Generation Forecasting

Demand and renewable energy generation in power systems are unsteady and affected by a high number of variable factors such as electricity prices and weather conditions, which further complicates the application of DR programs. Demand, price, and renewable energy generation forecasts are, therefore, crucial tools for effective DR applications. High-accuracy forecasts can reduce the operating costs and improve the power system reliability by adjusting the generation level of power plants at certain periods. These forecasts also help system operators adjust the electricity prices taking into account the correlation between demand and corresponding prices [5].

In order to achieve high accuracy forecasts, real-time exploitation of high amounts of data is required. For instance, the demand forecasts are acquired using historical demand data and various influencing factors including building specifications such as its location, size and type, end user's habits and usage patterns, appliance specifications, socio-economic factors such as income and education level, and weather characteristics such as daily and seasonal changes in temperature and humidity. The studies on demand forecasting have recently focused on the forecasts at the household level, even at the level of a specific appliance, aligned with the advancements in data collection for these small-scale units, which further increases the size of both input and output data. It is also noted that the application of forecasting methods is expected to become wider in the context of DR in the near future together with the emerging technologies enabled by smart grids. For instance, the forecast of plug-in hybrid electric vehicle charging load using the data of driver habits and travel pattern data analysis is considered as a promising research topic due to its potential effect on efficient DR. Regarding the renewable (i.e., wind and solar) generation forecasts, the inputs are generally composed of historical wind speed and solar irradiance data, other meteorological quantities such as wind direction, pressure and temperature, and local terrain structure [6].

Three types of data are generally used as input in the forecasting applications in smart grid environment: (i) historical data of the variable to be forecasted, (ii) historical data of different exogenous variables (e.g., electricity price information for demand forecasts and temperature for renewable energy generation forecasts), and (iii) the parameters that effect the variable to be forecasted (e.g., socio-economic factors for load forecasts and physical specifications of the related area for renewable energy generation forecasts) [7]. Furthermore, the data collected from the neighboring regions can also be incorporated into the input data sets in addition to the data collected from the point where the forecasts will be performed. The historical data from both target and exogenous variables generally consist of the recent measurements. Nevertheless, the data corresponding to the forecasted period in the previous years (called similar days) might be included in certain forecasting applications, particularly in the case of relatively steady patterns of energy consumption and weather conditions [7].

The use of all three types of data or at least one or two types from different locations converts the forecasting methods into multivariate models as shown in Eq. (1):

$$y_t^{r^*,v^*} = \sum_{\substack{r=1 \\ v=1}}^{R,V} \sum_{i=1}^{p} y_{t-i}^{r,v} \varnothing_i^{r,v} \tag{1}$$

where $y_t^{r^*,v^*}$ is the target variable for time t, region r, and variable v. R and V are the numbers of regions and variables, respectively. $\varnothing_i^{r,v}$ represents regression coefficients and p is model order. Eq. (1) can be given in an extended format as in Eq. (2) to clearly observe the amount of the data that might be used in demand and renewable energy forecasting:

$$\begin{bmatrix} y_{p+1}^{r^*,v^*} \\ y_{p+2}^{r^*,v^*} \\ \vdots \\ y_{p+N}^{r^*,v^*} \end{bmatrix} = \begin{bmatrix} y_p^{1,1} & \cdots & y_1^{1,1} \\ y_{p+1}^{1,1} & \ddots & \vdots \\ \vdots & \ddots & \vdots \\ y_{p+N-1}^{1,1} & \cdots & y_N^{1,1} \end{bmatrix} \cdots \cdots \begin{bmatrix} y_p^{R,V} & \cdots & y_1^{R,V} \\ y_{p+1}^{R,V} & \ddots & \vdots \\ \vdots & \ddots & \vdots \\ y_{p+N-1}^{R,V} & \cdots & y_N^{R,V} \end{bmatrix} \begin{bmatrix} \varnothing_1^{1,1} \\ \vdots \\ \varnothing_p^{1,1} \\ \vdots \\ \vdots \\ \varnothing_1^{R,V} \\ \vdots \\ \varnothing_p^{R,V} \end{bmatrix} \tag{2}$$

where $p+N$ represents the number of target variable datased in the training stage of the forecasting model. As seen from Eq. (2), the number of data used in the regression coefficient vector would be equal to the total number of the observations to be included for each variable from different regions, which implies a huge amount of data, especially with low time granularity, and hence long training times (up to a few hours) and inaccurate forecasts. It should be noted that Eqs. (1), (2) show the data used for autoregressive-based methods. The same input data sets can be also used in other data-driven forecasting methods such as other time series-based methods or ML-based methods.

The amount of data can be further increased when the decomposition methods are used for improved forecasts by dividing the relatively complex data into more meaningful components which are mostly easier to model [7]. For instance, wavelet transform (WT) decomposes the time series into several subseries depending on the decomposition level, which causes the input data to increase by several times. For another decomposition model, called empirical mode decomposition (EMD), the number of subseries might reach to a higher number compared to WT.

In order to deal with the problems caused by the use of a vast amount of data, while still ensuring a satisfactory level of accuracy, particularly in real-time forecasting applications, variable selection (aka feature selection) methods can be pointed out as one of the most effective methods. These methods basically select the most relevant variables among a set of variables (e.g., time factors such as the day of the week and the hour of the day, weather characteristics such as temperature and pressure, etc.) with the objective of reducing the number of variables to be used in the model construction. Regarding the type and characteristics of the data, different feature selection approaches such as statistical analysis, correlation analysis, principle component analysis, sensitivity analysis, and load curve analysis can be used. As one of these approaches, mutual information (MI) criterion examines the MI of each input with the variable to be forecasted and order them according to their information value. The values with higher information values are then used only in the forecasting process. Various optimization methods are also very effective in the selecting of the most beneficial features and removing the features with no or little influence on the forecasting performance. These methods can obtain the relation between different influential factors and the target variable depending on the time series data, which is of great importance in building the forecasting models. Particle swarm optimization model inspired by the flocking behavior of the birds, and ant colony optimization model inspired by the behavior of ants in finding the shortest path are the widely used methods in determining the influential factors. Information entropy theory is also widely employed in the literature for the purpose of reducing the irrelevant variables and thus improving convergent speed.

Preprocessing of historical time series by data clustering methods can also provide good results in forecasting applications. Each clustered group can then be modeled with a different method. This task is generally carried out using optimization algorithms mentioned above. Also, SOM network is mostly used for the partition of input data sets into a number of subsets with similar characteristics as these methods provide data clustering without prior knowledge about the classifying criteria (called unsupervised learning). Each partitioned subset is then applied to different forecasting methods, which are generally based on ML approaches such as ANN and SVM, considering their data characteristics.

The data collected from different sensors or meters might include some erroneous data due to various factors such as bad weather conditions, malfunctioning of these devices, and the problems in data transmission and storage. These data might considerably degrade the performance of the forecasting methods. The potential corrupted or noisy data are therefore filtered out before the forecasting process using different thresholds and criteria. In other words, when the magnitude of the data is over or under a predefined threshold, which is generally defined in the training stage, these data are removed from the input dataset. As one of these filtering methods, the use of a zero phase filter, which is a special case of a linear-phase filter, generally provides favorable results in removing the data that are highly close to the original data. It is noted that it might not be possible to differentiate the erroneous data from the original data for certain cases. In these cases, human intervention is generally required in addition to the filtering models used.

3.4   Dynamic Pricing

Providing electricity at different prices for each hour might lead to considerable increases and decreases in the total energy consumption in power systems due to the responses of consumers to this variable pricing. This method, called dynamic pricing or real-time pricing, can improve the reliability of power systems by changing the energy consumption behaviors of end users through the smart meters that can receive price information from the utilities. From the end-user perspective, the main benefit of dynamic pricing is that the prices of the energy used (i.e., the electricity bills) can be reduced considerably while controlling the demand or allowing system operators to control the demand in contracted periods. The end users can shift the operations of the certain loads to off-peak hours where the prices are relatively lower in order to realize this objective, which can be formulated as shown in Eq. (3):

$$minimize\, E = \sum_t P_{grid,t} \cdot \Delta T \cdot \lambda_{buy,t} \tag{3}$$

where $P_{grid,t}$ is the power drawn from the grid, ΔT is the time granularity, and $\lambda_{buy,t}$ is the price of the electricity within the corresponding time interval.

In Eq. (3), ΔT has a great impact on the amount of the data used in the analyses. When this value is on the scale of a few seconds, which is generally required by power system operational conditions, the amount of data will be highly large. On the contrary, a ΔT value of 10 or 15 min can be generally indicated as sufficient for residential applications. As shown in Eq. (4), the total load used by commercial, industrial, and domestic end users in power systems consists of flexible loads, which can be curtailed when required, and inflexible loads that are crucial for the operation of equipment in commercial/industrial buildings and for the end-user comfort in residential households. These inflexible loads can also be classified into two different classes according to their control availability: (i) interruptible loads that can be turned off for a certain time period (generally on the scale of minutes), and (ii) deferrable loads that can be scheduled to be used after a certain time period (generally on the scale of hours).

$$P_{\text{grid},t} = P_{\text{inflex\_load},t} + P_{\text{flex\_load},t} \tag{4}$$

The literature and real-world implementations on dynamic pricing-based programs have been mostly focused on the investigation of the benefits of these programs on commercial and industrial end users due to their relatively higher portion of demand and more predictable power profiles. The share of the electrical energy consumed in residential premises, however, has recently increased significantly in the total energy consumption, which has oriented the studies toward this field [8]. Compared to the studies on commercial/industrial buildings, the studies on the residential level contain a considerably higher amount of data as the measurement of different controllable appliances is included in the data sets. Besides, in order to accurately model the effects of dynamic pricing on the residential consumer energy consumption behavior, i.e., when and why they turn on or off the appliances or shift their use, interviews and questionnaires are also required in addition to the historical load data. The examination of such a huge data including responses to generally over 100 questions (e.g., number and specifications of the appliances, their usage frequencies and operation times, occupancy level of the house, etc.) from a few hundred thousand consumers might provide a valuable information that can be used effectively in regulating the hourly prices of energy supplied to the consumers. Furthermore, observing the responses of the end users to the changing prices and modeling these responses for later use in energy management has gained importance in the last years. All these data-intensive tasks are highly cumbersome and subject to challenges in dealing with the huge amount of data. It is therefore obvious that big data analytics will be a key role in effective dynamic pricing implementations in the near future. There are currently only a few studies in the literature on the use of big data analytics for dynamic pricing implementations; however, it can be indicated as a promising topic, which will be more common together with the widespread use of dynamic pricing programs all over the world.

4 REAL-WORLD APPLICATIONS AND RESEARCH ON BIG DATA-DRIVEN DEMAND RESPONSE

As explained above, data mining and ML techniques are essential for DR programs in processing and analyzing of immense amounts of data from smart meters and sensors. With this objective, various studies have been presented in the literature, particularly in the last decade on the resources of big data and relevant data management techniques in energy sector.

In Ref. [9], a cloud-based software program was developed for dynamic demand response (D^2R) concept that was applied in University of South California as a pilot microgrid project. Together with the data management techniques, the refined data are then used for demand and curtailment forecasting models in order to assess the DR potential in the pilot project in Ref. [9]. The increasing impact of machine-to-machine (M2M) communication concept on data amount in smart grid context was analyzed in Ref. [10] together with a discussion on the exact areas where data mining and ML techniques can play a vital role. Besides, a new technique for data management in M2M communication, specifically regarding smart meters, was also proposed in Ref. [10]. In order to reduce the smart meter data size and hence to provide an increase in computing speed, dimensionality reduction was used in Ref. [11] using random projection. For the classification of consumer load curves, an approach based on ANN was proposed in Ref. [12]. For effective data harvesting, an online clustering based on unsupervised learning techniques has been proposed in Ref. [13]. Distributed data analysis methods were used in Ref. [14] for energy demand forecasting. Several technical studies that cannot be all considered here also offered new technical developments in big data management in smart grid context including demand-side activities.

Furthermore several review papers have been also presented in the literature. In Ref. [15], a comprehensive literature survey on smart energy management based on big data was provided together with the detailed analysis on characteristics of big data and the relevant process methods for demand-side actions in energy sector. It was stated in Ref. [15] that the main resource of big data in DR applications is the AMI as mentioned previously. As the penetration of smart meters increases rapidly, the amount of data collected from smart meters can become significantly huge. As a sample analysis, a simple load demand data collection with 15 min resolution from 1 million smart meters can lead to nearly 3 k terabytes of data in a single year [15]. Besides, a comprehensive survey on literature as well as industrial companies that provides products and services on big data management was also provided in Ref. [15].

Another review paper in Ref. [16] investigated the contribution of energy big data on the analysis of various aspects of residential energy consumption behavior, which is of great importance particularly for DR applications. The social issues in energy consumption from information science point of view were given a specific importance in Ref. [16] in order to enhance the understanding of the big data resources in demand-side activities. In this context, big data issues and challenges faced in dynamic energy management within smart grid vision were also discussed in Ref. [17]. A general overview of data processing methods in this regard was also provided in Ref. [17] together with the discussion on possible upcoming concepts. Different and detailed survey studies on big data in smart grid vision can also be found in Refs. [18–20].

Together with the penetration of DR solution in the overall world [21], the applications of big data managements for smart power system operation have increased gradually in recent times. Several companies have announced solutions and software structures in this regard. IBM offers several customized infrastructure solutions for big data management specifically in power system operation. The provided solutions generally consider how the big data can be lessened and transformed into meaningful outputs for power system participants from operators to end users [22]. T-Systems provides several concepts for different big data imperatives of power system participants from big data security to rapid big data transition approaches for smart grid management, forecasting, etc. [23]. VPS presents a software namely VPS ICE for power system operation and discusses that the developed software can be combined with any commercial off-the-shelf hardware for use in energy management in any scale within power systems [24]. Siemens offers a big data analysis tool on smart metering, namely EnergyIP, for utilities and power system operators [25]. EnergyIP tool provides analysis on data patterns for notifying energy theft, identifying vulnerable or overloaded devices and plants and forecasting load in different levels. Even not specified for power system operation, several companies such as Cisco [26] and Hewlett Packard Enterprise [27] have recently provided software-based solutions for big data management for several areas of use. A different and also detailed survey on companies providing big data management solutions especially for power system operation can also be found in Ref. [15].

Many real-world applications also exist in this area. The Pacific Northwest Smart Grid Demonstration Project has been realized in United States with 60,000 participants in five different states, which is estimated to provide several terabytes of data over 2 years [28]. It was also discussed in this project that the direct load control (DLC) applications considering single appliances (air conditioners, clothing dryers, water heaters, etc.) would add significant additional burden from data point of view [28]. Electric Power Board (EPB) as the city-owned utility of Chattanooga, Tennessee, US and Oak Ridge National Laboratory collaborate on a project including big data management software for managing the aim of enhancing smart grid concept including demand-side actions for around 170,000 commercial and residential customers EPB serves [29]. Spanish power utility Viesgo and Siemens have collaborated on using Siemens'

EnergyIP tool for managing the big data from the smart meters which is aimed to cover around 700,000 power customers Viesgo serves [30]. EnergyIP tool has also been used in German Energy Market for different purposes under the collaboration of Siemens and German authorities [25]. "Pecan Street Smart Grid Project" has been developed in Austin, Texas, United States for the demonstration of demand-side activities and the project demands very detailed usage and therefore management of data from several renewable energy sources, loads in end-user premises, etc. [31]. Many more pilot and real-world applications of big data management concepts specifically developed for demand-side actions in smart grid environment exist also in different areas of the world and the need for big data management is expected to boost considerably in parallel to the development of more developed DR solutions even reaching to the plugs or single appliances of residential end users.

5 SUMMARY AND FUTURE PROSPECTS

Higher amounts of data have been increasingly accumulating within the smart grid environments together with wider applications of advanced measurement, communication, and control technologies. The emerging challenge of taking advantage of these data, called big data, has been investigated in two different areas: Generation side management (including renewable energy management) and DSM. Compared to the generation side management, it can be indicated that DSM includes significantly higher amounts of data due to the large number of end-use metering devices. DSM strategies have gained increasing interest in the last years since these programs have showed promising results in affecting the energy consumption behavior of end users, particularly at peak periods, through price- or incentive-based methods. These methods allow the system operators to apply various dynamic pricing schemes, and also induce end users to mitigate their consumption during peak periods, postpone the working times of certain appliances, and even replace the energy-inefficient appliances. These changes, therefore, satisfy the desired changes in demand level.

It can be indicated that the energy consumption data collected are one of the most important resources to apply the DR programs efficiently. These data, therefore, should be of high quality, that is, it should not contain any missing and erroneous data, which is almost not possible under real conditions due to the temporary faults in smart meters, sensors, and related communication tools. In this context, big data management techniques can be indicated as essential tools for exploiting the available data. These techniques can define the patterns of different data sets, determine the correlations among them (load classification), and complete and correct the required parts of data using prediction and imputation approaches (predictive analytics, bad data detection, and correction), each of which is vital importance for optimizing the DR implementations such as energy generation scheduling and setting real-time energy price.

Big data approaches have been effectively used in different areas; however, using them in smart grid environment, particularly in DR implementations, is a challenging task since it is required to select, classify, analyze, and forecast a massive amount of data in real time. In this chapter of the book, the challenges faced or to be faced in the DR implementations using big data have been investigated. It is noted that big data-driven DR causes also some problems in addition to its significant benefits aforementioned. For instance, the problems about the data security have become more challenging together with the increasing amount of data in DR applications. Therefore, traditional methods which have been widely used in different areas, such as encryption and data anonymization, have started to use in smart grid environments recently for the purpose of ensuring data confidentiality and achieving various services such as authentication and access control.

Acknowledgment

The author would like to thank Dr. Ozan Erdinç from Yildiz Technical University for his valuable suggestions.

References

[1] US Energy Information Administration, International Energy Outlook 2016, May 2016.

[2] A. Tascikaraoglu, B.M. Sanandaji, K. Poolla, P. Varaiya, Exploiting sparsity of interconnections in spatio-temporal wind speed forecasting using wavelet transform, Appl. Energy 165 (2016) 735–747.

[3] P. Siano, Demand response and smart grids—a survey, Renew. Sust. Energ. Rev. 30 (2014) 461–478.

[4] M. Kuzlu, M. Pipattanasomporn, S. Rahman, Communication network requirements for major smart grid applications in HAN, NAN and WAN, Comput. Netw. 67 (2014) 74–88.

[5] N.G. Paterakis, A. Taşcıkaraoğlu, O. Erdinc, A.G. Bakirtzis, J.P. Catalão, Assessment of demand-response-driven load pattern elasticity using a combined approach for smart households, IEEE Trans. Ind. Inf. 12 (4) (2016) 1529–1539.

[6] A. Tascikaraoglu, B. Sanandaji, G. Chicco, V. Cocina, F. Spertino, O. Erdinc, N. Paterakis, J.P. Catalao, Compressive Spatio-temporal forecasting of meteorological quantities and photovoltaic power, IEEE Transactions on Sustainable Energy 7 (3) (2016) 1295–1305.

[7] A. Tascikaraoglu, M. Uzunoglu, A review of combined approaches for prediction of short-term wind speed and power, Renew. Sust. Energ. Rev. 34 (2014) 243–254.

[8] O. Erdinç, A. Taşcıkaraoğlu, N.G. Paterakis, Y. Eren, J.P. Catalão, End-user comfort oriented day-ahead planning for responsive residential HVAC demand aggregation considering weather forecasts, IEEE Trans. Smart Grid 8 (1) (2017) 362–372.

[9] Y. Simmhan, S. Aman, A. Kumbhare, R. Liu, S. Stevens, Q. Zhou, V. Prasanna, Cloud-based software platform for big data analytics in smart grids, Comput. Sci. Eng. 15 (4) (2013) 38–47.

[10] Z. Fan, Q. Chen, G. Kalogridis, S. Tan, D. Kaleshi, in: The power of data: data analytics for M2M and smart grid, In 2012 3rd IEEE PES Innovative Smart Grid Technologies Europe (ISGT Europe), October, IEEE, 2012, pp. 1–8.

[11] A.D. Martins, E.C. Gurjão, in: Processing of smart meters data based on random projections, In Innovative Smart Grid Technologies Latin America (ISGT LA), 2013 IEEE PES Conference on, 2013, April, pp. 1–4.

[12] M.N.Q. Macedo, J.J.M. Galo, L.A.L. de Almeida, A.D.C. Lima, Demand side management using artificial neural networks in a smart grid environment, Renew. Sust. Energ. Rev. 41 (2015) 128–133.

[13] A. Monti, F. Ponci, in: Power grids of the future: why smart means complex, In Complexity in. Engineering, 2010. COMPENG'10, IEEE, February, 2010, pp. 7–11.

[14] R. Mallik, N. Sarda, H. Kargupta, S. Bandyopadhyay, Distributed data mining for sustainable smart grids, Proc. ACM SustKDD 11 (2011) 1–6.

[15] K. Zhou, C. Fu, S. Yang, Big data driven smart energy management: from big data to big insights, Renew. Sust. Energ. Rev. 56 (2016) 215–225.

[16] K. Zhou, S. Yang, Understanding household energy consumption behavior: the contribution of energy big data analytics, Renew. Sust. Energ. Rev. 56 (2016) 810–819.

[17] P.D. Diamantoulakis, V.M. Kapinas, G.K. Karagiannidis, Big data analytics for dynamic energy management in smart grids, Big Data Res. 2 (3) (2015) 94–101.

[18] D. Alahakoon, X. Yu, Smart electricity meter data intelligence for future energy systems: a survey, IEEE Trans. Ind. Inf. 12 (1) (2016) 425–436.

[19] A. Vasilakos, J. Hu, Energy big data analytics and security: challenges and opportunities, IEEE Trans. Smart Grid. 7 (5) (2016) 2423–2436.

[20] H. Jiang, K. Wang, Y. Wang, M. Gao, Y. Zhang, Energy big data: a survey, IEEE Access 4 (2016) 3844–3861.

[21] N.G. Paterakis, O. Erdinç, J.P. Catalão, An overview of demand response: key-elements and international experience, Renew. Sust. Energ. Rev. 69 (2017) 871–891.

[22] IBM Solutions for Big Data Analytics in Power Systems. http://www-03.ibm.com/systems/uk/power/solutions/bigdata-analytics/.

[23] T-Systems–Smarter Energy Management: Intelligent Monitoring of Power Usage, Big Data Analysis for Utilities. https://www.t-systems.com/blob/198454/6ab7d26105862dc24295d3fb92392b2c/dl-usecase-energy-data.pdf.

[24] VPS ICE Software Platform. http://virtualpowersystems.com/the-platform.

[25] Siemens website: Siemens expands data analysis tool for smart metering by adding big data option. http://www.siemens.com/press/en/pressrelease/?press=/en/pressrelease/2016/energymanagement/pr2016020154emen.htm&content=EM.

[26] Cisco: Big Data. http://www.cisco.com/c/en/us/solutions/data-center-virtualization/big-data/index.html.

[27] Hewlett Packard Enterprise: Big Data Solutions. https://www.hpe.com/us/en/solutions/big-data.html.

[28] FORBES: Big Data Meets The Smart Electrical Grid. http://www.forbes.com/sites/tomgroenfeldt/2012/05/09/big-data-meets-the-smart-electrical-grid/#2b14e6f91adc.

[29] E&E News: Big Data Means Big Challenges for Utilities. http://www.eenews.net/stories/1060018115.

[30] Siemens software will manage smart meter data for Spanish power utility. http://www.siemens.com/press/en/pressrelease/?press=/en/pressrelease/2016/energymanagement/pr2016020154emen.htm&content%5B%5D=EM.

[31] A. Tascikaraoglu, B.M. Sanandaji, Short-term residential electric load forecasting: a compressive spatio-temporal approach, Energy Build. 111 (2016) 380–392.

Topology Learning in Radial Distribution Grids

Deepjyoti Deka, Michael Chertkov

Los Alamos National Laboratory, Los Alamos, NM, United States

OVERVIEW CHAPTER

Accurate estimation of the state and topology of the distribution grid is hindered by the limited placement of real-time flow meters and breaker statuses at distribution grid lines. In recent years, increasing presence of smart devices and sensors at households have made measurements of consumption and voltages available at distribution buses. This chapter discusses greedy algorithms to learn the grid topology using voltage measurements collected at a subset of the buses in the distribution grid. The distribution grids are operated in a radial topology. This topological restriction leads to provable trends in voltage second moments (covariances) and enables the design of our learning algorithms. For the case where voltage measurements are available at all grid buses, our framework does not require any additional information related to line impedances of grid lines or consumption statistics at buses to estimate the operational topology. Further in presence of such information, we demonstrate guaranteed topology learning in scenarios with varying fraction of "missing" buses that have no voltage measurements. The efficiency of the algorithms is highlighted by their computational complexity that scales polynomially in the number of grid buses.

1 INTRODUCTION

Power grids are organized hierarchically into transmission and distribution grids. While transmission grids comprise of the high voltage lines, distribution grids consist of the low and medium voltage lines that link the distribution substations to the load buses. The majority of static and dynamic operations of the grid as well as power markets hinge on accurate estimation of the grid state (bus voltages, line flows, bus injections) and operational topology (breaker statuses). As the grid has redundant lines, estimation of the operational topology refers to determining the grid lines with breakers turned on. Traditionally, control and operation in the grid have emphasized on the transmission side as majority of the bulk generation is connected there. Measurement devices, real-time devices in particular, have been largely deployed in the transmission grid. The distribution grid has thus had sparse observability and real-time estimation [1]. In recent times a plethora of smart controllable devices like plug-in electric vehicles, smart air-conditioners, residential batteries as well as renewable resources like roof-top solar panels has increasingly penetrated the

Big Data Application in Power Systems. https://doi.org/10.1016/B978-0-12-811968-6.00012-7

distribution grid. To maximize the benefit derived from these resources and create new opportunities like distribution grid market, real-time state and topology estimation in the distribution grid has gained prominence. It is noteworthy that several of these new devices also provide local measurements of voltage and power injection at their resident buses. Further, expansion efforts of modern meters like phasor measurement units (PMUs) [2], micro-PMUs [3], frequency monitoring networks into the distribution grid are slowly getting traction. In this chapter, we study topology estimation in the distribution grid using high fidelity voltage measurements collected from smart meters placed at a subset of the grid buses. It is worth mentioning that these measurements are essentially nodal in nature. They do not include line measurements (line breakers, flows, etc.) that enable direct estimation of the topology.

Distribution grid topologies, in a majority of power grids, have one unifying feature. A typical distribution grid is operated as disjoint radial graphs, with a substation bus at the root node and customers (load buses) at the nonroot nodes of each tree. Switching from one radial layout to another is initiated through switching on and off breakers [4] in the available lines (see Fig. 1A for the illustration). The goal of topology estimation is to estimate the current tree structure in operation. Note that as the underlying graph of available lines in the grid is loopy, the number of candidate tree topologies that can be operational is large. Brute force (combinatorial) check of the true operational radial topology consistent with the measurements collected is thus computationally

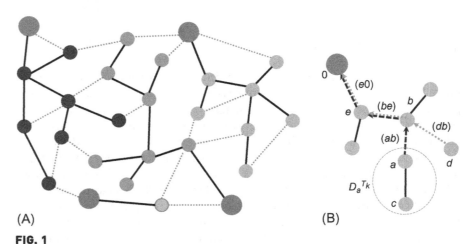

(A)   (B)

FIG. 1

(A) Distribution grid with four substations (*large* nodes). Operational lines are colored *solid black*, nonoperational lines are marked dotted. (B) Nodes in a radial distribution grid. Nodes a and c are descendants of node a. *Dotted lines* represent the paths from nodes a and d to the root node.

prohibitive. In this chapter we develop low complexity algorithms that circumvent this problem by exploring novel trends in second-order statistics of the collected nodal measurements. Further, such trends are able to estimate the grid topology accurately even when measurements are collected from a subset of the grid buses/nodes—a realistic feature as ubiquitous meter placement has still not been realized.

1.1 Prior Work

Learning the topology of power grids and distribution grids in particular is a growing area of research with several approaches being proposed in the past. In [5], a maximum likelihood estimator with sparsity promoting regularizers is used to recover the grid structure using locational marginal prices. In [6], a model using bus phase angles as a Markov random field for the DC power flow (PF) builds a dependency graph-based approach to detect faults in grids. In work specific to radial grids, Bolognani et al. [7] considers grids where transmission lines have constant resistance to reactance ratio and provides a learning algorithm that uses signs within the inverse covariance matrix of voltage measurements. In [8], topology identification with limited measurements in a distribution grid with Gaussian loads is used to design a machine learning estimate with approximate schemes. Cavraro et al. [9] compares available time-series measurements from smart meters with a database of permissible signatures to identify topology changes. Similar envelope-based comparison schemes have been used for parameter estimation [10, 11]. Available line flow measurements have been used for topology estimation using maximum likelihood tests in [12]. Conditional independence-based tests have been used to identify the radial topology in [13]. This has been extended to topology identification from samples collected from grid dynamics in [14].

1.2 Technical Contribution

In this chapter we consider a setting where the observer has access to voltage magnitude measurements at the grid nodes but no edge-based measurements. We utilize a linearized PF model [4, 15–17] and demonstrate that under uncorrelated nodal injections, the variance of *voltage magnitude differences* increase along paths in the operational grid topology. Thus we present a computationally fast spanning tree-based learning algorithm (originally outlined in [18]) for the operational tree using only voltage measurements at all nodes. Crucially, the algorithm is agnostic to the individual nodal injections and parameters of lines (resistances or reactances). Further, we extend our algorithm to the case with missing nodes/buses with no measurements, where the missing nodes are separated by at least three hops from each other and covariances of nodal power consumption are available. We consider topology learning in another setting where voltage measurements are limited to the terminal nodes/leaves (end-users) alone. All intermediate nodes are unobserved and hence assumed to be missing nodes. Learning in this regime relies on functions of voltages at pair or triplets of terminal nodes that enable the construction of the operational topology iteratively from the leaves onward to the substation node. Parts of the results in this chapter are compiled from [4, 18, 19].

Our algorithm shows similar aspects as learning of tree-structured graphical models in [20] using information distances derived from multivariate probability distributions. However, our approach relies on the Kirchhoff's laws of physical network flows that relate nodal voltages and injections that, to the best of our knowledge, do not have an analog in graphical model learning literature [20, 21]. Further, voltage magnitude-based weights used in our work are not restricted to satisfy graph additivity unlike information distances in graphical models.

The rest of this chapter is organized as follows. Section 2 introduces nomenclature and PF relations in the distribution grids. Section 3 describes key features (equalities and inequalities) of the statistics nodal voltage magnitudes that provide the machinery for our learning algorithms. Algorithm reconstructing operational spanning tree in the case of complete visibility (voltage magnitudes are observed at all nodes) is discussed in Section 4. Modification of the algorithm for missing data (missing nodes separated by at least three hopes) is described in Section 5. We also discuss topology learning when the available measurements are limited to leaves (end-users) with detailed examples. Simulation of our learning algorithm on test radial networks is presented in Section 6. Finally, Section 7 contains conclusions and discussion of future work.

2   DISTRIBUTION GRID: STRUCTURE AND POWER FLOWS

Radial Structure

We represent the distribution grid by the graph $\mathcal{G} = (\mathcal{V}, \mathcal{E})$, where $\mathcal{V}$ is the set of buses/nodes of the graph and $\mathcal{E}$ is the set of all undirected lines/edges (open or operational). We denote nodes by alphabets (a, b, ...) and the edge connecting nodes a and b by (ab). The operational grid has a "radial" structure as shown in Fig. 1A. In general the operational grid is a collection of K disjoint trees.

In this chapter, we will focus on grids where the operational structure consists of only one tree $\mathcal{T}$ with nodes $\mathcal{V}_{\mathcal{T}}$ and operational edge set $\mathcal{E}_{\mathcal{T}} \subset \mathcal{E}$. The tree's root node has degree one (connected by one edge) and represents a substation as shown in Fig. 1B. Let $\mathcal{P}_T^a$ denote the set of edges in the unique path from node a to the root node (reference bus) in tree $\mathcal{T}$. A node b is termed as a *descendant* of node a if $\mathcal{P}_T^b$ includes some edge (ac) connected to node a. We use D_T^a to denote the set of descendants of a and include node a in it by definition. If b is an immediate descendant of a ($(ab) \in \mathcal{E}_T$), we term a as parent and b as its child. Nodes that do not have any children, that is, $D_T^a = \{a\}$ are termed leaves.

PF Model

Let $z_{ab} = r_{ab} + ix_{ab}$ denote the complex impedances of a line (ab) ($i^2 = -1$). Here r_{ab} and x_{ab} are line resistance and reactance, respectively. The a.c. power injection at node a is given by Kirchhoff's laws as

$$P_a = p_a + iq_a = \sum_{b:(ab)\in\mathcal{E}_T} \frac{v_a^2 - v_a v_b \exp(i\theta_a - i\theta_b)}{z_{ab}^*} \tag{1}$$

where the real-valued scalars, v_a, θ_a, p_a, and q_a denote the voltage magnitude, voltage phase, active, and reactive power injection, respectively, at node a. $V_a (= v_a \exp(i\theta_a))$ and P_a denote the nodal complex voltage and injection, respectively. Note that Eq. (1) is nonlinear and nonconvex. Under realistic assumption that losses of both active and reactive power losses on each line of tree $\mathcal{T}$ is small, we ignore second-order terms in Eq. (1) to achieve the following linearization [4, 7, 18]:

$$p_a = \sum_{b:(ab)\in\mathcal{E}^T} (\beta_{ab}(\theta_a - \theta_b) + g_{ab}(v_a - v_b)), \quad q_a = \sum_{b:(ab)\in\mathcal{E}^T} (-g_{ab}(\theta_a - \theta_b) + \beta_{ab}(v_a - v_b)) \tag{2}$$

where

$$g_{ab} \doteq \frac{r_{ab}}{x_{ab}^2 + r_{ab}^2}, \quad \beta_{ab} \doteq \frac{x_{ab}}{x_{ab}^2 + r_{ab}^2} \tag{3}$$

This linearization assumes that phase difference between neighboring nodes and magnitude deviations from the reference voltage are small. Deka et al. [4] show that Eqs. (2) are equivalent to the LinDistFlow model [15, 16, 22], if deviations in voltage magnitude are assumed to be small. Further, if line resistances are equated to zero, they reduce to the DC PF model [23] used for transmission grids. Similar to LinDistFlow model, Eqs. (2) are lossless with sum power equal to zero ($\sum_{a\in V_T} P_a = 0$). Note that in our linearized model, active and reactive power injections are functions of difference in voltage magnitudes and phases of neighboring nodes. Thus the analysis of the system is reduced by measuring the voltage magnitude and phase at all buses relative to one specific bus termed as reference bus/node with voltage magnitude 1 p.u. and phase 0. As per convention, the substation is taken as the reference bus since its injection also balances the power injections in the remaining network. Inverting Eqs. (2) for the reduced system (without the reference node), we express voltages as a function of nodal power injections in the following vector form:

$$v = H_{1/r}^{-1}p + H_{1/x}^{-1}q \quad \theta = H_{1/x}^{-1}p - H_{1/r}^{-1}q \tag{4}$$

We term this as the *linear coupled power flow (LC-PF) model* where p, q, v, and θ are the vectors of real power, reactive power injections, relative voltage magnitudes, and phase angles, respectively, at the nonsubstation nodes. $H_{1/r}$ and $H_{1/x}$ are the reduced weighted Laplacian matrices for tree $\mathcal{T}$ where reciprocal of resistances and reactances are used, respectively, as edge weights. The reduction removes the row and column corresponding to the reference bus in the original weighted Laplacian matrix. Due to the radial topology, the inverse of the reduced weighted graph Laplacian matrix $H_{1/r}$ has the following structure (see Deka et al. [4] for the derivation).

$$H_{1/r}^{-1}(a,b) = \sum_{(cd) \in \mathcal{P}_{\mathcal{T}}^a \cap \mathcal{P}_{\mathcal{T}}^b} r_{cd} \tag{5}$$

Thus, the (a, b)th entry in $H_{1/r}^{-1}$ is given by the sum of line resistances of edges that are included in the path to the root from either node. For node a and its parent b in tree $\mathcal{T}$ (see Fig. 1B), it follows from Eq. (5) that

$$H_{1/r}^{-1}(a,c) - H_{1/r}^{-1}(b,c) = \begin{cases} r_{ab} & \text{if node } c \in D_{\mathcal{T}}^a \\ 0 & \text{otherwise} \end{cases} \tag{6}$$

We denote the mean of a random vector X by $\mu_X = \mathbb{E}[X]$. For two random vectors X and Y, the covariance matrix is denoted by $\Omega_{XY} = \mathbb{E}[(X - \mu_X)(Y - \mu_Y)^T]$. Using the LC-PF model, the means and covariances of voltage magnitudes can be related with those of active and reactive injection as follows:

$$\mu_v = H_{1/r}^{-1}\mu_p + H_{1/x}^{-1}\mu_q, \quad \Omega_v = H_{1/r}^{-1}\Omega_p H_{1/r}^{-1} + H_{1/x}^{-1}\Omega_q H_{1/x}^{-1} + H_{1/r}^{-1}\Omega_{pq} H_{1/x}^{-1} + H_{1/x}^{-1}\Omega_{qp} H_{1/r}^{-1} \tag{7}$$

In the next section we derive key results for functions of nodal voltages in a radial distribution grid that will subsequently be used in the topology learning algorithm.

3   PROPERTIES OF VOLTAGE MAGNITUDES IN RADIAL GRIDS

First, we make the following assumption regarding statistics of power injections at the nonsubstation grid nodes, under which our results hold.

Assumption 1. Power injections at different nodes are not correlated, while active and reactive injections at the same node are positively correlated. Mathematically, $\forall a, b$ nonsubstation nodes

$$\Omega_{qp}(a,a) > 0, \quad \Omega_p(a,b) = \Omega_q(a,b) = \Omega_{qp}(a,b) = 0$$

This is a valid assumption for many distribution grids due to independence between different nodal load fluctuations and alignment/correlations between same node's active and reactive power usage. Note that Assumption 1 is applicable when nodal injections are negative (loads), positive (due to local generation) or a mixture of both. Now we consider the quantity $\phi_{ab} = \mathbb{E}[(v_a - \mu_{v_a}) - (v_b - \mu_{v_b})]^2$, which measures the variance of the difference in voltage magnitudes between nodes a and b. Using Eq. (7), we have

$$\phi_{ab} = \Omega_v(a,a) - 2\Omega_v(a,b) + \Omega_v(b,b) \tag{8}$$

Expressing Eq. (8) in terms of the four matrices that constitute Ω_v leads to the following

$$\begin{aligned}
\phi_{ab} &= \sum_{d \in \mathcal{T}} \left(H_{1/r}^{-1}(a,d) - H_{1/r}^{-1}(b,d) \right)^2 \Omega_p(d,d) + \left(H_{1/x}^{-1}(a,d) - H_{1/x}^{-1}(b,d) \right)^2 \Omega_q(d,d) \\
&\quad + 2 \left(H_{1/r}^{-1}(a,d) - H_{1/r}^{-1}(b,d) \right) \left(H_{1/x}^{-1}(a,d) - H_{1/x}^{-1}(b,d) \right) \Omega_{pq}(d,d)
\end{aligned} \tag{9}$$

This relation is important as we use it to identify inequality trends and equality relations for ϕ_{ab} that we use in our learning algorithms. The next result identifies order in ϕ_{ab} along the radial grid.

Theorem 1. *For three nodes $a \neq b \neq c$ in grid tree $\mathcal{T}$, $\phi_{ab} < \phi_{ac}$ holds for the following cases:*

1. *Node a is a descendant of node b and node b is a descendant of node c (see Fig. 2A).*
2. *Nodes a and c are descendants of node b and the path from a to c passes through node b (see Fig. 2B).*
3. *Node c is a descendant of node b and node b is a descendant of node a (see Fig. 2C).*

The proofs for the first two cases can be found in [4], while that of the third case is presented in [18]. Specifically for adjacent nodes in $\mathcal{T}$, the following results show equality expressions involving ϕ.

Theorem 2. *Let (ab) and (bc) be operational edges in $\mathcal{T}$*

1. *If node a is the parent of node b (see Fig. 2C) then*
   $$\phi_{ab} = \sum_{d \in D_{\mathcal{T}}^b} r_{ab}^2 \Omega_p(d,d) + x_{ab}^2 \Omega_q(d,d) + 2 r_{ab} x_{ab} \Omega_{pq}(d,d)$$

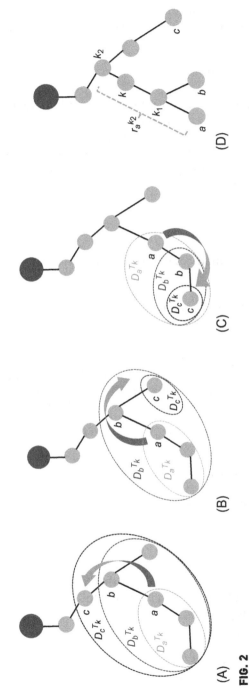

FIG. 2

Distribution grid tree with substation/root node represented by *large* node. (A) Node *a* is a descendant of node *b*, node *b* is a descendant of node *c*. (B) Nodes *a* and *c* are descendants of node *b* along disjoint subtrees. (C) Node *c* is a descendant of node *b*, node *b* is a descendant of node *a*. (D) Nodes *a* and *b* are leaf nodes with common parent k_1. $r_a^{k_2}$ is the sum of resistances on path from *a* to k_2.

2. *If node* b *is the parent of node* c *and child of node* a *(see Fig. 2C), then*

$$\phi_{ac} = \sum_{d \in D_T^c} (r_{ab} + r_{bc})^2 \Omega_p(d, d) + (x_{ab} + x_{bc})^2 \Omega_q(d, d) + 2(r_{ab} + r_{bc})(x_{ab} + x_{bc})\Omega_{pq}(d, d)$$
$$+ \sum_{d \in D_T^b - D_T^c} r_{ab}^2 \Omega_p(d, d) + x_{ab}^2 \Omega_q(d, d) + 2r_{ab}x_{ab}\Omega_{pq}(d, d) \tag{10}$$

3. *If node* b *is the parent of both nodes* a *and* c *(see Fig. 2B), then*

$$\phi_{ac} = \sum_{d \in D_T^a} r_{ab}^2 \Omega_p(d, d) + x_{ab}^2 \Omega_q(d, d) + 2r_{ab}x_{ab}\Omega_{pq}(d, d)$$
$$+ \sum_{d \in D_T^c} r_{bc}^2 \Omega_p(d, d) + x_{bc}^2 \Omega_q(d, d) + 2r_{bc}x_{bc}\Omega_{pq}(d, d) \tag{11}$$

Proof

1. Use Eq. (6) in Eq. (9) as (ab) is an edge.
2. Use Eq. (6) in Eq. (9) as (ab), (bc) are edges and consider different descendant sets.
3. As (ab) and (bc) are operational edges, the only node d such that $(H_{1/r}^{-1}(a, d) - H_{1/r}^{-1}(c, d)) \neq 0$ are either descendants of a (set $\mathcal{D}_a$) or of c (set $\mathcal{D}_c$) that are disjoint. Using this in the formula for ϕ_{ac} in Eq. (9) gives us the relation. □

It is worth mentioning that all three statements in Theorem 2 involve line impedances corresponding to edges (ab) and (bc) only and injections at their descendants. This is critical in the search for missing nodes in our topology learning algorithms. Finally we discuss another result that relates the variance of voltage magnitude differences at groups of three nodes that are leaves but possibly separated from each other.

Theorem 3. *Let terminal nodes* a *and* b *have common parent node* k_1. *Let* c *be another terminal node such that* $c, k_1 \in \mathcal{D}_{k_2}$ *and* $\mathcal{P}_{k_1} \cap \mathcal{P}_c = \mathcal{P}_{k_2}$ *for some intermediate node* k_2 *(see Fig. 2D). Let* $r_a^{k_2}$ *and* $x_a^{k_2}$ *denote the sum of resistance and reactance, respectively, on lines on the path from node* a *to node* k_2, *that is,* $r_a^{k_2} = \sum_{(ef) \in \mathcal{P}_a - \mathcal{P}_{k_2}} r_{ef}$, $x_a^{k_2} = \sum_{(ef) \in \mathcal{P}_a - \mathcal{P}_{k_2}} x_{ef}$. *Define* $r_b^{k_2}$, $r_{k_1}^{k_2}$, *etc., in the same way.*

$$\phi_{ac} - \phi_{bc} = \Omega_p(a, a)((r_a^{k_2})^2 - (r_{k_1}^{k_2})^2) + \Omega_q(a, a)((x_a^{k_2})^2 - (x_{k_1}^{k_2})^2)$$
$$+ \Omega_{pq}(a, a)(r_a^{k_2}x_a^{k_2} - r_{k_1}^{k_2}x_{k_1}^{k_2}) - \Omega_p(b, b)((r_b^{k_2})^2 - (r_{k_1}^{k_2})^2) \tag{12}$$
$$+ \Omega_q(b, b)((x_b^{k_2})^2 - (x_{k_1}^{k_2})^2) + \Omega_{pq}(b, b)(r_b^{k_2}x_b^{k_2} - r_{k_1}^{k_2}x_{k_1}^{k_2})$$

The proof for this result relies on expansion of the expression on the left and using similar techniques as the previous theorem. Check Deka et al. [19] for exact details. Theorem 3 shows that the difference in ϕ between a leaf node (node c here) and two sibling leaves (nodes a and b) depends on injections only at nodes a and b. Further the lines whose impedances appear on the right-hand side of Eq. (12) are (ak_1), (bk_1) and the ones on the path from node k_1 to k_2. This will be used in learning the path from terminal pairs with common parent (here a and b) to the root iteratively, when missing nodes are present.

4 TOPOLOGY LEARNING WITH FULL OBSERVATION

Given an underlying loopy graph with edge set $\mathcal{E}$ (possibly complete), the goal of topology learning is to estimate the operational edge set $\mathcal{E}_T$ in radial grid $\mathcal{T}$. We first discuss the case where voltage magnitude measurements are available at all nodes.

Theorem 4. *Let the weight of each permissible edge* $(ab) \in \mathcal{E}$ *of the underlying loopy graph be* $\phi_{ab} = \mathbb{E}[(v_a - \mu_{v_a}) - (v_b - \mu_{v_b})]^2$. *Then operational edge set* $\mathcal{E}_T$ *in radial grid* $\mathcal{T}$ *forms the minimum weight spanning tree in the underlying graph.*

Proof. From Theorem 1, it is clear that for each node a, the minimum value of ϕ_{ab} along any path in $\mathcal{T}$ (toward or away from the root node) is attained at its immediate neighbor b on that path, connected by edge $(ab) \in \mathcal{E}_T$. The minimum spanning tree for the original loopy graph with ϕ's as edge weights is thus given by the operational edges in the radial tree. □

Note that if node a is taken as the substation/root node ($v_a = 1$), the weight of any edge (ab) is given by $\phi_{ab} = \Omega_v(b, b)$. In the spanning tree construction, the root is thus connected to the node with lowest variance of voltage magnitude.

Algorithm 1. We consider input as voltage magnitude readings for all nonsubstation buses in the system and compute ϕ_{ab} for all permissible edges $(ab) \in \mathcal{E}$.

Algorithm 1 determines the operational edge-set $\mathcal{E}_T$ by the minimum spanning tree. Note that Algorithm 1 does not need any information on line parameters (resistances and reactances) or on statistics of active and reactive nodal power consumption. If impedances of lines in $\mathcal{E}$ and phase angle measurements at all nodes are known, Eqs. (2), (7) can subsequently estimate means and covariances of each node's power injections as well.

ALGORITHM 1 MINIMUM WEIGHT SPANNING TREE-BASED TOPOLOGY LEARNING

Input: m voltage magnitudes v for all nodes, set of all edges $\mathcal{E}$.

Output: Operational edge set $\mathcal{E}_T$.

1: $\forall (ab) \in \mathcal{E}$, compute $\phi_{ab} = \mathbb{E}[(v_a - \mu_{v_a}) - (v_b - \mu_{v_b})]^2$

2: Find minimum weight spanning tree from $\mathcal{E}$ with ϕ_{ab} as edge weights.

3: $\mathcal{E}_T \leftarrow$ edges in spanning tree

Algorithm Complexity

Using Kruskal's algorithm [24, 25], the minimum spanning tree from $\mathcal{E}$ edges can be computed in $O(|\mathcal{E}| \log |\mathcal{E}|)$ operations. If $\mathcal{E}$ is not known, we consider a complete graph where all edges are permissible. Algorithm 1's complexity in that case is $O(N^2 \log N)$.

Extension to Multiple Trees

Note that each tree has a reference bus with reference voltage known. The voltage magnitudes at nodes a and b belonging to disjoint trees will be uncorrelated and $\phi_{ab} = \Omega_v(a, a) + \Omega_v(b, b)$. This result can be used to separate nodes into disjoint subsets before running Algorithm 1 to generate each operational tree. The same technique can be used to extend algorithms in later sections to the case with multiple trees.

In the next section, we use our spanning tree-based algorithm to consider two cases with missing nodes.

5 TOPOLOGY LEARNING WITH MISSING DATA

In a realistic power grid, communication packet drops or random noise events may erase voltage magnitude measurements for a missing node set $\mathcal{M}$ in $\mathcal{T}$. We consider two cases of missing nodes, one with missing nodes separated by greater than two hops in the operational tree and the other where all nodes other than the terminal nodes/leaves are missing. We assume that the observer estimates or has access to historical information for the values of Ω_p, Ω_q, and Ω_{pq} covariance matrices for all nodal injections and impedances R, X of all lines in permissible edge set $\mathcal{E}$. We consider the first case now.

5.1   Missing Nodes Separated by Three or More Hops

The setting here is highlighted in the following assumption.

Assumption 2. Missing nodes $\mathcal{M}$ are separated by greater than two hops in the grid tree $\mathcal{T}$.

Note that under Assumption 2, an observable node cannot be connected to two or more unobserved nodes.

To reconstruct the operational topology with missing nodes, we first construct the minimum weight spanning tree $\mathcal{T}_{\mathcal{M}}$ between observable nodes using ϕ as edge weights. Note that edges between neighbors in $\mathcal{T}$ also appear in $\mathcal{T}_{\mathcal{M}}$ but not vice versa. We then analyze edges in tree $\mathcal{T}_{\mathcal{M}}$ and detect unobserved node locations. First, consider the case where the missing node is a leaf node l (see Fig. 3A). By Assumption 2, information from its parent (q) and grandparent (w) are observed. Note that ϕ_{qw} satisfies Statement 1 in Theorem 2. If all other descendants of q are known, statement 1 of the theorem can be used to identify the existence of unobserved node l by checking for equality.

We now discuss the identification of a nonleaf missing node b (see Fig. 3A). b's parent a and children node set $C = \{c_1, c_2, c_3, c_4\}$ comprise its one-hop neighborhood in operational tree $\mathcal{T}$, and are observable under Assumption 2. Using statements 1 and 3 in Theorem 1, it is clear that descendants of b (set C) are connected to the rest of $\mathcal{T}_{\mathcal{M}}$ through node a. However, edges in $\mathcal{T}_{\mathcal{M}}$ between nodes in C and a can exist in either of the configurations shown in Fig. 3B and C. In either configuration, a parent-child pair in $\mathcal{T}_{\mathcal{M}}$ may represent a grandparent-child pair or even sibling pair in the true grid $\mathcal{T}$. To identify the existence of the missing node, we use Theorem 2 on either configuration.

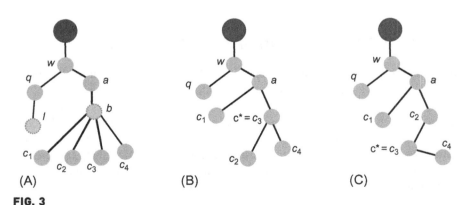

(A)   (B)   (C)

FIG. 3

(A) Distribution grid tree $\mathcal{T}$ with unobserved leaf node l nonleaf unobserved node b. Node a is b's parent while nodes c_1, c_2, c_3, c_4 are its children. The spanning tree $\mathcal{T}_{\mathcal{M}}$ of observed nodes exists in either (B) configuration A or (C) configuration B.

Algorithm 2. Given missing set $\mathcal{M}$, $\mathcal{V}_T - \mathcal{M}$ is the observed set. Algorithm 2, first, constructs spanning tree $\mathcal{T}_\mathcal{M}$ for observed nodes using edge weights given by ϕ. Observed nodes in $\mathcal{T}_\mathcal{M}$ are then arranged in decreasing depth from root node to check the location of missing nodes. For each leaf b with parent a, Steps 5 to 10 checks if edge $(ab) \in \mathcal{E}_T$ with or without some unobserved leaf node h connected to b. C denotes the set of undecided children of node a. Step 13 determines if nodes in C are connected to grandparent node a via a missing node h using Statement 2 in Theorem 2, while Step 16 checks if nodes in C and node a are siblings with a common missing node as parent using Statement 3 in Theorem 2. In each iteration, nodes (both missing and observed) are removed from tree $\mathcal{T}_\mathcal{M}$ or missing set $\mathcal{M}$ if their parents are identified, and discovered edges are added to $\mathcal{E}_T$. This process is iterated by picking a new node a with all children as leaf nodes until no missing nodes remain to be discovered.

Complexity

For an N node system with $\mathcal{M}$ set of missing nodes, Deka et al. [18] prove that the overall complexity of Algorithm 2 is $O((N - |\mathcal{M}|)^2 \log(N - |\mathcal{M}|) + (N - |\mathcal{M}|)|\mathcal{M}|)$, which is $O(N^2 \log N)$ in the worst case. Note that this is also the worst-case complexity of Algorithm 1.

ALGORITHM 2 MINIMUM WEIGHT SPANNING TREE-BASED TOPOLOGY LEARNING WITH MISSING DATA

Input: Injection covariances Ω_p, Ω_q, Ω_{pq} of all nodes, missing nodes set $\mathcal{M}$, m voltage observations v for nodes in $\mathcal{V}_T - \mathcal{M}$, set of all edges T with line impedances. **Output:** Operational edge set $\mathcal{E}_T$.

1: $\forall$ observable nodes a, b, compute ϕ_{ab} and find minimum weight spanning tree $\mathcal{T}_\mathcal{M}$ with ϕ_{ab} as edge weights. Sort nodes in $\mathcal{T}_\mathcal{M}$ in reserve topological order.

2: **while** $|\mathcal{M}| > 0$ **do**

3:     Select node a with children set C in $\mathcal{T}_\mathcal{M}$ consisting only of leaf nodes

4:     **for all** $b \in C$ **do**

5:         **if** ϕ_{ab} satisfies Statement 1 in Theorem 2 with $D_T^b = \{b\}$ **then**

6:             $\mathcal{E}_T \leftarrow \mathcal{E}_T \cup \{(ab)\}$, $C \leftarrow C - \{b\}$, add injection covariance of b to a. Remove node b from $\mathcal{T}_\mathcal{M}$.

7:         **end if**

8:         **if** $\exists h \in \mathcal{M}$ s.t. ϕ_{ab} satisfies Statement 1 in Theorem 2 with $D_T^b = \{b,h\}$ **then**

9:             $\mathcal{E}_T \leftarrow \mathcal{E}_T \cup \{(ab),(bh)\}$, $\mathcal{M} \leftarrow \mathcal{M} - \{h\}$, $C \leftarrow C - \{b\}$, $\mathcal{T}_\mathcal{M} \leftarrow \mathcal{T}_\mathcal{M} - \{b\}$. Add injection covariances of b, h to a.

10:         **end if**

11:     **end for**

12:     **if** $|C| > 0$ **then**

13:         **if** $\exists c \in C, h \in \mathcal{M}$ s.t. ϕ_{ab} satisfies Statement 2 in Theorem 2 with $D_T^c = \{c\}$ and $D_T^h = \{h\} \cup C$ **then**

14:             $\mathcal{E}_T \leftarrow \mathcal{E}_T \cup \{(ah)\} \cup \{(ch)\forall c \in C\}$, $\mathcal{M} \leftarrow \mathcal{M} - \{h\}$. Add injection covariances $\forall c \in C$, h to a, $\mathcal{T}_\mathcal{M} \leftarrow \mathcal{T}_\mathcal{M} - C$

15:         **else**

16:             Pick $b \in C$. Find $h \in \mathcal{M}$ s.t. ϕ_{ab} satisfies Statement 3 in Theorem 2 with h as parent and $D_T^b = \{b\}$, $D_T^a = \{a\}$.

17:             $\mathcal{E}_T \leftarrow \mathcal{E}_T \cup \{(ah)\} \cup \{(ch)\forall c \in C\}$. Add injection covariances of a, $\forall c \in C$ to h, $\mathcal{T}_\mathcal{M} \leftarrow \mathcal{T}_\mathcal{M} - \{a\} \cup C$.

18:         **end if**

19:     **end if**

20: **end while**

Note

Empirically computed second moments of voltages may differ from their true values. Hence we use tolerances to check the correctness all equality relations. Similar tolerances are used in the next section as well.

5.2 All Nonleaf Nodes Are Missing

We now consider the case where all observed nodes are limited to terminal nodes or leaves. Thus, all intermediate nodes that are neighbors are missing. As before, we assume knowledge of line parameters at all permissible lines in set $\mathcal{E}$. However, in this setting, we assume injection statistics to be available only at the terminal nodes (not at missing nodes). Further we assume the following structural constraint.

Assumption 3. All missing intermediate nodes are assumed to have a degree greater than two.

This assumption is necessary as without it, the solution to the topology learning problem may not be unique for any learning algorithm. An example for this is given in [19]. Similar assumptions for uniqueness in learning general graphical models are mentioned in [26]. Note that under Assumption 3, the sibling of a leaf node may be another leaf node or a missing intermediate node.

To learn the topology given only measurements at leaf nodes, we present Algorithm 3. The topology learning is done in three major steps: (a) identifying parent of sibling leaves, (b) building path from siblings to the root iteratively, and (c) identifying location of leaves with no leaf siblings. As the injection covariances at terminal nodes are known, Statement 4 in Theorem 2 shows that ϕ_{ab} for two sibling nodes depends on the impedances on the lines to their parent. We use this to determine the parent of all leaf pairs that are siblings. Theorem 3 shows that for a leaf node c and two sibling leaves a and b, $\phi_{ac} - \phi_{bc}$ depend on the impedances of edges in the paths from a and b to the root. This result can thus be used iteratively to identify missing nodes on the path from a and b to the root. Finally the location of leaves that do not have other leaves as siblings is determined through Theorem 3. The correct location is ensured by checking at all candidate nodes before checking at their parents. See Deka et al. [19] for a detailed explanation.

Computational Complexity

As detailed in [19], the overall complexity of the algorithm is $O(N^3)$ in the worst case. Further it can be shown that for specific configurations, Algorithm 3 can learn the grid with phase measurements limited to only 50% of the grid nodes, which is the lower limit for exact reconstruction for any algorithm.

ALGORITHM 3 TOPOLOGY LEARNING USING TERMINAL NODE DATA

Input: Injection covariances Ω_p, Ω_q, Ω_{pq} at terminal nodes $\mathcal{L}$, missing node set $\mathcal{M} = \mathcal{V}_T - \mathcal{L}$, m voltage magnitude observations v for nodes in $\mathcal{L}$, set of all edges $\mathcal{E}$ with line impedances. **Output:** Operational edge set $\mathcal{E}_T$.

1: $\forall$ nodes $a, c \in \mathcal{L}$, compute ϕ_{ac}. $\forall a \in \mathcal{V}_T$, define $par_a \leftarrow \Phi$, $des_a \leftarrow \Phi$
2: **for all** $a, c \in \mathcal{L}$ **do**
3:     **if** $b \in \mathcal{M}$ s.t. ϕ_{ac}, b satisfy Statement 3 in Theorem 2 **then**
4:         $\mathcal{E}_T \leftarrow \mathcal{E}_T \cup \{(ab), (bc)\}$, par_a, $par_c \leftarrow b$, $des_b \leftarrow a, c$, $tp \leftarrow 1$
5:     **end if**
6: **end for**
7: **while** $tp > 0$ **do**
8:     $tp \leftarrow 0$
9:     **for all** $k \in \mathcal{M}$ with some $a, b \in des_k$, $par_k = \Phi$ **do**
10:         **if** $k_2 \in \mathcal{M}$, $c \in \mathcal{L}$, s.t. $\phi_{ac} - \phi_{bc}$ satisfy Theorem 3 **then**
11:             $\mathcal{E}_T \leftarrow \mathcal{E}_T \cup \{(kk_2)\}$, $par_k \leftarrow k_2$, $des_{k_2} \leftarrow des_k$, $tp \leftarrow 1$
12:         **end if**
13:     **end for**
14: **end while**
15: If one missing node has unidentified parent, join it to root. Form postorder traversal set $\mathcal{W}$ for missing nodes with known parents
16: **for all** $c \in \mathcal{L}$, $par_c = \Phi$ **do**
17:     **for** $j \leftarrow 1$ to $|\mathcal{W}|$ **do**
18:         $k_2 \leftarrow \mathcal{W}(j)$ with $a, b \in des_{k_2}$
19:         **if** $\phi_{ac} - \phi_{bc}$ satisfy Eq. (12) **then**
20:             $\mathcal{E}_T \leftarrow \mathcal{E}_T \cup \{(ck_2)\}$, $\mathcal{W} \leftarrow \mathcal{W} - \{k_2\}$, $j \leftarrow |\mathcal{W}|$
21:         **end if**
22:     **end for**
23: **end for**

6 EXPERIMENTS

In this section, we present simulation results that highlight the performance of our three learning algorithms. First, we consider Algorithm 1 that learns the operational edge set $\mathcal{E}_T$ using voltage magnitude measurements at all nodes. We consider a radial network [27, 28] with 29 load nodes and 1 substation as shown in Fig. 4A. In each of our simulation runs, we first collect complex

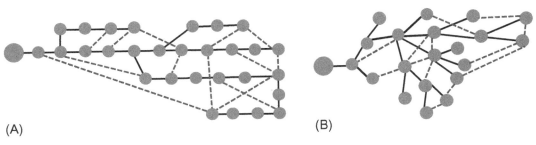

(A)   (B)

FIG. 4

Layouts of the grids tested. The *large circle* represents substation (marked as *S*). The *smaller circles* represent numbered load nodes. *Black lines* represent operational edges. Some of the additional open lines are represented by *dotted lines*. (A) Twenty-nine-bus system for Algorithm 2. (B) Twenty-bus system for Algorithm 3.

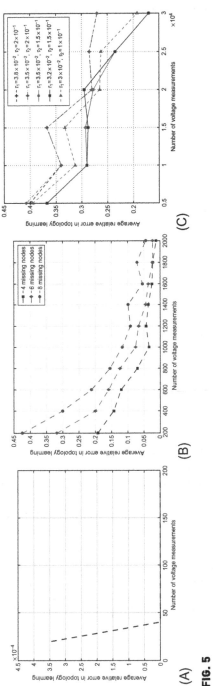

FIG. 5

Average fractional errors in learning operational edges versus number of samples used. (A) Algorithm 1, (B) Algorithm 2 with 4, 6, and 8 missing nodes, and (C) Algorithm 3 with data at terminal nodes.

power injection samples at the nonsubstation nodes from a uncorrelated multivariate Gaussian distribution that is uncorrelated between different nodes and generate voltage magnitude samples from the LC-PF model. We also introduce 30 additional edges (at random) forming the loopy edge set $\mathcal{E}$. The results for topology learning for this case are presented in Fig. 5A. Note that the estimation is extremely accurate and average errors expressed relative to the size of the operational edge set decay to zero at sample sizes less than 50.

Next we present simulations for Algorithm 2 where the operational grid structure is reconstructed in the presence of unobserved nodes separated by three or more hops. We consider three cases with four, six, and eight missing nodes. The locations of the unobserved nodes are selected at random in accordance with Assumption 2. The average number of errors shown in Fig. 5B decreases with an increase in the number of samples for all the cases considered. This tendency is seen clearly for all the cases of the unobserved node sets. Further, the average errors increase with increase in the number of unobserved nodes for a fixed number of measurement samples.

Finally we present results for Algorithm 3 that learns the topology with voltage measurements only at terminal nodes. For that we consider a 20-bus radial grid satisfying Assumption 3 depicted in Fig. 4B. As before we introduce 30 additional edges (at random) to construct the loopy edge set $\mathcal{E}$. In Fig. 5C, we plot the average fractional errors in Algorithm 3 that decrease steadily with increase in the number of voltage magnitude measurements used for the different values of tolerances considered. The values of tolerance to achieve the most accurate results are selected manually.

7 CONCLUSIONS

This chapter highlights different schemes to estimate the operational radial topology of distribution grids from a dense underlying loopy graph using statistics of voltage magnitudes. For the case where voltage magnitudes are available at all nodes, we show that if the variance of voltage differences is used as edge weights in the underlying graph, the operational grid is given by the minimum weight spanning tree. In fact, no additional information is necessary for the algorithm to operate. We also show two extensions of our learning framework to cases where available voltage measurements are limited to a subset of the grid nodes. For unobserved nodes separated by greater than three hops, we show that exact reconstruction is possible by verifying equality relations satisfied by the edge weights. Similarly, such equality relations enable the exact reconstruction when all voltage measurements are limited to the leaf nodes and all intermediate nodes are missing. Future directions of using voltage moment-based learning include relaxation of the assumptions used, for example, allowing unobserved nodes to be separated by less than three hops and without available historical injection statistics. Further inclusion of lossy PF models and analysis of effect of measurement noise will extend the practical usefulness of the learning algorithms proposed.

References

[1] R. Hoffman, Practical state estimation for electric distribution networks, in: IEEE PES Power Systems Conference and Exposition, IEEE, 2006, pp. 510–517.

[2] A.G. Phadke, Synchronized phasor measurements in power systems, IEEE Comput. Appl. Power 6 (2) (1993) 10–15.

[3] A. von Meier, D. Culler, A. McEachern, R. Arghandeh, Micro-synchrophasors for distribution systems, in: 2014 IEEE PES Innovative Smart Grid Technologies Conference (ISGT), 2014, pp. 1–5.

[4] D. Deka, M. Chertkov, S. Backhaus, Structure learning in power distribution networks, in: IEEE Transactions on Control of Network Systems, IEEE, 2017.

[5] V. Kekatos, G.B. Giannakis, R. Baldick, Grid topology identification using electricity prices, in: 2014 IEEE PES General Meeting | Conference & Exposition, IEEE, July, 2014, pp. 1–5.

[6] M. He, J. Zhang, A dependency graph approach for fault detection and localization towards secure smart grid, IEEE Trans. Smart Grid 2 (2) (2011) 342–351.

[7] S. Bolognani, N. Bof, D. Michelotti, R. Muraro, L. Schenato, Identification of power distribution network topology via voltage correlation analysis, in: 2013 IEEE 52nd Annual Conference on Decision and Control (CDC), IEEE, 2013, pp. 1659–1664.

[8] Y. Sharon, A.M. Annaswamy, A.L. Motto, A. Chakraborty, Topology identification in distribution network with limited measurements, in: 2012 IEEE PES Innovative Smart Grid Technologies (ISGT), IEEE, 2012, pp. 1–6.

[9] G. Cavraro, R. Arghandeh, K. Poolla, A. Von Meier, Data-driven approach for distribution network topology detection, in: 2015 IEEE Power & Energy Society General Meeting, July, 2015, pp. 1–5.

[10] J. Peppanen, J. Grimaldo, M.J. Reno, S. Grijalva, R.G. Harley, Increasing distribution system model accuracy with extensive deployment of smart meters, in: 2014 IEEE PES General Meeting Conference & Exposition, IEEE, 2014, pp. 1–5.

[11] J. Peppanen, M.J. Reno, M. Thakkar, S. Grijalva, R.G. Harley, Leveraging AMI data for distribution system model calibration and situational awareness, IEEE Trans. Smart Grid 6 (4) (2015) 2050–2059.

[12] R. Sevlian, R. Rajagopal, Feeder topology identification, 2015 (arXiv:1503.07224).

[13] D. Deka, S. Backhaus, M. Chertkov, Estimating distribution grid topologies: a graphical learning based approach, in: Power Systems Computation Conference (PSCC), IEEE, 2016, pp. 1–7.

[14] S. Talukdar, D. Deka, D. Materassi, M.V. Salapaka, Exact Topology Reconstruction of Radial Dynamical Systems with Applications to Distribution System of the Power Grid, in: American Control Conference (ACC), 2017 (accepted).

[15] M. Baran, F.F. Wu, Optimal sizing of capacitors placed on a radial distribution system, IEEE Trans. Power Delivery 4 (1) (1989) 735–743.

[16] M.E. Baran, F.F. Wu, Optimal capacitor placement on radial distribution systems, IEEE Trans. Power Delivery 4 (1) (1989) 725–734.

[17] S. Bolognani, S. Zampieri, On the existence and linear approximation of the power flow solution in power distribution networks, IEEE Trans. Power Syst. 31 (1) (2016) 163–172.

[18] D. Deka, S. Backhaus, M. Chertkov, Learning topology of the power distribution grid with and without missing data, in: 2016 European Control Conference (ECC), IEEE, 2016, pp. 313–320.

[19] D. Deka, S. Backhaus, M. Chertkov, Learning topology of distribution grids using only terminal node measurements, in: IEEE Smartgridcomm, 2016.

[20] M.J. Choi, V.Y.F. Tan, A. Anandkumar, A.S. Willsky, Learning latent tree graphical models, J. Mach. Learn. Res. 12 (2011) 1771–1812.

[21] C.K. Chow, C.N. Liu, Approximating discrete probability distributions with dependence trees, IEEE Trans. Inf. Theory 14 (3) (1968) 462–467.

[22] M.E. Baran, F.F. Wu, Network reconfiguration in distribution systems for loss reduction and load balancing, IEEE Trans. Power Delivery 4 (2) (1989) 1401–1407.

[23] A. Abur, A.G. Exposito, Power System State Estimation: Theory and Implementation, CRC Press, Boca Raton, FL, 2004.

[24] J.B. Kruskal, On the shortest spanning subtree of a graph and the traveling salesman problem, Proc. Am. Math. Soc. 7 (1) (1956) 48–50.

[25] T.H. Cormen, C.E. Leiserson, R.L. Rivest, C. Stein, Introduction to Algorithms, MIT Press, Cambridge, MA, 2001.

[26] J. Pearl, Probabilistic Reasoning in Intelligent Systems: Networks of Plausible Inference, Morgan Kaufmann, San Mateo, CA, 2014.

[27] U. Eminoglu, M.H. Hocaoglu, A new power flow method for radial distribution systems including voltage dependent load models, Electr. Power Syst. Res. 76 (1–3) (2005) 106–114.

[28] Available at http://www.dejazzer.com/reds.html.

CHAPTER 13

Grid Topology Identification via Distributed Statistical Hypothesis Testing

Saverio Bolognani
Automatic Control Laboratory ETH Zürich, Zürich, Switzerland

CHAPTER OVERVIEW

We consider the problem of automatically identifying the topology of a power distribution network, based on data measurements collected on the grid. Possible applications include the detection of changes in the operational topology, the deployment and tuning of plug-and-play volt-VAR regulators, and the implementation of active management strategies for congestion relief.

We first show, by using a first-order model of the grid, that voltage measurements exhibit some specific correlation properties, that can be described via a sparse Markov random field. By specializing the tools available for the identification of graphical models, we propose a centralized algorithm for the reconstruction of the grid topology.

We then show how it is possible to formulate the grid topology identification task as a series of distributed statistical tests that agents need to perform on their measurements. As the number of collected samples increases, agents can answer the test with increasing confidence, select the correct hypothesis (e.g., a switch being open or close), and ultimately infer the grid topology. The computational complexity of each of these tests is independent from the grid size.

The effectiveness of both the centralized and the distributed approach is tested in simulations, based on household power demand measurements from a real distribution feeder.

1 INTRODUCTION

Power distribution networks, compared to transmission systems, have been historically planned and designed according to a *fit-and-forget* approach. During operation, these grids remain mostly unmonitored, with minimal sensing and actuation, and often no communication infrastructure available to collect measurements or dispatch real-time commands.

Different challenges are now emerging in the power distribution networks, and are motivating a much deeper integration of information, communication, and control technology in this realm. One example is the large-scale penetration of microgenerators from fluctuating energy sources. Distributed power generation, especially inside the highly resistive, radial, low voltage networks, can cause local overvoltage and power line congestion issues [1]. Another example is the connection of dispatchable loads to the power distribution network (e.g., plugin electric vehicles and smart buildings). Today's power distribution

281

Big Data Application in Power Systems. https://doi.org/10.1016/B978-0-12-811968-6.00013-9

grids will face major congestion issues if proper scheduling and coordination protocols will not be enforced to these consumers [2, 3].

In the last few years, international research projects have been funded in order to develop and engineer solutions to these challenges, while maintaining grid efficiency and reliability [4–7]. Many of these solutions require that the topology of the power distribution grid is known, which is not always true. In many cases, the deployment of intelligence in the power distribution grid will consist of retro-fitting an existing infrastructure via the installation of new devices. A *plug-and-play* approach is often considered and may constitute in some cases the only viable solution. According to this approach, the devices must identify the physical system in which they operate, starting from the topology of the grid, and reconfigure the communication and control infrastructure in order to being able to perform the assigned tasks. Even after deployment, the topology of the grid may be subject to changes, via the operation of dedicated switches to achieve higher efficiency or better quality of the service. Such changes of topology need to be detected by the controllers in the grid, which in turn need to be promptly reconfigured.

In this chapter, we consider the problem of identifying the grid topology from field measurements that can be performed in the grid, and in particular from voltage magnitude measurements at the buses. The approach proposed in this paper is closely related to the methods for the identification of Markov random fields (graphical models) [8], and is based on some conditional correlation properties that characterize voltage measurements in a radial grid. These properties are reported in Section 3, where they are also used to derive a centralized topology identification algorithm. In Section 4, we show how the same reasoning can be used to design distributed tests that involve only three nodes, and return elementary bits of information regarding the topology of the grid. These tests require minimal sensing and computational resources, and return a reliable response after an extremely limited number of measurement samples. Both the correlation analysis and the distributed identification algorithms are tested in Section 5 on power measurements data obtained from a real distribution network.

1.1   Related Works

The literature on grid topology identification is mostly divided into two areas: works that consider the problem of determining the position of a limited number of grid switches, and therefore test a limited number of hypotheses, and works that aim at identifying the entire topology of a feeder. We review these two areas separately in the next two paragraphs.

The idea of using field measurements to detection of unmonitored switching of circuit breakers in the reconfiguration of the power distribution grid has been presented for example in [9], where the task has been formulated as a classification problem. More recently, correlation analysis methods have been proposed to verify whether the topology available in a geographic information system is correct [10], tackling the problem as a hypothesis testing problem. An algorithm for real-time detection of topology changes, based on PMU measurements and on the identification of patterns in the time series following a switching event, has been proposed in [11]. The Markov random field nature of voltage phasor measurements has been recognized in [12], and used as a test for fault detection.

Algorithm like the one proposed in [13], on the other hand, aim at reconstructing the entire topology, without assuming any specific library of possible configurations. The mathematical analysis presented in this chapter is adopted (and extended) from there. More recently, the same idea of constructing a minimum-cost spanning tree for the identification of the power distribution network, which is closely related to the Chow-Liu algorithm for graphical models, has been used in [14] on the mutual information matrix, yielding better performance but still requiring very large number of samples. Topology learning tools based on the trends in second-order moments of voltage measurements have been proposed in [15, 16]. Interestingly, the same authors also considered the case in which some measurements are missing, a challenging problem in graphical learning [17]. In [18], the bus connectivity and topology estimation problems are formulated as a linear regression problem with least absolute shrinkage on grouped variables (Group Lasso).

In all these works synthetic data are used rather than real measurements. A notable exception is [19], where, however, linear voltage sensitivity coefficients are computed, without enforcing sparsity or tree structure.

2 POWER DISTRIBUTION GRID MODEL

We model a power distribution grid as a radial graph $\mathcal{G}$, in which edges represent the power lines, and nodes $\mathcal{V} = \{0, \ldots, n\}$ represent the buses of the grid (including the substation, indexed as 0).

We limit the study to a grid-connected distribution feeder, and we therefore assume that the voltage at node 0 is constant (i.e., it does not depend on the power demands in the feeder) and that all other nodes $h \in \mathcal{L} := \{1, \ldots, n\}$ can be modeled as PQ buses (i.e., with an active and reactive power demand that does not depend on the bus voltage).

In this framework, the steady state of the grid is described by the following nodal quantities, for each node $h \in \mathcal{V}$:

- complex voltage $u_h = v_h e^{j\varphi_h}$
- complex power injection $s_h = p_h + jq_h$
- complex current injection i_h

Based on this notation, we write the nonlinear power flow equations of the grid as $n + 1$ complex-valued equations of the form

$$v_h e^{j\varphi_h} \sum_{k \in \mathcal{V}} \overline{y}_{hk} v_k e^{-j\varphi_k} = p_h + jq_h, \quad \forall h \in \mathcal{V} \tag{1}$$

where $\overline{y}_{hk}$ is the complex conjugate of the admittance of the line connecting h to k, and $y_{hh} = -\sum_{k \neq h} y_{hk}$ (assuming negligible shunt admittances at the buses).

For the subsequent analysis, it is convenient to introduce the following vectorial notation. We denote by v, s, p, q the vectors of dimension n, obtained by stacking the scalar quantities v_h, s_h, p_h, q_h, respectively, for all $h \in \mathcal{L}$. Similarly, we partition the bus admittance matrix Y, whose elements are the scalars y_{hk}, as

$$Y = \begin{bmatrix} Y_{00} & Y_{0\mathcal{L}} \\ Y_{\mathcal{L}0} & Y_{\mathcal{L}\mathcal{L}} \end{bmatrix}$$

where $Y_{\mathcal{L}\mathcal{L}}$ has dimensions $n \times n$.

For the subsequent analysis, we consider the linearization of the nonlinear power flow equations (1) around the flat voltage profile ($v_h = v_0$, $\forall h \in \mathcal{V}$), corresponding to the *linear coupled power flow model*. We refer to [20, Section V] for the derivation of the model, and to [21] for a geometric interpretation of this linearization. Based on this approximation, voltage magnitudes at the buses in $\mathcal{L}$ can be expressed as

$$v \approx \mathbf{1} v_0 + \frac{1}{v_0} \operatorname{Re}(Z\overline{s}) \tag{2}$$

where the bus impedance matrix $Z \in \mathbb{C}^{n \times n}$ is defined as $Z = Y_{\mathcal{L}\mathcal{L}}^{-1}$.

We recall that the elements of the bus impedance matrix have the following well-known interpretation in the case of radial networks.

Definition 1 (Shortest Electric Path). Given two buses $h, k \in \mathcal{V}$, we define the electric path $\mathcal{P}_{hk}$ as the smallest connected subset of nodes of $\mathcal{V}$ such that $h, k \in \mathcal{P}_{hk}$.

Lemma 1. *Let* h, k *be two buses in* $\mathcal{V}$, *and let* $\mathcal{P}_{0h}$ *and* $\mathcal{P}_{0k}$ *be the shortest electric paths that connect them to node 0. Then* Z_{hk} *is the sum of the impedances of the edges connecting the nodes in the intersection* $\mathcal{P}_{0h} \cap \mathcal{P}_{0k}$.

Without loss of generality, in the following, we assume $v_0 = 1$. We also consider the approximation error in Eq. (2) negligible, and therefore adopt the grid model

$$v = 1 + Rp + Xq \tag{3}$$

where $R = \operatorname{Re}(Z)$ and $X = \operatorname{Im}(Z)$ are the reduced bus resistance and reactance matrices, respectively.

3 VOLTAGE CORRELATION ANALYSIS

In this section, we review and extend the voltage correlation analysis proposed for the first time in [13], as it lays the groundwork for the derivation of the proposed distributed identification strategy.

Based on Eq. (3), the covariance matrix of the bus voltages can be directly expressed as

$$\mathrm{cov}(v) = \mathbb{E}(v - \mathbb{E}v)(v - \mathbb{E}v)^T$$

$$= \mathbb{E}(R(p - \mathbb{E}p) + X(q - \mathbb{E}q))(R(p - \mathbb{E}p) + X(q - \mathbb{E}q))^T \tag{4}$$

$$= R\Sigma_{pp}R + X\Sigma_{qp}R + R\Sigma_{pq}X + X\Sigma_{qq}X$$

where

$$\begin{bmatrix} \Sigma_{pp} & \Sigma_{pq} \\ \Sigma_{qp} & \Sigma_{qq} \end{bmatrix}$$

is the positive definite covariance matrix of the bus power injection vector $[p^T q^T]^T$.

The covariance matrix (4) clearly contains information regarding the topology of the grid, which is encoded in the matrices R and X. For this information to be reconstructable, we need the following assumption.

Assumption 1 (Uncorrelated Power Demands). Active and reactive power injections at different buses are mutually uncorrelated. Therefore Σ_{pp}, Σ_{pq}, Σ_{qp}, and Σ_{qq} are all diagonal matrices.

Assumption 1 is a critical step in the derivation of correlation-based topology identification methods, and has been adopted throughout the literature reviewed in Section 1. In Section 5 we verify whether this assumption holds on the power measurements from a real distribution feeder, and we discuss how it depends on the time scale under consideration.

We then introduce two other assumptions that will be later employed in the analysis.

Assumption 2 (Uniform X/R Ratio). All power lines in the distribution grid have the same inductance/resistance ratio, that is

$$\gamma_{hk} = e^{j\theta}|\gamma_{hk}|, \quad \forall h, k \in \mathcal{V}$$

where θ is fixed across the network.

Assumption 2 is satisfied when the grid is relatively homogeneous, and is reasonable in most practical cases, including the IEEE test feeder considered in the numerical experiments of Section 5.

Assumption 3 (Uniform Power Factor). All loads in the distribution feeder have the same power factor, that is

$$q_h = \kappa p_h, \quad \forall h \in \mathcal{L}$$

where κ is fixed across the network.

Notice that Assumption 3 is trivially verified if the loads are perfectly compensated, in which case $\kappa = 0$ and thus $q_h = 0$ for all $h \in \mathcal{L}$.

We can therefore state the following result, which shows how it is possible to compute the voltage covariance matrix a completely equivalent purely resistive grid.

Lemma 2 (Equivalent Resistive Grid). *Let Assumption 1 hold, together with either Assumption 2 or 3. Then there exists a purely resistive network, with the same topology of $\mathcal{G}$, bus conductance matrix G, and uncorrelated active power injections with covariance matrix Σ, which yields the same voltage covariance matrix cov(v) as the original grid.*

Moreover, the voltage covariance matrix can be explicitly expressed as

$$\mathrm{cov}(v) = M\Sigma M$$

where $M = G_{\mathcal{L}\mathcal{L}}^{-1}$.

Proof. Let us first consider the case in which Assumption 2 holds. Then $Y = e^{j\theta}Y'$, where $Y' \in \mathbb{R}^{(n+1)\times(n+1)}$. It follows that $R = \mathrm{Re}\left(Y_{\mathcal{L}\mathcal{L}}^{-1}\right) = \cos\theta\left(Y_{\mathcal{L}\mathcal{L}}'\right)^{-1}$ and $X = \mathrm{Im}\left(Y_{\mathcal{L}\mathcal{L}}^{-1}\right) = -\sin\theta\left(Y_{\mathcal{L}\mathcal{L}}'\right)^{-1}$. Therefore $X = -\tan\theta R$, and expression (4) can be rewritten as

$$\mathrm{cov}(v) = R\left(\Sigma_{pp} - 2\tan\theta\Sigma_{qp} + \tan^2\theta\Sigma_{qq}\right)R$$

The statement of the lemma is therefore verified by defining the positive definite matrix

$$\Sigma = \Sigma_{pp} - 2\tan\theta\Sigma_{qp} + \tan^2\theta\Sigma_{qq} = \begin{bmatrix} I & -\tan\theta I \end{bmatrix} \begin{bmatrix} \Sigma_{pp} & \Sigma_{pq} \\ \Sigma_{qp} & \Sigma_{qq} \end{bmatrix} \begin{bmatrix} I \\ -\tan\theta I \end{bmatrix}$$

and by considering the graph Laplacian $G = \mathrm{Re}(Y)$.

Let us now consider the case in which Assumption 3 holds. Then $\Sigma_{pq} = \Sigma_{qp} = \kappa\Sigma_{pp}$ and $\Sigma_{qq} = \kappa^2\Sigma_{pp}$. Expression (4) can be manipulated to obtain

$$\mathrm{cov}(v) = (R + \kappa X)\Sigma_{pp}(R + \kappa X)$$

The statement of the lemma is therefore verified by taking $\Sigma = \Sigma_{pp}$ (which is positive definite) and by considering the graph Laplacian G obtained by weighting each edge hk as $\mathrm{Re}(y_{hk}) - \kappa\mathrm{Im}(y_{hk})$, where y_{hk} is the admittance of the corresponding power line. $\square$

Lemma 2 is instrumental to prove the following result, which shows how the inverse of the covariance matrix cov(v) is sparse and has a specific sign pattern.

Theorem 1. *Let Assumption 1 hold, together with either Assumption 2 or 3. Let* $K = cov(v)^{-1}$. *Then for any pair* h, k *in* $\mathcal{L}$ *we have that*

$$
K_{hk} \begin{cases} > 0 & \text{if } h = k \\ < 0 & \text{if } h \sim k \\ > 0 & \text{if } \exists \ell \in \mathcal{L} \text{ such that } h \sim \ell \text{ and } \ell \sim k \\ 0 & \text{otherwise} \end{cases}
$$

where the $\sim$ *sign indicates neighbors in the electric topology (i.e., there exists an edge of the graph connecting them).*

Proof. Based on Lemma 2, the matrix K can be explicitly expressed as

$$
K = cov(v)^{-1} = (M\Sigma M)^{-1} = G_{\mathcal{L}\mathcal{L}} \Sigma^{-1} G_{\mathcal{L}\mathcal{L}}
$$

for some positively weighted Laplacian L and some positive definite diagonal matrix Σ^{-1}. We introduce the notation

$$
\Sigma^\dagger = \begin{bmatrix} 0 & 0 \\ 0 & \Sigma^{-1} \end{bmatrix} \in \mathbb{R}^{(n+1)\times(n+1)}
$$

By defining $\mathcal{N}(h)$ as the set of neighbors of node h, and by using the fact that G is a positively weighted Laplacian and therefore

$$
G_{hk} \begin{cases} > 0 & \text{if } h = k \\ < 0 & \text{if } h \sim k \\ 0 & \text{otherwise} \end{cases}
$$

we have that

$$
\mathbf{1}_h^T G \Sigma^\dagger G \mathbf{1}_k = \left(G_{hh} \mathbf{1}_h^T + \sum_{h' \in \mathcal{N}(h)} G_{hh'} \mathbf{1}_{h'}^T \right) \Sigma^\dagger \left(\mathbf{1}_k G_{kk} + \sum_{k' \in \mathcal{N}(k)} \mathbf{1}_{k'} G_{k'k} \right)
$$

Now, using the fact that

$$
\mathbf{1}_v^T \Sigma^\dagger \mathbf{1}_w = \begin{cases} (\Sigma_{vv})^{-1} > 0 & \text{if } v = w \neq 0 \\ 0 & \text{otherwise} \end{cases}
$$

we have that, in $\mathcal{L}$,

- for all h,

$$
K_{hh} = \mathbf{1}_h^T G \Sigma^\dagger G \mathbf{1}_h = G_{hh} (\Sigma_{hh})^{-1} G_{hh} + \sum_{\ell \in \mathcal{N}(h)} G_{h\ell} (\Sigma_{\ell\ell})^{-1} G_{\ell h} > 0;
$$

- if $h \sim k$ then

$$K_{hk} = 1_h^T G \Sigma^\dagger G 1_k = G_{hh} (\Sigma_{hh})^{-1} G_{hk} + G_{hk} (\Sigma_{kk})^{-1} G_{kk} < 0;$$

- if $h \neq k$ and $\exists \ell$ such that $\ell \sim h$ and $\ell \sim k$, then

$$K_{hk} = 1_h^T G \Sigma^\dagger G 1_k = G_{h\ell} (\Sigma_{\ell\ell})^{-1} G_{\ell k} > 0;$$

- $K_{hk} = 1_h^T G \Sigma^\dagger G 1_k = 0$ otherwise.   □

Remark 1. The matrix K is known as *concentration matrix*, and has an interesting and well-known interpretation in terms of conditional correlation. In particular, $K_{hk} = 0$ if and only if v_h and v_k are conditionally uncorrelated given all other voltages v_ℓ, $\ell \neq h, k$. According to Theorem 1, the sparsity pattern of K is the same of $Y_{\mathcal{L}\mathcal{L}}^2$. In the terminology of Markov random fields [8], this means that the corresponding graphical model is an undirected graph in which nodes are connected by an edge (and therefore are conditionally correlated) if they are 1-hop or 2-hop neighbors in the graph describing the power distribution grid topology.

Theorem 1 also shows that the strictly negative elements of K have the sparsity pattern of $Y_{\mathcal{L}\mathcal{L}}$ and can therefore be directly used to reconstruct the electrical topology of the grid.

Based on these observations, we propose the following steps for the identification of the grid topology, given a sequence of T voltage measurements at all buses of the grid, $v^{(t)}$, $t = 1, ..., T$.

1. Compute the sample covariance matrix $\hat{\Sigma} = \text{cov}(v^{(t)}, t = 1, ..., T)$.
2. Compute the sample concentration matrix $\hat{K}$ as $\hat{\Sigma}^{-1}$.
3. Consider the complete graph $\mathcal{C}$ defined on the nodes $\mathcal{L}$, with edge weights corresponding to the elements of $\hat{K}$. Compute the *minimum spanning tree* on $\mathcal{C}$, that is, the subgraph of $\mathcal{C}$ that is a tree, connects all the nodes, and whose total edge cost is less than or equal to any other spanning tree.

Notice that the minimum spanning tree can be computed in polynomial time by greedy algorithms like the Prim's algorithm [22].

The proposed algorithm resembles, in some sense, the well-known Chow-Liu algorithm [23] for graphical model identification, in which, however, the choice of the best spanning tree is motivated by the search for the closest approximation of the actual distribution in an information-theoretic sense. In our scenario, on the other hand, we know in advance that there exists a tree which is the *root* (in the graph-theory sense) of the graph that describes the actual distribution (i.e., the actual graph connects nodes that can be reached in one or two hops in such tree), and we make explicit use of this additional information. Because of this a priori knowledge, we also do not need the tools that have been developed for model selection [24] (i.e., to tune the sparsity of the estimated graph).

In Section 5 we illustrate the voltage correlation analysis presented in this section, by considering a dataset of real power demand measurements and by implementing the proposed identification algorithm. We show how this algorithm typically requires a large number of samples, and how its performance deteriorates when the assumptions are not verified.

4 A DISTRIBUTED TOPOLOGY TEST

As reviewed in Section 1, different approaches based on similar statistical analysis of voltage measurements have been proposed, improving the topology detection rate and in some cases reducing the number of samples needed [14–16]. These works, however, consider the same centralized scenario introduced in Section 3. Voltages at all nodes are supposed to be measured (with the possible exception of few nodes, as suggested in [17]), making the implementation of these schemes impractical in poorly monitored networks.

In this section, the same correlation-based approach will be employed to derive a set of distributed topology tests, where small clusters of three buses will have to communicate and share their voltage measurements in order to provide elementary bits of information regarding the topology of the grid. This approach better suits the typical needs of distribution network operators: it requires a minimal amount of sensing (three bus voltage sensors) and it allows to discern simple (but relevant) hypotheses on the grid topology, such as

- whether a switch is open or closed;
- which of the three sensor lies closest to the substation, in an electrical sense; and
- what is the relative position of a newly connected sensor with respect to other sensors.

In order to present the details of this approach, we introduce the following definitions, which are also illustrated in Fig. 1 and apply only to radial graphs.

Definition 2 (Triad). A set of three buses $\mathcal{T} \subset \mathcal{L}$ is a triad, if one of the three buses belongs to the shortest electric path that connects the other two.

Definition 3 (Minimal Interleaved Graph). A graph is a minimal interleaved graph connecting a triad $\mathcal{T}$ if it is obtained by interleaving $\mathcal{T}$ with another triad $\mathcal{T}' = \{0, v, w\}$, as shown in Fig. 1.

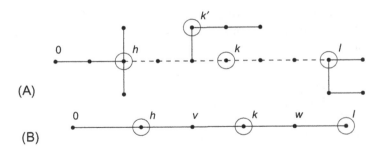

FIG. 1
Schematic representation of the definitions introduces in Section 4. (A) The subset of buses $\mathcal{T} = \{h, k, \ell\}$ is a *triad* because k belongs to the *shortest-electric path* $\mathcal{P}_{h\ell}$ (the *thick-dashed path*). The subset $\{h, k', \ell\}$ is not a triad. (B) The *minimal interleaved graph* that connects the triad $\mathcal{T}$.

Definition 4 (Node Depth). Given a bus $h \in \mathcal{V}$, and a weighted Laplacian G, we define the node depth x_h as the sum of the weights of the edges that connect the nodes in the shortest electric path $\mathcal{P}_{0h}$.

Based on these definitions, we can state the following result, which shows that the voltage covariance at the buses of a triad is identical to the voltage covariance of a triad in a much smaller purely resistive grid, with properly chosen line parameters and active power injection covariance.

Lemma 3 (Minimal Equivalent Resistive Grid). *Let Assumption 1 hold, together with either Assumption 2 or 3. Consider a triad $\mathcal{T} = \{h, k, \ell\}$. Then there exists a minimal equivalent resistive grid, whose graph is a minimal interleaved graph connecting $\mathcal{T}$, with conductance matrix $\tilde{G}$ and diagonal power covariance $\tilde{\Sigma}$, which yields the same covariance $\mathrm{cov}(v_\mathcal{T})$ of the voltages at the nodes h, k, ℓ.*

Proof. The proof is constructive, and follows these steps, which are also shown in Fig. 2.

- Based on Lemma 2, we consider the equivalent resistive grid with positively weighted Laplacian G and power covariance Σ (Step (A)).
- Without loss of generality, we lump all the power demands in each *lateral* (i.e., branches that do not belong to the path $\mathcal{P}_{0\ell}$) to the node where the lateral connects (Step (B)).
- We denote by $\mathcal{L}_1$, $\mathcal{L}_2$, and $\mathcal{L}_3$, the three subsets of nodes indicated in Step (B).
- We consider the minimal interleaved graph in Step (C), with positively weighted Laplacian $\tilde{G}$ such that nodes h, k, ℓ have the same depths x_h, x_k, x_ℓ as in the original graph with Laplacian G, while nodes v and w have depths

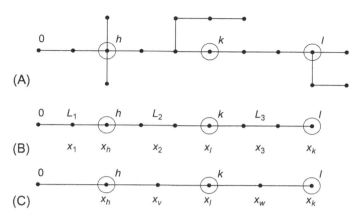

FIG. 2
Schematic representation of the steps of the constructive proof of Lemma 3.

$$x_v = \frac{\sum_{i \in \mathcal{L}_2} x_i \sigma_i (x_i - x_h)}{\sum_{i \in \mathcal{L}_2} \sigma_i (x_i - x_h)}$$

$$x_w = \frac{\sum_{i \in \mathcal{L}_3} x_i \sigma_i (x_i - x_k)}{\sum_{i \in \mathcal{L}_3} \sigma_i (x_i - x_k)}$$

where σ_i is the diagonal element of Σ corresponding to node i.

- We construct the diagonal power covariance matrix $\widetilde{\Sigma}$ with diagonal elements

$$\widetilde{\sigma}_v = \frac{\sum_{i \in \mathcal{L}_2} \sigma_i (x_i - x_h)}{x_v - x_h}$$

$$\widetilde{\sigma}_w = \frac{\sum_{i \in \mathcal{L}_3} \sigma_i (x_i - x_k)}{x_w - x_k}$$

$$\widetilde{\sigma}_h = \sigma_h + \frac{\sum_{i \in \mathcal{L}_1} x_i^2 \sigma_i}{x_h^2} + \sum_{i \in \mathcal{L}_2} \sigma_i - \widetilde{\sigma}_v$$

$$\widetilde{\sigma}_k = \sigma_k + \sum_{i \in \mathcal{L}_3} \sigma_i - \widetilde{\sigma}_w$$

$$\widetilde{\sigma}_\ell = \sigma_\ell$$

The voltage covariance matrix for the minimal equivalent resistive grid can be computed as $\widetilde{M}\widetilde{\Sigma}\widetilde{M}$. Using the fact that, for line graphs, $M_{ij} = \min\{x_i, x_k\}$ (which follows directly from Lemma 1), via lengthy but otherwise standard computations, it is possible to show that the covariance of the triad voltages v_T is the same in the original equivalent resistive grid and in the minimal equivalent resistive grid. Via Lemma 2, the same covariance matrix is also equal to the voltage covariance $\text{cov}(v_T)$ in the original grid. □

The purpose of the equivalence introduced by Lemma 3 is to allow a much simpler study of the voltage covariance matrix $\text{cov}(v_T)$. In fact, for a minimal interleaved graph, it is possible to explicitly write $\text{cov}(v_T)$ as a function of the parameters $\widetilde{G}$ and $\widetilde{\Sigma}$. This fact is used to obtain the following result.

Theorem 2. *Let Assumption 1 hold, together with either Assumption 2 or 3.*

Consider a triad $T = \{h, k, \ell\}$, in which $k \in \mathcal{P}_{h\ell}$, and let $\text{cov}(v_T)$ be the covariance of the voltages at the buses h, k, ℓ. *Then the matrix $K = [\text{cov}(v_T)]^{-1}$ has sign pattern*

$$\text{sign}(K) = \begin{bmatrix} +1 & -1 & +1 \\ -1 & +1 & -1 \\ +1 & -1 & +1 \end{bmatrix}$$

Proof. We first construct, via Lemma 2, an equivalent resistive network. Then, via Lemma 3, we consider a minimal equivalent resistive network which is guaranteed to yield the same voltage covariance matrix at the nodes of the triad.

The voltage covariance matrix $\text{cov}(v_T)$ can be then obtained by selecting the rows and columns of the full voltage covariance matrix $\text{cov}(v) = \widetilde{M}\widetilde{\Sigma}\widetilde{M}$. Using Lemma 1, we can write

$$\text{cov}(v_T) = N \begin{bmatrix} \tilde{\sigma}_h & & & & \\ & \tilde{\sigma}_v & & & \\ & & \tilde{\sigma}_k & & \\ & & & \tilde{\sigma}_w & \\ & & & & \tilde{\sigma}_\ell \end{bmatrix} N^T \text{ where } N = \begin{bmatrix} \tilde{x}_h & \tilde{x}_h & \tilde{x}_h & \tilde{x}_h & \tilde{x}_h \\ \tilde{x}_h & \tilde{x}_v & \tilde{x}_k & \tilde{x}_k & \tilde{x}_h \\ \tilde{x}_h & \tilde{x}_v & \tilde{x}_k & \tilde{x}_w & \tilde{x}_\ell \end{bmatrix}$$

As the determinant of K is positive, the sign pattern of K is the same sign pattern of $\text{adjoint}(\text{cov}(v_T))$. The matrix $\text{adjoint}(\text{cov}(v_T))$ can be evaluated via standard symbolic math software. It is convenient to operate a change of variable, introducing the positive depth differences $\delta_{h0} = \tilde{x}_h$, $\delta_{vh} = \tilde{x}_v - \tilde{x}_h$, $\delta_{kv} = \tilde{x}_k - \tilde{x}_v$, $\delta_{wk} = \tilde{x}_w - \tilde{x}_k$, and $\delta_{\ell w} = \tilde{x}_\ell - \tilde{x}_w$. The elements of $\text{adjoint}(\text{cov}(v_T))$ are fourth-order polynomials in these quantities. For example, we have that

$$[\text{adjoint}(\text{cov}(v_T))]_{\ell\ell} = \tilde{\sigma}_h(\tilde{\sigma}_k + \tilde{\sigma}_\ell + \tilde{\sigma}_v + \tilde{\sigma}_w)\delta_{h0}^2\delta_{vh}^2$$
$$+ 2\tilde{\sigma}_h(\tilde{\sigma}_k + \tilde{\sigma}_\ell + \tilde{\sigma}_w)\delta_{h0}^2\delta_{vh}\delta_{kv} + (\tilde{\sigma}_h + \tilde{\sigma}_v)(\tilde{\sigma}_k + \tilde{\sigma}_\ell + \tilde{\sigma}_w)\delta_{h0}^2\delta_{kv}^2$$

We do not report them here for space reasons. The sign of all the elements of $\text{adjoint}(\text{cov}(v_T))$ is then apparent by inspection, and the matrix K exhibits the sign pattern

$$\text{sign}(K) = \begin{bmatrix} +1 & -1 & +1 \\ -1 & +1 & -1 \\ +1 & -1 & +1 \end{bmatrix}$$

□

Based on Theorem 2, we can then propose the following steps as a distributed hypothesis test. Given a triad $\mathcal{T}$, and a sequence of T voltage measurements at the three buses in $\mathcal{T}$, $v_{\mathcal{T}}^{(t)}$, $t = 1, ..., T$, this test determines which node of the triad belongs to the shortest-electric path that connects the other two nodes (in other words, the "order" of the nodes in the triad). It consists of the following steps.

1. Compute the 3×3 sample covariance matrix $\hat{\Sigma}_{\mathcal{T}} = \text{cov}(v_{\mathcal{T}}^{(t)}, t = 1, ..., T)$.
2. Compute the 3×3 sample concentration matrix $\hat{K}$ as $\hat{\Sigma}_{\mathcal{T}}^{-1}$.
3. Consider the complete graph $\mathcal{C}$ defined on the nodes $\mathcal{T}$, with edge weights corresponding to the elements of $\hat{K}$. Compute the *minimum spanning tree* on $\mathcal{C}$, i.e., the line subgraph of $\mathcal{C}$ that connects the three nodes, and whose total edge cost is less than or equal to any other spanning tree.

The advantage of this approach is that the test has to identify the most likely topology among only three possible topologies of the triad, resulting in a much smaller search space. For example, in the scenario represented in Fig. 3, a distribution grid operator may be interested in knowing whether switch S1 or S2 is closed. By selecting the triad $\{h, k, \ell\}$, this binary question is cast into the form of a hypothesis test on the relative topology of the nodes in the triad: if $k \in \mathcal{P}_{h\ell}$, then S1 must be open while S2 must be closed; on the other hand, if $h \in \mathcal{P}_{k\ell}$, then S2 must be open, and S1 must be closed. We show in Section 5 that very few samples are needed in order to identify the correct hypothesis in similar configurations.

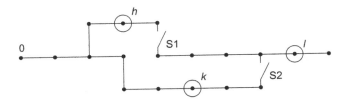

FIG. 3
An example of how the problem of determining the position of a pair of grid switches can be cast into the problem of determining the relative position of the nodes of a triad $\{h, k, \ell\}$.

5   NUMERICAL EXPERIMENTS

In this section, we use a dataset of power demand measurements from a real power distribution grid to validate both the correlation analysis presented in Section 3 and the distributed algorithm proposed in Section 4.

The dataset is provided as part of the DiSC simulation framework [25], and has been obtained as anonymized data from the Danish DSO NRGi. It represents the power consumption of about 1200 individual households from the area around the Danish city Horsens. Each profile has a temporal resolution of 15 min, and spans more than a year. In order to recreate the sub-15 min variability of loads, we superimposed power demand fluctuations obtained via a simple generative model similar to the one proposed in [26].

As a test feeder, we adopted the test feeder proposed in [20], available as an online repository [27], and consisting in the three-phase backbone of the standard IEEE 123 distribution test feeder [28]. At each bus of the feeder, we considered an aggregation of power demand profiles proportional to the nominal power demand of the bus. Some examples of power profiles are plotted in Fig. 4.

The entire analysis presented in this chapter, and similar approaches proposed in the literature, requires that Assumption 1 is verified, that is, that power demands at the different buses are uncorrelated. Fig. 5 shows the covariance matrix computed on the bus active power injections, and shows how correlation is in fact present. This fact was also observed in [25], and is mostly due to the fact that different households follow similar hourly patterns, and are exposed to the same weather conditions. On the other hand, the right panel of Fig. 5 shows that the mutual correlation vanishes at high frequencies (shorter

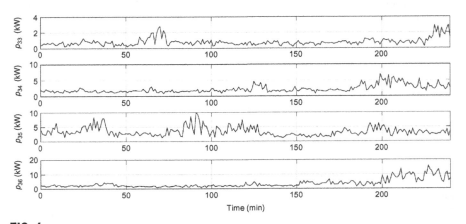

FIG. 4

Example of demand profiles for four buses of the test case. Each bus demand is obtained as the aggregation of multiple real data measurements from individual buildings.

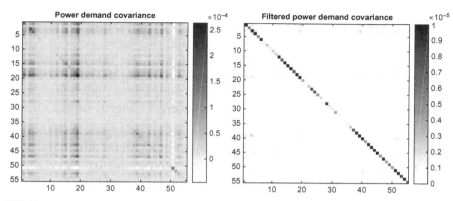

FIG. 5

Sample covariance matrix obtained from a 12-h dataset of real data. The *left panel* represents the covariance matrix obtained from the raw data, and exhibits relevant interbus correlation. The *right panel* shows that the covariance matrix obtained after a high-pass filter has been applied (stop-band frequency: $\frac{1}{8 \text{ min}}$, pass-band frequency $\frac{1}{0.8 \text{ min}}$). Above this cut-out frequency, power demands are practically uncorrelated.

time scales). Once measurement is preprocessed through a high-pass filter, it is therefore reasonable to assume uncorrelation. This recommendation is clearly valid not only for the approach proposed in this chapter, but for the other similar approaches reviewed in the literature as well.

In Fig. 6, we compute the concentration matrix K in order to verify the sparsity pattern predicted in Theorem 1, and thus apply the centralized identification

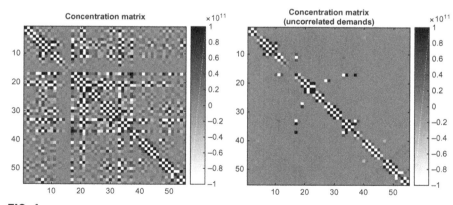

FIG. 6

The *left panel* represents the sample concentration matrix obtained from the real data measurements. The concentration matrix should exhibit the same sparsity pattern of the squared Laplacian. Spurious elements are mainly caused by the correlation between power demands, as shown in the *right panel*, where the same matrix is computed based on synthetic uncorrelated power demands, and the sparsity patter is practically exact.

algorithm proposed in Section 3. The left panel shows how the concentration matrix presents many spurious elements, due to nonlinearity, nonuniform X/R ratio of the lines, nonuniform P/Q ratio, and correlation between power demands. Further numerical experiments confirmed that this latter cause is predominant. The right panel, obtained by performing the same analysis, on the same test grid, but with synthetic uncorrelated power demands, shows that the sign patter predicted in Theorem 1 emerges correctly in this case.

Interestingly, the identification algorithm proposed in Section 3, based on the construction of a minimum spanning tree, is quite robust also in the presence of spurious elements in the sample concentration matrix. Fig. 7 shows the result of a typical execution of the algorithm (which in general depends

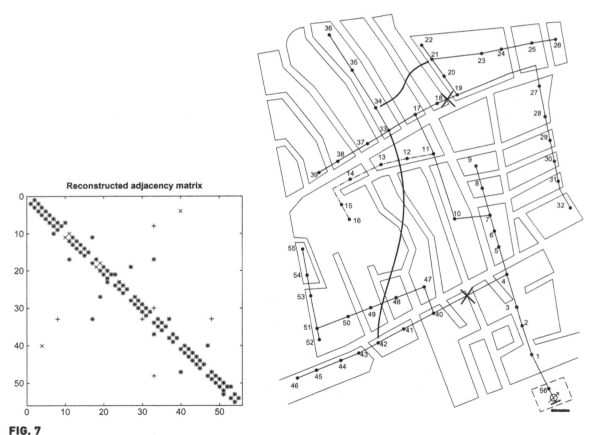

FIG. 7

Typical results of the application of the approach presented in Section 3 to the modified IEEE 123 test case, with real power measurement data. The *left panel* shows how the true Laplacian (×) is correctly identified (+) except for two misidentified edges (*highlighted* also in the map).

on the specific samples that are measured). The grid topology is reconstructed almost correctly, with the exception of two edges, highlighted in the right panel.

Finally, we consider the distributed statistical hypothesis testing proposed in Section 4. We consider the triad $\mathcal{T} = \{6, 7, 10\}$. Fig. 8 shows the concentration matrices K, computed for increasing number of samples and different sets of measurements. On the matrices, we overlay (as black dots) the result of the identification algorithm proposed in Section 4. With as few as 30 samples, the sign pattern predicted by Theorem 2 emerges correctly, and allows the correct reconstruction of the relative position of the three nodes. As the number of samples increases, the concentration matrices become identical.

To illustrate the performance of this distributed approach, in Fig. 9 we considered six different triads, and for each one of them we plotted the error rate in the reconstruction of the relative position of the three nodes, for increasing number of samples. For some triads, the error rate decreases extremely fast, and very few samples are needed in order to successfully identify the topology of the triad. In few other cases, slightly more samples are needed.

A natural question, which we have not addressed in this chapter, is about the optimal placement of the sensors (i.e., selection of the triad) in order to

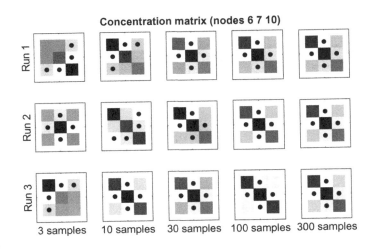

FIG. 8

Sample concentration matrix for the voltage measurements collected at three nodes (6, 7, 10). Each *column* corresponds to a different number of samples. Each *row* corresponds to a different dataset (corresponding to different starting times during the day). The *black dots* correspond to the topology identified via the minimum-spanning-tree algorithm. As the number of samples increases, the sample concentration matrix converges to its true value (allowing the identification of the correct topology). For very small sample sets, different realization can yield different results.

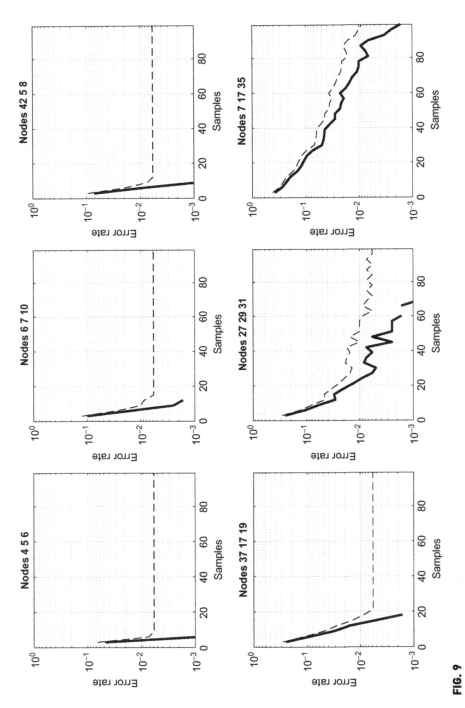

FIG. 9

Error rate of the proposed hypothesis-testing algorithm for the identification of the relative connection of different triads of nodes. The *thick line* represents the unbiased estimate of the error rate computed on 1200 realizations. The *dashed line* is the 95% confidence interval computed via the Wilson score interval [29], and settles at 0.58% when no errors are observed in the 1200 realizations.

maximize the efficiency of the algorithm in selecting the right hypotheses between those available (e.g., which bus voltage to measure in order to determine the position of the switches in Fig. 3).

6 CONCLUSIONS

In this chapter we presented an analysis of the correlation of voltage magnitude measurements in a radial distribution feeder, showing how, under some assumptions, such correlation encodes information about the topology of the grid.

An immediate application of this result consists in a centralized algorithm for the identification of the topology of the grid. The applicability of this algorithm, and of many other similar algorithms in the literature, is limited by the fact that all voltage buses need to be measured, and many samples are required in order to converge to the correct topology.

In most cases, the identification of the full topology is not even needed. Most topology identification problems can be cast in the form of an hypothesis test: which network switch is closed, which node is closer to the substation, etc. In order to address this need, a distributed statistical test has been proposed. Based on the voltage measurements of a set of only three nodes, it is possible to conclude, after very few samples, on the relative position and interconnection of these nodes. The resulting algorithm is very robust with respect to weakly correlated power demands, and extremely lightweight to implement.

The optimal placement of these triads of sensors, given a specific topology hypothesis to test, is an open problem with relevant practical implications.

References

[1] Q. Zhou, J.W. Bialek, Generation curtailment to manage voltage constraints in distribution networks, IET Gener. Transm. Distrib. 1 (3) (2007) 492–498.

[2] K. Clement-Nyns, E. Haesen, J.L.J. Driesen, The impact of charging plug-in hybrid electric vehicles on a residential distribution grid, IEEE Trans. Power Syst. 25 (1) (2010) 371–380.

[3] J.A.P. Lopes, F.J. Soares, P.M.R. Almeida, Integration of electric vehicles in the electric power system, Proc. IEEE 99 (1) (2011) 168–183.

[4] ADDRESS, Active distribution networks with full integration of demand and distributed energy RESourceS (EU FP7-ENERGY project), (2008). http://www.addressfp7.org. (accessed 8.10.17).

[5] DREAM, Distributed renewable resources exploitation in electric grids through advanced heterarchical management (EU FP7-ENERGY project), (2013). http://www.dream-smartgrid.eu. (accessed 8.10.17).

[6] EvolvDSO, Development of methodologies and tools for new and evolving DSO roles for efficient DRES integration in distribution networks (EU FP7-ENERGY project), (2013). http://www.evolvdso.eu. (accessed 10.01.17).

[7] PlanGridEV, Distribution grid planning and operational principles for EV mass roll-out while enabling DER integration (EU FP7-ENERGY project), 2013, http://www.plangridev.eu. (accessed 8.10.17).

[8] M.J. Wainwright, M.I. Jordan, Graphical models, exponential families, and variational inference, Found. Trends Mach. Learn. 1 (1–2) (2008) 1–305.

[9] Y. Sharon, A.M. Annaswamy, A.L. Motto, A. Chakraborty, Topology identification in distribution network with limited measurements, in: IEEE Innovative Smart Grid Tech. Conf. (ISGT), 2012.

[10] W. Luan, J. Peng, M. Maras, J. Lo, B. Harapnuk, Smart meter data analytics for distribution network connectivity verification, IEEE Trans. Smart Grid 6 (4) (2015) 1964–1971.

[11] G. Cavraro, R. Arghandeh, K. Poolla, A. von Meier, Data-driven approach for distribution network topology detection, in: Proc. IEEE PES General Meeting, 2015.

[12] M. He, J. Zhang, A dependency graph approach for fault detection and localization towards secure smart grid, IEEE Trans. Smart Grid 2 (2) (2011) 342–351.

[13] S. Bolognani, N. Bof, D. Michelotti, R. Muraro, L. Schenato, Identification of power distribution network topology via voltage correlation analysis. in: Proc. IEEE 52nd Annual Conference on Decision and Control (CDC), 2013. https://doi.org/10.1109/CDC.2013.6760120.

[14] Y. Weng, Y. Liao, R. Rajagopal, Distributed energy resources topology identification via graphical modeling, IEEE Trans. Power Syst. 2016, https://doi.org/10.1109/TPWRS.2016.2628876.

[15] D. Deka, S. Backhaus, M. Chertkov, Structure learning and statistical estimation in distribution networks—part I, 2015 (arXiv:1501.04131v2 [math.OC]).

[16] D. Deka, S. Backhaus, M. Chertkov, Estimating distribution grid topologies: a graphical learning based approach, in: Proc. Power Systems Computation Conference (PSCC), 2016, https://doi.org/10.1109/PSCC.2016.7541005.

[17] D. Deka, S. Backhaus, M. Chertkov, Learning topology of distribution grids using only terminal node measurements, in: Proc. IEEE International Conference on Smart Grid Communications (SmartGridComm), 2016, https://doi.org/10.1109/SmartGridComm.2016.7778762.

[18] Y. Liao, Y. Weng, R. Rajagopal, Urban distribution grid topology reconstruction via Lasso. in: Proc. IEEE PES General Meeting, 2016, https://doi.org/10.1109/PESGM.2016.7741545.

[19] S. Weckx, R. D'Hulst, J. Driesen, Voltage sensitivity analysis of a laboratory distribution grid with incomplete data, IEEE Trans. Smart Grid 6 (3) (2015) 1271–1280, https://doi.org/10.1109/TSG.2014.2380642.

[20] S. Bolognani, S. Zampieri, On the existence and linear approximation of the power flow solution in power distribution networks, IEEE Trans. Power Syst. 31 (1) (2016) 163–172.

[21] S. Bolognani, F. Dörfler, Fast power system analysis via implicit linearization of the power flow manifold, in: Proc. 53rd Annual Allerton Conference on Communication, Control, and Computing, 2015, https://doi.org/10.1109/ALLERTON.2015.7447032.

[22] D. Cheriton, R.E. Tarjan, Finding minimum spanning trees, SIAM J. Comput. 5 (4) (1976) 724–742.

[23] C.K. Chow, C.N. Liu, Approximating discrete probability distributions with dependence trees, IEEE Trans. Inf. Theory 14 (3) (1968) 462–467.

[24] M. Yuan, Y. Lin, Model selection and estimation in the Gaussian graphical model, Biometrika 94 (1) (2007) 19–35.

[25] R. Pedersen, C. Sloth, G.B. Andresen, R. Wisniewski, DiSC: a simulation framework for distribution system voltage control, in: Proc. European Control Conference, 2015.

[26] A. Pohl, J. Johnson, S. Sena, R. Broderick, J. Quiroz, High-resolution residential feeder load characterization and variability modelling, in: Proc. 40th IEEE Photovoltaic Specialist Conference (PVSC), 2014.

[27] S. Bolognani, Approx-PF—approximate linear solution of power flow equations in power distribution networks, (accessed 8.10.17). 2014, http://github.com/saveriob/approx-pf. (accessed 8.10.17).

[28] W.H. Kersting, Radial distribution test feeders, in: IEEE Power Engineering Society Winter Meeting, vol. 2, 2001, pp. 908–912, https://doi.org/10.1109/PESW.2001.916993.

[29] R.V. Hogg, E.A. Tanis, Probability and Statistical Inference, sixth ed., Prentice Hall, Upper Saddle River, NJ, 2001.

CHAPTER 14

Supervised Learning-Based Fault Location in Power Grids

Hanif Livani

University of Nevada Reno, Reno, NV, United States

CHAPTER OVERVIEW

In modern societies and with the introduction of "smart power grids," customers are more sensitive to power outages. There are also more complex power transmission configurations to integrate renewable energy-based power generation at remote locations. Therefore, more efficient and accurate methods of fault location along these complex configurations are required, which target improving power supply restoration process, reducing the overall power outages time and costs, and enhancing end-users satisfaction. The availability of high-resolution/high-volume data, due to the proliferation of intelligent electronic devices in smart grids, paves ground to implement more accurate and intelligent fault location methods. This chapter presents a supervised-learning fault location method for complex power transmission lines by using high-resolution voltage and current measurements data. The fault location methods are developed for two complex high-voltage AC transmission systems, (1) three-terminal transmission lines, (2) hybrid transmission lines. The presented methodologies utilize discrete wavelet transform and support vector machine (SVM) as a supervised learning algorithm where the power system operating and fault conditions are taken into account through the learning steps of the SVM classifiers.

In modern societies and with the introduction of "smart power grids," customers are more sensitive to power outages. There are also more complex power transmission configurations to integrate renewable energy-based power generation at remote locations. Therefore, more efficient and accurate methods of fault location along these complex configurations are required, which target improving power supply restoration process, reducing the overall power outages time and costs, and enhancing end-users satisfaction. The availability of high-resolution/high-volume data, due to the proliferation of intelligent electronic devices in smart grids, paves ground to implement more accurate and intelligent fault location methods. This chapter presents a supervised-learning fault location method for complex power transmission lines by using high-resolution voltage and current measurements data. The fault location methods are developed for two complex high-voltage AC (HVAC) transmission systems, (1) three-terminal transmission lines, (2) hybrid transmission lines. The presented methodologies utilize discrete wavelet transform (DWT) and support vector machine (SVM) as a supervised learning algorithm where the power

Big Data Application in Power Systems. https://doi.org/10.1016/B978-0-12-811968-6.00014-0

system operating and fault conditions are taken into account through the learning steps of the SVM classifiers.

1 FUNDAMENTALS OF SVM

SVM was first introduced by Vapnik as a binary linear supervised classification algorithm [1]. The original SVM classifier separates data sets with two binary classes ($\{+1, -1\}$), by finding an optimal hyperplane. The linear hyperplane is defined as

$$W^T x + b = \begin{cases} \geq 1, & \text{class} + 1 \\ \leq -1, & \text{class} - 1 \end{cases} \tag{1}$$

where $x \in R^{n \times 1}$ is the input vector with n features, $W \in R^{n \times 1}$ is a weight vector, and b is a bias term. The linear separating hyperplane for a binary linear SVM in a two-dimensional space (i.e., $n = 2$) is shown in Fig. 1. The separation margin (m) between two classes is calculated as

$$m = \frac{2}{\|W\|} \tag{2}$$

where $\|W\|$ is the 2-norm of the weight vector. In order to maximize the separation margin, $\|W\|$ is minimized. Therefore, the maximum margin can be established by solving the following quadratic optimization problem

$$\min \frac{1}{2} \|W\|^2 \tag{3}$$

subject to $y_i \left(W^T x_i + b \right) \geq 1$ \hfill (4)

where x_i is the ith input vector and $y_i \in \{+1, -1\}$ is the corresponding label for x_i.

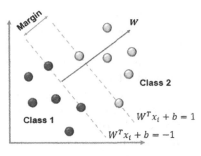

FIG. 1
The two-dimensional feature space with optimal separating hyperplane.

The solution of the optimization problem provides the values of W and b such that the separation margin between the two classes is maximum. In order to solve the above optimization problem, the duality theorem is used and the SVM parameters are obtained by solving the following dual problem

$$\max L(\alpha) = \sum_{i=1}^{N} \alpha_i - 2^{-1} \sum_{i=1}^{N} \sum_{j=1}^{N} \alpha_i \alpha_j \gamma_i \gamma_j x_i x_j \tag{5}$$

$$\text{subject to } \sum_{i=1}^{N} \alpha_i \gamma_i = 0 \tag{6}$$

$$\alpha_i \geq 0 \tag{7}$$

where α_i is the Lagrangian multiplier and N is the number of labeled training data set.

Once the dual optimization problem is solved, the training points with $\alpha_i^* > 0$ are identified as the support vectors (SVs), and W^* and b^* are then calculated as

$$W^* = \sum_{i=1}^{N_{SV}} \alpha_i^* \gamma_i x_i \tag{8}$$

$$b^* = \frac{1}{N_{SV}} \left(\sum_{i=1}^{N_{SV}} \gamma_i - W^* x_i \right) \tag{9}$$

where N_{SV} is the number of obtained SVs.

In most data-driven power systems application, the input feature vectors are not linearly separable in the original input space. Therefore, they can be mapped into a higher dimensional feature space using a nonlinear function Φ to obtain a linearly separable data set. As the calculation of the inner product of Φ in higher dimensional feature space is computationally complex, kernel function k is utilized to calculate the inner product directly as a function of the original input vectors. Thus, the SVMs are obtained by solving the following optimization problem

$$\max L(\alpha) = \sum_{i=1}^{N} \alpha_i - 2^{-1} \sum_{i=1}^{N} \sum_{j=1}^{N} \alpha_i \alpha_j \gamma_i \gamma_j k(x_i, x_j) \tag{10}$$

$$\text{subject to } \sum_{i=1}^{N} \alpha_i \gamma_i = 0 \tag{11}$$

where $k(x_i, x_j)$ is the kernel function.

Thus, the optimization problem is solved and the training points with $\alpha_i^* > 0$ are the SVs. The optimal decision function is then expressed as

$$sign\left(\sum_{i \in SV} \alpha_i^* y_i k(x, x_i) + b^*\right) = \begin{cases} > 0, & \text{class} + 1 \\ < 0, & \text{class} - 1 \end{cases} \tag{12}$$

$$b^* = \frac{1}{N_{SV}}\left(\sum_{i=1}^{N_{SV}} y_i - \sum_{j=1}^{N_{SV}} \alpha_j^* y_j k\left(x_i, x_j\right)\right) \tag{13}$$

The kernel functions such as linear, sigmoidal, and Gaussian radial basis function (RBF) are the most commonly used in power systems application. The Gaussian RBF is chosen for supervised learning-based fault location due to its better performance. The Gaussian RBF kernel function is given as

$$k\left(x_i, x_j\right) = \exp\left(-\left(||x_i - x_j||^2\right)/\gamma\right) \tag{14}$$

where x_i and x_j are n-dimensional input vectors, $\gamma = 2\sigma^2$, σ is the standard deviation of the Gaussian function. The kernel function parameter (γ) needs to be tuned only once in order to achieve sufficient classification accuracy.

2 POWER SYSTEM APPLICATIONS OF SVM

SVM has been extensively used for numerous power system applications that are categorized as:

- Power quality (PQ) analysis such as disturbance classification.
- Power system protection.
- Voltage and rotor angle stability prediction.
- Energy price forecasting and load forecasting.
- Fault classification and fault location.

PQ study has become an important issue in recent years due to large applications of inverter-based energy resources and transactive-based load controllers. Harmonics, voltage swell, voltage sag, and the power interruption can downgrade the power supply quality. The detection of power disturbances is an important tool to ensure the PQ of supply and to detect location and type of disturbances. The authors in Ref. [2] propose a new method of PQ classification based on SVM and neural network. Space phasor is used for feature extraction from three-phase measurements to create suitable patterns for classifiers. The trained classifier is utilized for different disturbance classification including voltage sags, voltage fluctuations, and voltage transients. Ref. [3] presents an integrated model for PQ disturbances recognition using a novel wavelet multiclass SVM. It combines linear SVMs and the disturbances-versus-normal approach to form the multiclass SVM, which is capable of processing multiple classification problems.

Protective relays may mal-operate or the required information may be missed for a proper relay action. Supportive protection systems are required to aid the conventional protection by providing selective and secure coordination. SVMs have considerable potential as zone classifiers for distance relay coordination. This typically requires a multiclass SVM classifier to effectively analyze and to build the underlying concept between the reach of different zones and the apparent impedance trajectory during a fault. Several methods have been proposed for multiclass classification where typically several binary SVM classifiers are combined together. In Ref. [4], one-step multiclass classification, one-against-all, and one-against-one multiclass methods are compared for their performance with respect to accuracy, training, and testing time. Ref. [5] proposes a new machine learning approach for protective relays based on binary SVMs, and communications between the protective relays and the supervisory control and data acquisition, which is called smart protective relays. The goal of smart relays is to classify and discriminate the normal conditions from fault conditions using local measurements. It is shown that the proposed SVM-based smart relays can detect the location of an initial fault using local current, voltage, real power, and reactive power measurements. Smart relays can make a correct decision even when the state of the system changes after some equipment failure.

Real-time monitoring of power system stability is an essential task to prevent blackouts. In case of a disturbance leading to transient instability, fast recognition of the instability conditions is crucial for allowing sufficient time to take emergency control actions. In Ref. [6], a new method for rotor angle stability prediction in a power system immediately after a large disturbance is presented. The proposed two-stage method first estimates the similarity of post-fault voltage trajectories of the generator buses after the disturbance to some preidentified templates. The stability status prediction is then carried out using SVM classifier which takes the similarity values calculated at the different generator buses as the inputs. In Ref. [7], a method based on SVM classifier is presented for rotor angle stability prediction. Generator voltages, frequencies, measured by phasor measurements units immediately after the fault clearance, are used as the inputs for the SVM classifier.

Electricity price forecasting is a difficult and essential task for market participants in a deregulated electricity market. Market participants are sometimes more interested in forecasting the prediction interval of the electricity price, rather than forecasting the value [8]. The prediction interval forecasting is essential to estimate the uncertainty involved in the price. Thus, it is useful to make generation bidding strategies and investment decisions. In Ref. [8], a novel data mining-based algorithm is proposed to achieve two major objectives: to accurately forecast the value of the electricity price series, which is widely accepted as a nonlinear time series, and to accurately estimate the prediction interval of the electricity price series. In the proposed method, SVM is used to forecast the value of the price.

Short-term load forecasting (STLF) is the basis for power system planning and operation. Many power system operations, such as unit commitment, economic dispatch, maintenance scheduling, and planning, are performed effectively with accurate STLF results. The use of SVM for STLF was initially introduced in Ref. [9]. Another STFT method, based on an adaptive two-stage hybrid network with self-organized map (SOM) and SVM, is presented in Ref. [10]. In the first stage, the SOM network is applied to cluster the input data set into several subsets in an unsupervised manner. As the second stage, SVMs are used to fit the training data of each subset.

3 FAULT CLASSIFICATION AND LOCATION FOR THREE-TERMINAL TRANSMISSION LINES

Three-terminal systems are used in power transmission networks to connect three power sources, *A*, *B*, and *C*. The power sources are either generators or Thevenin equivalent of a connected network. As shown in Fig. 2, the three terminals are connected through a *T*-point which does not have any measurements or protection devices. This section presents SVM-based fault classification and location algorithms. The main contributions of the presented methods over the state-of-the-art algorithms are

- The proposed method uses SVM for fault type and faulty line identification based on unsynchronized high-frequency measurements from only two substations.
- The proposed faulty-half identification is based on SVM classifier. The state-of-the-art algorithms are based on the time delay between the arrival time of the initial traveling waves in ground mode and aerial mode. Therefore, the SVM-based approach reduces the sensitivity of traveling

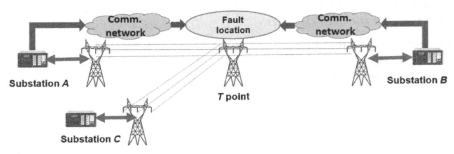

FIG. 2

Three-terminal transmission lines with two unsynchronized high-resolution measurements.

wave-based fault location to the possible errors resulting from time delay calculation, especially for the faults close to the middle of the lines.
- The proposed methodologies need a smaller set of input feature vector for the SVM classifiers compared to the state-of-the-art supervised learning-based algorithms.

3.1 SVM-Based Fault Classification

In this section, an SVM-based fault type classification algorithm is presented. Four binary support vector machines, SVM_i ($i = 1,...,4$), are used to classify the type of fault. The labeled fault data is used to train the SVMs to detect the fault at phases a, b, c and ground. The output of each SVM_i ($i = 1,..,4$) is either $+1$ or -1 which implicates if the fault happens in the corresponding phase or not. As an example if the outputs of SVM_1 and SVM_4 are $+1$ and the rest are -1, the fault is classified as a phase-a-to-ground fault.

In order to train the SVM classifiers, different labeled fault scenarios in a given topology are used. The performance of the trained SVM classifiers is validated by utilizing other fault scenarios. The input features to each binary SVM classifier are the normalized wavelet energies of post-fault three-phase and ground-mode high-resolution voltages from substations A and B. In order to verify the performance of different wavelets, the SVM classifiers are trained and evaluated using three commonly used wavelets: Daubechies-4 (db-4), db-8, and Meyer. The accuracy of the fault location algorithm remains unchanged for the three utilized wavelets. This section presented the results with db-4 mother wavelet as it is one of the widely adopted wavelets in power system applications. The fault type classification algorithm requires the following steps to obtain the input features for the SVMs:

1. Unsynchronized three-phase high-resolution voltage measurement data is captured in substations A and B. The aerial and ground-mode high-resolution voltages are obtained using Clarke transformation as

$$\begin{bmatrix} V_0 \\ V_1 \\ V_2 \end{bmatrix} = \frac{2}{3} \begin{bmatrix} \frac{1}{2} & \frac{1}{2} & \frac{1}{2} \\ 1 & -\frac{1}{2} & -\frac{1}{2} \\ 0 & \frac{\sqrt{3}}{2} & -\frac{\sqrt{3}}{2} \end{bmatrix} \begin{bmatrix} V_a \\ V_b \\ V_c \end{bmatrix} \tag{15}$$

where V_0, V_1, and V_2 are the modal voltages, and V_a, V_b, and V_c are three-phase voltages. In the case of untransposed lines, the modal transformation matrix is used instead of Clarke transformation that can be obtained by any EMTP software.

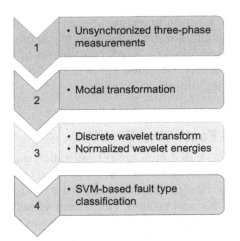

FIG. 3

The flowchart of the SVM-based fault type classification.

2. DWT is applied to the measured three-phase high-resolution voltages (V_a, V_b, V_c) and the calculated ground-mode voltage (V_0) for duration of 40 ms after the fault initiation at substations A and B. The wavelet transformation coefficients (WTCs) are calculated in scale 2, and then squared denoted as WTC^2s. The wavelet energies of the high-resolution voltage measurements, E_{Vk} ($k \in \{a,b,c \text{ and } 0\}$), are calculated for one cycle after the fault initiation as

$$E_V^i = \sum_{m=0}^{M-1} WTC_i^2(m), \quad \text{for } i \in \{a, b, c \text{ and } 0\} \tag{16}$$

where M is the number of samples in one cycle.

3. The normalized wavelet energies are calculated at each substations as

$$E_{Nv}{}^i = \frac{E_V^i}{E_{Va} + E_{Vb} + E_{Vc} + E_{V0}}, \quad \text{for } i \in \{a, b, c \text{ and } 0\} \tag{17}$$

4. The calculated normalized wavelet energies at substations A and B are used as the 8×1 input feature vector to the SVM-based fault type classifiers (Fig. 3).

3.2   Single-Ended Traveling Wave-Based Fault Location

The essential steps in traveling wave-based fault location algorithm in three-terminal transmission lines are faulty line and fault half identification. SVM-based algorithms are used to identify the faulty line among the three transmission lines, and then the faulty half in the identified line. Two new SVM classifiers are first trained using the input feature vectors based on the unsynchronized high-resolution voltage measurements at substations A and B, for faulty line identification. Three separate SVM classifiers associated with each individual line are then trained for faulty half identification in the corresponding faulty line.

Fault or switching initiated transients are composed of forward and backward traveling waves. While these waves are traveling along the lines, reflections occur due to the discontinuities such as the fault point, receiving or sending end terminals of a line. These traveling waves continue to bounce back and forth between the fault point and the terminals until a post-fault steady state is reached. Traveling wave's behavior can be understood better by using the Lattice diagram method. Fig. 4 shows a single-phase line with a fault at F. The Lattice diagram shows multiple reflections and refractions initiated by the fault. τ is the travel time associated with the total length of the line.

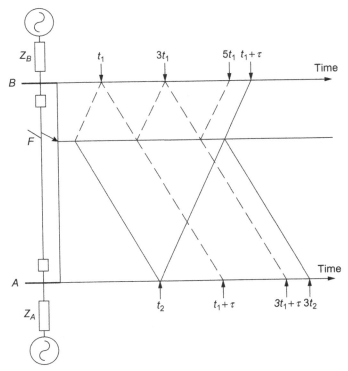

FIG. 4
Lattice diagram for a fault at point F.

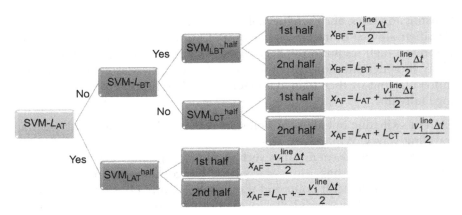

FIG. 5
Hierarchical fault section and fault location algorithms for three-terminal transmission lines.

The arrival times of the backward and forward traveling waves are indicated in Fig. 4. If the fault occurs at x miles away from bus A, the arrival time of the forward traveling wave at bus B is $t_1 = \frac{l-x}{v}$, and the arrival time of the backward traveling wave at bus A is $t_2 = \frac{x}{v}$, where l is the total length of the line and v is the traveling wave velocity. This information is utilized to locate the fault.

In three-phase transmission lines, there are three modes of traveling wave propagation. Therefore, the traveling wave calculations have to be done in the modal domain for each mode separately. Once the faulty line and half are identified based on SVM classifiers, the single-ended traveling wave-based fault location is carried out using the aerial mode (or mode-1) voltages at substation A or B. The hierarchical fault section identification and location algorithm is shown in Fig. 5. It needs to be noted that for a fault identified in line AT or CT, the mode-1 voltage at substation A is used to observe the traveling wave arrival times, and calculate the fault distance with respect to substation A. On the other hand, for a fault identified line BT, the mode-1 voltage at substation B is used to calculate the fault distance with respect to substation B. In Fig. 5, Δt (s) is the time difference between the first and the second peaks of WTC^{2}s at substation A or B corresponding to the backward traveling wave and the reflected backward traveling wave from the fault point respectively, and v_1^{line} (mi/s) is the traveling wave velocity on the identified faulty line.

3.3   Results and Discussion

In order to validate the performance of the proposed fault classification and location algorithms, a 230-kV 60-Hz three-terminal transmission systems with $L_{AT} = 200$ mi, $L_{BT} = 180$ mi, and $L_{CT} = 170$ mi is simulated using an open-source Electromagnetic Transient Program called ATP [11]. Numerous scenarios are simulated with respect to fault type, location, resistance, inception angle, and lines loading to resemble all possible fault conditions. Gaussian noises with the mean equal to zero and standard deviation (σ) equal to 1% of the sampled measurements are added to the high-resolution voltage data at substations A and B. The fault type classification is carried out using four trained SVMs with Gaussian RBF kernel function, associated with each phase and ground. Fig. 6 shows the classification accuracies for phases a, b, c, and ground.

The faulty line identification is then executed using two trained SVMs with Gaussian RBF kernel function. The average accuracy of faulty line identification using two SVMs with kernel parameters, $\gamma_1 = 0.9$ and $\gamma_2 = 1.1$, is 97.4%. Three SVM-based faulty half classifiers are then trained and evaluated using the created fault scenarios. The obtained accuracies for $\text{SVM}^{half}_{LAT}(\gamma = 1.3)$, $\text{SVM}^{half}_{LBT}(\gamma = 1.1)$, and $\text{SVM}^{half}_{LCT}(\gamma = 1.5)$ are 99%, 99%, and 98%, respectively.

Illustrative test case: As an illustrative test case, a phase-a-to-ground fault is assumed to occur in line AT at 60 mi from substation A. Once the fault type, faulty line, and half are identified using trained SVM classifiers, the obtained wavelet transformation coefficient squared (WTC^2) at bus A is observed to identify the first and second traveling waves as shown in Fig. 7. The time difference between the first and second traveling waves is $\Delta t = 0.00065$ s, and the aerial-mode traveling wave velocity is 1.85×10^5 mil/s. Therefore, the fault location is calculated as

$$x = \frac{65 \times 10^{-5} \times 1.85 \times 10^5}{2} = 60.125 \text{ mi}$$

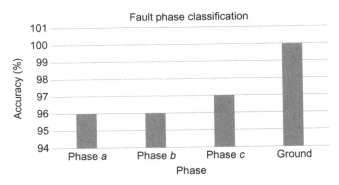

FIG. 6

Fault type classification accuracies using four SVM-based classifiers.

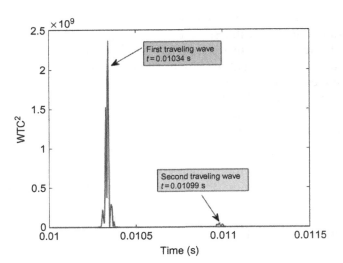

FIG. 7
WTC^2 at bus A for a phase-a-to-ground fault at 60 mi from substation A in line AT.

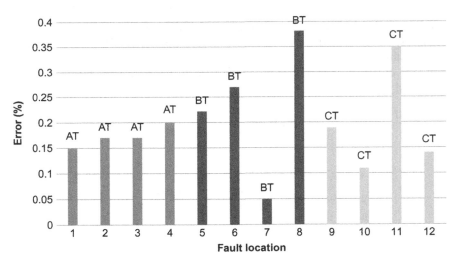

FIG. 8
Fault location errors in line AT, BT, and CT.

The fault location is carried out for a wide range of actual fault location in lines AT, BT, and CT and the errors are calculated using error $(\%) = \frac{\text{AFD} - \text{CFD}}{\text{Total section length}} 100$, where AFD is the actual fault distance and CFD is the calculated fault distance. Fig. 8 shows the fault location error for

12 different faults in lines AT, BT, and CT ranging from 4 mi from substation *A* to 3 mi from substation *B* and 9 mi from substation *C*.

Effect of fault parameters: In order to assess the performance of the proposed fault classification and location algorithm, fault parameters such as fault inception angle, fault resistance, and nonideal faults such as nonlinear high-impedance fault are considered through numerous simulation results. Faults are simulated for a wide range of resistance from 0.01 to 90 Ω, fault inception angles varying between 5 and 350 degrees. Dynamic time-varying nonlinear high impedance and inductive faults are studied to validate the performance of the proposed techniques [11].

4 FAULT LOCATION FOR HYBRID HVAC TRANSMISSION LINES

Hybrid HVAC transmission lines are composed of underground cables combined with overhead lines, and are used when right-of-way issues arise or to connect offshore wind farms to the grid. The proliferation of such complex systems poses difficulties for post-fault analysis such as fault location for system operators and maintenance crews. After an unprecedented event in Denmark in February 2015, when Dong Energy's 400-MW Anholt offshore wind farm in Danish waters had not transmitted electricity to the onshore grid for at least 3 weeks because of a subsea cable fault that occurred, it has been accepted that accurate identification of faulty segment is required. Repairing sea cables is far more difficult and time consuming than repairing cables on land. Thus, accurate fault location is needed to exactly pinpoint the location of a cable failure in a timely manner which results in power system reliability improvement, quick restoration of the power service, and reduction in outage time.

In this section, a single-ended traveling wave-based fault location algorithm is proposed as shown in Fig. 9. The state-of-the-art traveling wave fault location is improved by using SVM classifier for faulty line and faulty half identification. The proposed SVM classifiers are independent of the fault type.

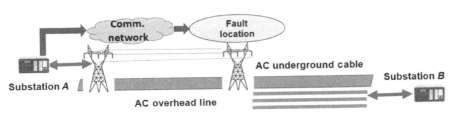

FIG. 9

Hybrid HVAC transmission line.

The input features for the faulty line and half SVM-based classifiers are obtained as:

1. Three-phase high-resolution voltage and current measurement data is captured in substation A, using optical voltage and current transducers. The aerial and ground-mode voltages and current are obtained using Eq. (15).
2. DWT is applied to the measured three-phase voltages, currents, and the calculated ground-mode voltage and current for duration of 40 ms after the fault initiation at substation A. The wavelet transformation coefficients (WTCs) are calculated in scale 2, and then squared denoted as WTC^2s. The wavelet energies are calculated for one cycle after the fault initiation using Eq. (16), and then normalized using Eq. (17).
3. The calculated normalized wavelet energies are used as the 8×1 input feature vector to the SVM-based classifiers.

4.1 Single-Ended Traveling Wave-Based Fault Location

The single-ended fault location algorithm is based on SVM-based faulty line and fault half identification. The SVM classifiers are first trained using the feature vectors that are calculated using high-resolution voltage and current measurements at only substation A. One SVM classifier is trained for faulty line identification and two SVM classifiers are used for faulty half identification associated with the faulty line.

Once the faulty line and half are identified based on SVM classifiers, the single-ended traveling wave-based fault location is carried out using the aerial-mode (or mode-1) voltages at substation A. The hierarchical fault section identification and location algorithm is shown in Fig. 10. In Fig. 10, Δt (s) is the time difference between the first and the second peaks of WTC^2s at substation A corresponding to the backward traveling wave and the reflected backward traveling

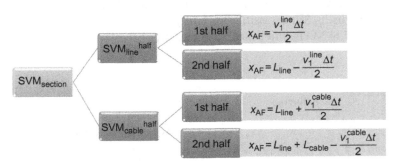

FIG. 10

Hierarchical faulty section and fault location algorithm for hybrid HVAC transmission lines.

wave from the fault point respectively, v_1^{line} (mi/s) is the traveling wave velocity on the overhead line, and v_1^{cable} (mi/s) is the traveling wave velocity on the underground cable.

4.2 Results and Discussion

The validation of the performance of the proposed fault location algorithm is carried out through simulation of a 230-kV 60-Hz hybrid transmission systems with $L_{line} = 100$ mi, $L_{cable} = 20$ mi [12]. Several fault scenarios are simulated with respect to fault type, location, resistance, inception angle, and system loading to resemble all possible fault and system conditions. Gaussian noises with mean of equal to zero and a standard deviation (σ) equal to 1% of the sampled measurements are added to the high-resolution voltage and current measurements. The faulty line and half identification are carried out using three trained SVMs. Fig. 11 shows the identification accuracies using three kernel functions, Gaussian RBF, polynomial, and linear.

The fault location is carried out for a wide range of fault in the overhead line or underground cable. The errors are calculated using $$error\,(\%) = \frac{AFD - CFD}{Total\ section\ length} 100,$$ where AFD is the actual fault distance and CFD is the calculated fault distance. Fig. 12 shows the fault location error with Gaussian RBF kernel function for 12 different faults in the overhead line

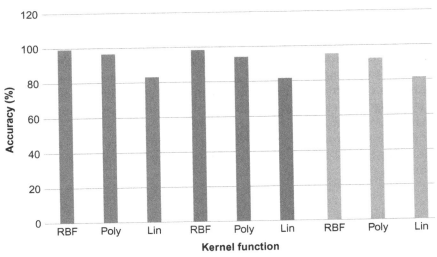

FIG. 11

Faulty section identification errors using three kernel functions, SVM$_{section}$ (*blue*), SVM$_{line}^{half}$ (*red*), and SVM$_{cable}^{half}$ (*green*).

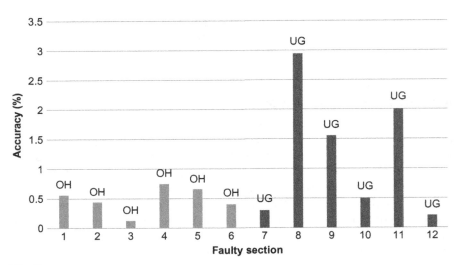

FIG. 12
Faulty location errors in overhead line and underground cable.

or underground cable ranging from 5 mi from substation *A* to 2 mi from substation *B*.

Effect of fault parameters and sensitivity analysis: In order to analyze the performance of the proposed fault location method and assess the sensitivity of the algorithm, the fault parameters, i.e., fault type, fault inception angle, fault resistance, nonideal faults such as nonlinear high-impedance fault are simulated. All 10 types of faults with a wide range of resistance from 0.01 to 100 Ω are considered. Fault inception angles changes between 5 and 350 degrees, and dynamic time-varying nonlinear high impedance and inductive faults are studied to validate the performance of the proposed techniques. Furthermore, the effects of cable aging on the accuracy of the single-ended fault location algorithm are assessed. Fault location accuracy degradation over time can be addressed by introducing a correction factor, which translates the change in cable parameters to a change in velocity. The correction factor is determined by carrying out site tests in certain time intervals or by employing a parameter estimation tool.

5 SUMMARY

In this chapter, supervised-learning SVM-based fault location methods are discussed for complex power transmission lines that are based on high-resolution voltage and current measurements data. The high-resolution measurements are obtained using high-precision optical voltage and current transducers. The

SVM-based fault location algorithms are discussed for two complex HVAC transmission systems, (1) three-terminal transmission lines, (2) hybrid transmission lines. The presented methodologies are developed based on DWT and SVM as a supervised learning algorithm. The power system operating and fault conditions are taken into account through the learning steps of the SVM classifiers.

References

[1] V. Vapnik, Statistical Learning Theory, John Wiley & Sons, New York, NY, 1998.

[2] P. Janik, T. Lobos, Automated classification of power-quality disturbances using SVM and RBF networks, IEEE Trans. Power Del. 21 (3) (2006) 1663–1669.

[3] W.M. Lin, C.H. Wu, C.H. Lin, F.S. Cheng, Detection and classification of multiple power quality disturbances with wavelet multiclass SVM, IEEE Trans. Power Del. 23 (4) (2008) 2575–2582.

[4] B. Ravikumar, D. Thukaram, H.P. Khincha, Comparison of multiclass SVM classification methods to use in a supportive system for distance relay coordination, IEEE Trans. Power Del. 25 (3) (2010) 1296–1305.

[5] Y. Zhang, M.D. Ilic, O.K. Tonguz, Mitigating blackouts via smart relays: a machine learning approach, IEEE Proc. 99 (1) (2011) 94–118.

[6] A.D. Rajapakse, F. Gomez, K. Nanayakkara, P.A. Crossley, V.V. Terzija, Rotor angle instability prediction using post-disturbance voltage trajectories, IEEE Trans. Power Syst. 25 (2) (2010) 947–956.

[7] F.R. Gomez, A.D. Rajapakse, U.D. Annakkage, I.T. Fernando, Support vector machine based algorithm for post-fault transient stability status prediction using synchronized measurements, IEEE Trans. Power Syst. 26 (3) (2011) 1474–1483.

[8] J. Hua Zhao, Z.Y. Dong, Z. Xu, K.P. Wong, A statistical approach for interval forecasting of the electricity price, IEEE Trans. Power Syst. 23 (2) (2008) 267–276.

[9] B.J. Chen, M.W. Chang, C.J. Lin, Load forecasting using support vector machines: a study on EUNITE competition 2001, IEEE Trans. Power Syst. 19 (4) (2004) 1821–1830.

[10] Y. Wang, Q. Xia, C. Kang, Secondary forecasting based on deviation analysis for short-term load forecasting, IEEE Trans. Power Syst. 26 (2) (2011) 500–507.

[11] H. Livani, C.Y. Evrenosoglu, A fault classification and location method for three-terminal circuits using machine learning, IEEE Trans. Power Del. 28 (4) (2013) 2282–2290.

[12] H. Livani, C.Y. Evrenosoglu, A machine learning and wavelet-based fault location method for hybrid transmission lines, IEEE Trans. Power Del. 5 (1) (2014) 51–59.

CHAPTER 15

Data-Driven Voltage Unbalance Analysis in Power Distribution Networks

Matthias Stifter*, Ingo Nader†

*AIT Austrian Institute of Technology, Center of Energy, Vienna, Austria, †Unbelievable Machine, Vienna, Austria

CHAPTER OVERVIEW

More data from various sources enable in-depth analysis of various network parameters. Efficient analysis requires the use and adaption of methods developed for big data applications, like MapReduce for parallel in-database processing. This chapter investigates the applicability of data-driven methods and interactive data visualization to discover new insight into low-voltage network states. The use of MapReduce functions based on Open Source Software like R or Java is demonstrated in combination with a commercial distributed analytics database. These customized functions are applied to analyze unbalance voltage conditions in low-voltage networks and discover and explore the reasons by relating it to other events in the network. The discovery process is supported by interactive visualization methods, like affinity graphs for representing collaborative filters. Performance comparisons to conventional database concepts are discussed at the end of the chapter.

CHAPTER POINTS

Data, methods, and technologies for data analysis
Adaption of MapReduce functions based on open source programming languages in combination with commercial analytical database
Discovery based on interactive visualization
Performance comparison

1   INTRODUCTION

With the increase of sensor and metering devices, like phasor measurement units and rollout of smart meters, better knowledge about system behavior and network states will be available. But these benefits come together with increasing amount of data, which has to be communicated, processed, stored, and analyzed. It turns out that conventional applications are not capable of handling the data and utility IT systems have been not designed with respect to integrating large amounts of data.

321

Big Data Application in Power Systems. https://doi.org/10.1016/B978-0-12-811968-6.00015-2

Many different research questions and applications have been addressed based on analytics of sensor and meter data. Applications and needs for analyzing data have been successfully demonstrated. They reach from better understanding of the network [1] and loads [2], improving forecasts and prediction [3] to business-driven objectives [4] and high-resolution low-voltage monitoring frameworks [5]. Clustering-based methods of voltages have been investigated in [6] to identify topology and connectivity. The authors in [7] present an approach assign sensors to their feeder location based on covariance clustering of voltages.

A number of companies have business and data analytics solutions in their portfolio, offering computational efficient and state-of-the-art data processing frameworks. Opposed to available commercial products, top languages for statistics, data analytics, and data science are dominated by open source software, like Gnu R, Python, or the Hadoop ecosystem, as several rankings and polls on the Internet report.

Beside an efficient data analytics programming language, a high-performance parallel processing system is necessary to cope with the amount of data. Clusters are evolving into cloud concepts, where computational power can be aggregated and managed to perform the required work load. An overview about concepts of cloud computing for smart grid applications can be found in [8]. Technology providers announce "big data" solutions for the energy utilities, where technology often has its origin from other domains, like telecommunication. Main differences of the processing requirements are volumes, velocity, and variety, where velocity regards to speed for collecting and processing data, often within real-time. Current trends and recommendations favor Hadoop-based technology and utilization of MapReduce technology for meeting future requirements on storing and analyzing massive data [9]. One example of data model integration is the design of Common Information Model [10]-based model databases on HBase—open source, nonrelational, distributed database—with a Hadoop-based query technology presented in [11].

2 PROBLEM STATEMENT

A factor for limiting hosting capacity of renewable generators in low-voltage distribution networks is the unbalance due to single phase loads and generators. With the ability to monitor, identify, and reduce high unbalances, thus increasing the available voltage band, networks can be better utilized and operated more cost effectively. High-resolution (and time-synchronized) measurements from smart meters make it possible to discovery various effects and relations which are not observable by existing conventional methods of monitoring.

2.1   Unbalance in Low-Voltage Distribution Networks

Unbalanced loads cause a neutral point displacement in three-phase/four-wire low-voltage distribution networks due to additional voltage drop in the neutral line (see Fig. 1). For this reasons some grid codes assess the voltage rise due to single-phase generators (e.g., inverters of photovoltaic systems) with the threefold increase, caused by the zero-sequence component. Unbalance is also harmful for many motors and cause additional losses in lines and transformers.

In general, low-voltage distribution grids have a star topology and feeders are operated in open meshes which can be manually connected in case of maintenance or outages. Therefore, the unbalance effects are feeder dependent. Fig. 2 shows the spread of the voltages between three phases of one feeder.

Investigations have shown that the used voltage band can be decreased if the unbalancing is reduced by switching of two phases—keeping the same rotation—in cable boxes [12].

2.2   Utilized Voltage Band

The highest and lowest voltage in the network has to be between the allowed voltage limits. Beside unbalance in the phases, loads with high-power consumption can cause high-voltage drops along the line. In case of generation the voltage can increase due to reverse of power flow and therefore increase the used voltage band in the network (decrease voltage band reserve). The used voltage band should be narrow in order to allow, for example, on-load tap changers to operate.

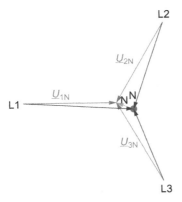

FIG. 1

Unsymmetrical voltages and neutral point displacement due to unbalanced loads.

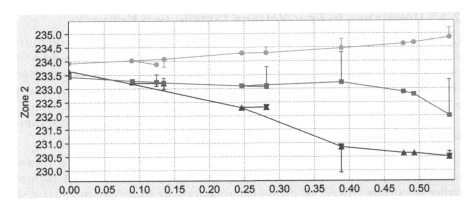

FIG. 2

Voltage drop diagram showing the spread between the phases (*red*, *blue*, *green*) along the distance from the substation.

3 DATA ACQUISITION AND STORAGE

3.1 Smart Meter Data Acquisition System

The so-called Power Snap Shot Analysis (PSSA) method [13] has been developed based on time-synchronous measurements of voltage, angle, active, and reactive power of all three phases. During the measurement snapshot cycle of 15 min every meter records the mentioned measurements for every second, thus resulting in 900 × 18 measurements plus additional 10 and 15 min means of the voltage. Due to the limited bandwidth in power line carrier (PLC)-based smart meter systems, only a selected subset of the meters (trigger meters) sending their three most relevant events, for example, highest or lowest voltage, or highest unbalance, to the data concentrator (DC). The DC decides what are the most interesting timestamps of the measurement cycle. Up to ten synchronous snapshots are then requested from every meter in the network, resulting in approx. 3000 data points for every snapshot.

The snapshots are transmitted over a central server to the analysis framework, the so-called PSS Host [14] where the XML-based files are parsed and loaded into a PostgreSQL relational database [15]. The PSS Analyzer is a graphical user front end, which can provide statistical analysis and visualization of the snapshots. In addition, measurement data can be used to feed and run a network simulator. The network model is also used to visualize the voltage drop diagram (Fig. 2) and other diagrams based on the geographical information.

Currently the snapshot datasets contain more than 1 million snapshots (3 billion measurements) from about 35 low-voltage networks. The number of networks is steadily increasing depending on the demand for analysis. In the area of Upper

Table 1 Measurement Data Set

Total Rows	Voltage Measurements[a]	Date
100 millions	14.1 millions	June 2014
800 millions	97 millions	August 2015

[a]Each measurement/row contains a vector of three phases.

Table 2 Measurement Data Set

Characteristics	Network A	Network B	Network C
Transformer (kVA)	630	400	800
Feeders	9	8	8
Customers	145	193	271
Max. feeder length	2307 m	1079 m	447 mm

Austria alone there are more than 8000 low-voltage networks. Table 1 shows the total number and voltage measurements.

Table 2 shows the characteristic of three exemplary low-voltage networks.

3.2 Distributed Database Storage

For analysis the data are loaded into the Teradata Aster Discovery Platform, which is a PostgreSQL-based, fully parallel database with additional functionality for data preparation and data analysis. This functionality includes computations performed in the MapReduce framework [16] that has been integrated into the SQL framework to make it easily accessible. The discovery platform allows performing the analysis without data movement (in-database processing), using predefined MapReduce functionality as well as custom functions which can be coded in various programming languages (Java and R were used for the present analyses).

4 DISTRIBUTED DATA PROCESSING

4.1 Statistical Method

Data discovery was performed using the statistical programming language R (Version 3.0.2) [17], Java, and the Teradata Aster Discovery Platform [18] using standard functionality as well as additionally implemented code for event generation (described later) developed by the authors. For analyzing the connection between meters, the power snapshot data have been processed by a custom MapReduce function to generate distinct events. These events were used as inputs to the collaborative filtering algorithm as implemented in the Teradata Aster Analytics Foundation, release number 5.11.

Collaborative filtering (`CFilter`) operates on defined events and determines how frequently they are happening together. A popular example is recommendations on books according to other customers preferences. A statistical confidence for the probability that they are happening together is calculated. The implemented Aster SQL-MR function provides figures for support, confidence, lift, z-score, and raw score probability [19]. `CFilter` provides the following calculation on $event_i$ and $event_j$:

Cooccurrence: Count of cooccurrence of both events

$$N_{i \cap j} = \sum_{event_i} event_j$$

Score: Product of two conditional probabilities

$$S_{i \cap j} = P(event_i | event_j) \cdot P(event_j | event_i)$$

Support: Percentage among all events, when the two events cooccur

$$Sp_{i \cap j} = N_{i \cap j} / N_{total}$$

Confidence: Percentage of $event_j$ occurrences in all events in which $event_i$ occurs

$$C_{i \cap j} = N_{i \cap j} / \sum event_j.$$

Lift: Ratio of the observed to the expected support value

$$Lift_{i \cap j} = \frac{Sp_{i \cap j}}{\sum event_i / N_{total} \times \sum event_j / N_{total}}$$

where Lift > 1 expresses a positive effect of $event_i$ and $event_j$ on the occurrence of the other events, Lift < 1 a negative effect, and Lift = 0 no effect. Z_{Score} measure of the significance of the cooccurrence

$$Z_{Score} = \overline{N}_{i \cap j} / \sigma(N_{i \cap j})$$

It can be visualized as a "sigma" graph with Aster Lens, showing nodes and their relationship as edges between them. For meters to be connected, they need to have the same event (strong asymmetry) at the same snapshot. The more often this connection exists, over the whole set of power snapshots, the stronger is the link between the meters.

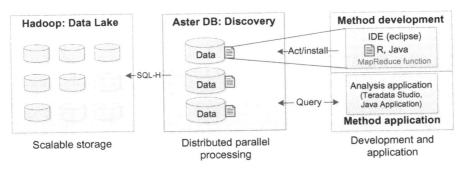

FIG. 3
Developing and deploying MapReduce functions into the distributed database.

4.2 Distributed Queries and Functions

The method of developing and injecting map/reduce functions—written for instance in R or Java and which are directly executed in the distributed database—are shown in Fig. 3. The Hadoop-based scalable storage is optional and was not used for this analysis.

5 DATA DISCOVERY

5.1 Distribution of Voltages

Initially, the histogram of measured voltages per phase indicates that there is a certain number of unbalanced voltages occurrences. Even by finding a point in time which corresponds to a power snapshot in Fig. 4, it is not obvious how this unbalanced voltage state relates to any causal network condition.

5.2 Relation Between Meters With High Unbalance Events

For the discovery of unbalanced network situations the MapReduce-based analysis has been applied to all available snapshots for different networks. The objective is to find and quantify a common event, which happens at the same time at other nodes as well (e.g., strong asymmetry). The unbalance between the phases is discretized by the MapReduce function according to Table 3.

Next, the dependency of an event happening at the same time in other nodes is investigated. Fig. 5 visualizes events as edges which link nodes where the event has happened at the same time.

In network A, events of unbalanced voltages happening at the same time, are shown in Fig. 5. By interactive selecting nodes in the graph it is seen that the events in feeder 1 are not happening together with events in other feeders (e.g., feeder 3, feeder 4). But events from feeder 1 happen together with events

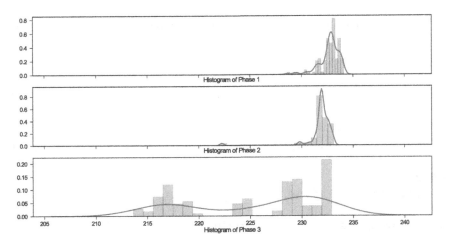

FIG. 4

Histogram of the voltages per phase of one snapshot, showing strong asymmetric voltages. *Note*: The *vertical dashed line* marks the trigger of this snapshot (determined by the lowest voltage of all of the three phases).

Table 3 Definition of Unbalance Events

Residuum	Description	Abbreviation
0–2	No asymmetry	no_asym
2–5	Slight asymmetry	slight_asym
5–9	Medium asymmetry	meda_sym
>9	Strong asymmetry	strong_aysm

in feeder 8, since they are connected by two red nodes (indicating a very high number of events). It can be misleading that these events from feeder 1 are caused by the same phenomena (e.g., high single-phase load). The network model reveals that these two—highly represented nodes—are at the very end of the feeder, connected to the rest over a 400 m line. It is highly probable that events of unbalance happen very often at the same time, but not physically related to events in feeder 1.

In network B (Fig. 6), simultaneous events are concentrated in feeder 6, feeder 4, and feeder 7. The interactive exploration shows that these events are concentrated on the feeder level, since affinities disappear when selecting the centers of the clusters (red nodes and/or red edges).

Visualization of network C (Fig. 7) reveals events at the bottom of the figure. These are unrelated to other rather global, events of unbalance in the network and linking only a smaller number of nodes. A zoom on these isolated events in

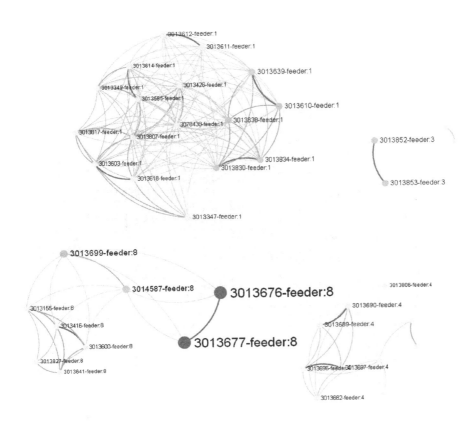

FIG. 5
Network A: Visualization of relations of number of voltage unbalance events which are occurring at the same time at meters. *Note*: A connection (edge) means only that there is also an event in the other node at the same time, but not necessarily the same event.

Fig. 8 shows that these events are concentrated on one feeder. A closer look to the network model and additional geographic data shows that they are all located in a residential building, equipped with ripple controlled warm water boilers.

Analyzing the phenomena in depth reveals that the asymmetry event happens during night time, at about 1 o'clock in the morning. Usually three phases are connected in random order at every household, only preserving the right rotation of the electric field. Unfortunately, in this case all single-phased load devices are connected to the same phase, thus resulting in a very high unbalance when all of them are switched on at the same time. To solve this issue load switch devices have to be connected to arbitrary phases, so that they are statistically even distributed among the phases again.

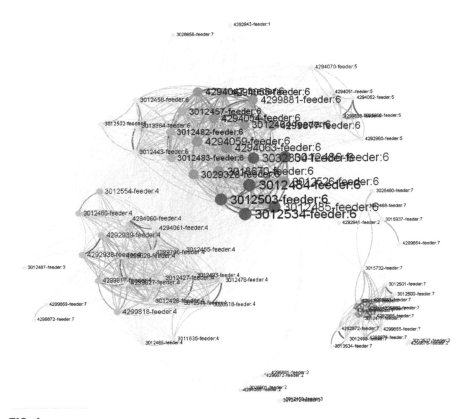

FIG. 6

Network B: Visualization of relations of number of voltage unbalance events (color and width of edges) which are occurring at the same time at meters (color and size of nodes).

5.3   Timely Distribution of High Unbalance Events

To narrow the reason of events further investigation on the distribution of event times are made, to find additional information on inadvertent network states.

Data analysis in network A of day and hour (Figs. 9 and 10) events are taking place, reveal no evidence of unintended conditions. It is highly probable that events are happening during the noon hours of the day time, where customers are active. A similar distribution of events is found in Network B (not shown).

In network C, plotting the day of the events on the timeline give evidence that this events are comparable rare and happening with month of distance in between (Fig. 11). The histogram of the events per hour of the day shows that they are concentrated at night time around 1 o'clock (Fig. 12). This could be taken as important criteria for discovering phenomena, since it gives evidence that events are not happening randomly, but by determined cause.

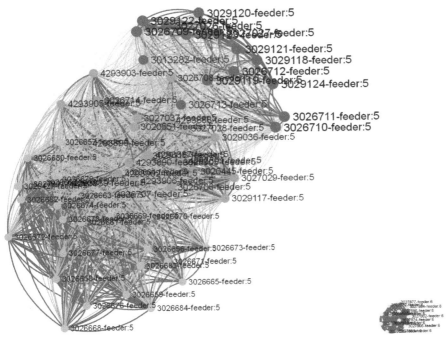

FIG. 7
Network C: Visualizations of the number of voltage unbalance events at the same time (color of edges) related to other meters. Note that the isolated events on the bottom are happening unrelated to all other events in the network.

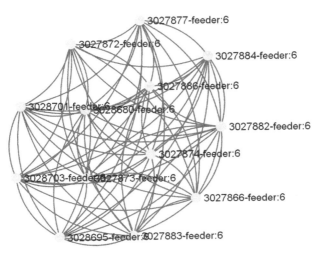

FIG. 8
Network C: Affinity graph. If an event of unbalance happens at one of the meters, an unbalance event happens at the same time as well, but only at these meters. *Note*: This is a zoom on the lower right part of the graph presented in Fig. 7.

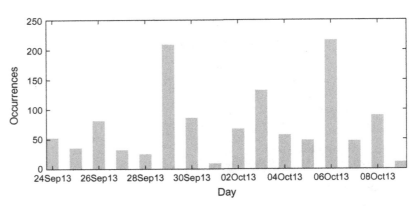

FIG. 9

Network A: Distribution of unbalanced voltage events during the evaluation period.

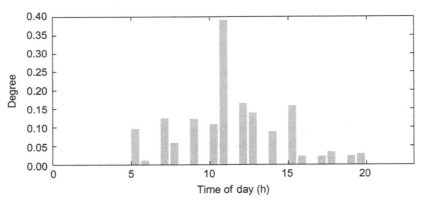

FIG. 10

Network A: Distribution of unbalanced voltage events for the hour of the day.

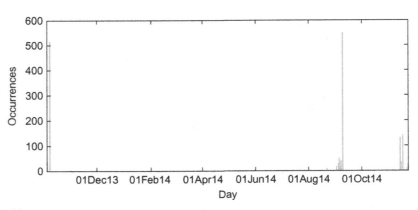

FIG. 11

Network C: Distribution of unbalanced voltage events during the evaluation period. Note that it points out to be very rare events.

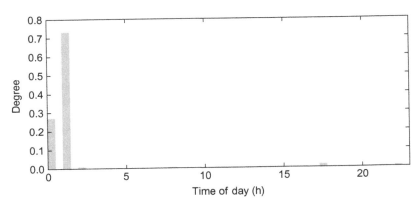

FIG. 12
Network C: Distribution of unbalanced voltage events for the hour of the day. Note that most of the events are taking place at 1 o'clock.

5.4   Relation Between Different Network States

Opposed to the previous analysis, now events are defined differently. Depending on the criteria of a network state, for example, single/three-phase asymmetry, single-phase high voltages, single/three-phase high active and reactive powers, the following definitions are used:

$$U_{max} = \max U_{iN} \quad i \in 1,2,3 \tag{1}$$

$$U_{min} = \min U_{iN} \quad i \in 1,2,3 \tag{2}$$

$$\Delta U = U_{max} - U_{min} \tag{3}$$

$$\bar{U} = \frac{1}{3}\sum_{i=1}^{3} U_{iN} \tag{4}$$

$$P_{sum} = \sum_{i=1}^{3} P_i \quad \Delta P = P_{max} - P_{min} \tag{5}$$

$$Q_{sum} = \sum_{i=1}^{3} Q_i \quad \Delta Q = Q_{max} - Q_{min} \tag{6}$$

In Table 4 the voltage, active, and reactive power criteria are listed with define the events calculated by the MapReduce function.

The earlier-defined events are then calculated for the networks under two different conditions:

1. Events which happen at the *same moment in time within a snapshot*, meaning all events which occur during the same moment in time in the whole network (network perspective).

Table 4 Definition of Unsymmetry Events

Voltage	
$\Delta U > 9$ V	asym_9v_voltage
$\Delta U > 6$ V	asym_6v_voltage
$\Delta U > 3$ V	asym_3v_voltage
Voltage peaks (for single phases: max of all phases)	
$U_{max} > 253$ V	high_single_voltage_253v
$U_{max} > 246.1$ V	high_single_voltage_246v
Voltage dips (for single phases: min of all phases)	
$U_{min} < 207$ V	low_single_voltage_207v
$U_{min} < 221$ V	low_single_voltage_221v
Voltage peaks and dips (for all phases: mean over three phases)	
$\bar{U} > 253$ V	high_mean_voltage_253v
$\bar{U} > 246.1$ V	high_mean_voltage_246v
$\bar{U} > 220$ V	(No event)
$\bar{U} > 207$ V	low_mean_voltage_221v
$\bar{U} < 207$ V	low_mean_voltage_207v
Active power	
$P_{sum} > 40$ kW	peak_40kw_act_power
$P_{sum} > 25$ kW	peak_25kw_act_power
Active power asymmetry	
$\Delta P > 20$ kW	asym_20kw_act_power
$\Delta P > 12$ kW	asym_12kw_act_power
$\Delta P > 7$ kW	asym_7kw_act_power
$\Delta P > 4$ kW	asym_4kw_act_power
Single-phase feed in	
$P_{min} < -10$ W and $P_{max} > 10$ W	single_phase_feed_in
$P_{max} < -10$ W	multi_phase_feed_in
Active power asymmetry	
$Q_{sum} > 4$ kW	peak_4kw_react_power
$\Delta Q > 2$ kW	asym_2000w_react_power
$\Delta Q > 1$ kW	asym_1000w_react_power
$\Delta Q > 0.5$ kW	asym_500w_react_power

2. Events which happen at the *same time at the same meter* (meter perspective).

The resulting computed figures represent events which happen together at the same moment in time (meaning they are in the same snapshot) once from the network perspective and once from the meter perspective.

To exemplify the difference between simultaneous events for the whole network B (Fig. 13) and events occurring simultaneous in 1 m (Fig. 14), the focus

FIG. 13

Network B (events in the network): Interactive selection of event low_single_voltage_207v (black frame), revealing other events taking place simultaneously in the whole network.

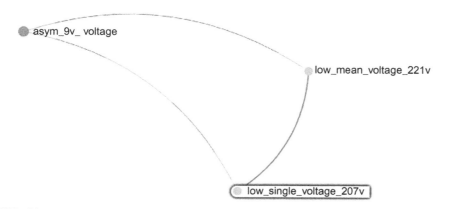

FIG. 14

Network B (events in meters): Interactive selection of event low_single_voltage_207v (black frame) revealing other events taking place simultaneously in the meters.

on the event `low_single_voltage_207v` is displayed. While other events happen simultaneously in the network B (Fig. 13), events at the meters occur only with two other events: `low_mean_voltage_221v` and `asym_9v_voltage` (Fig. 14).

Fig. 15 shows the isolated occurrence of `high_single_voltage_253v` events in network C. In Fig. 16 the simultaneity of single feed-in with asymmetric active and reactive power events is shown. This could be interpreted that a slight voltage unbalance is a frequent network situation, with the strongest relation to single-phase feed-in and asymmetric reactive power (1000 Var and above) (Fig. 17).

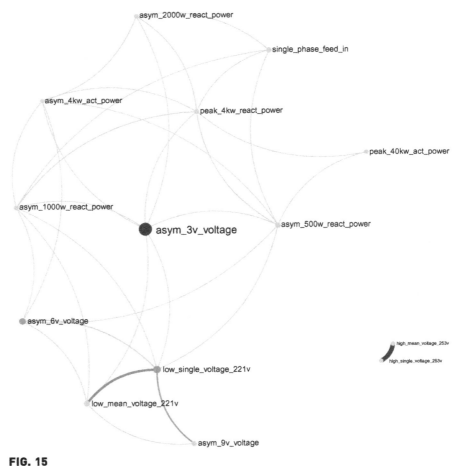

FIG. 15

Network C: Relation of events, which happen at the same moment in time, for individual meters. Events of `high_single_voltage_253v` are independent from feed-in and asymmetry events.

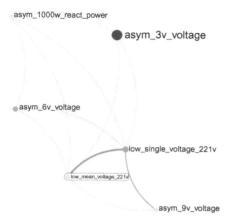

FIG. 16
Network C (selection): Relations between different events, represented by links and colors. Single-phase feed in of 4 kW correlates with asymmetric reactive power of 1 kW.

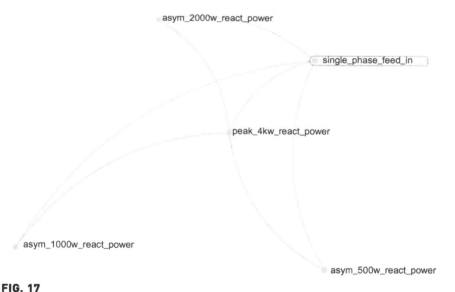

FIG. 17
Network C (selection): Interactive visualization of event `single_phase_feed_in` that occurs together with events of asymmetric reactive power consumption. It is not related to events of `low_single_voltage_221v`.

5.5   Maximum and Minimum Voltage: Voltage Spread

In the next example, a MapReduce function `MeterMinMax` calculates maximum and minimum voltage measurements of the three phases separately for every meter. For every measurement (row) which is processed, the minimal

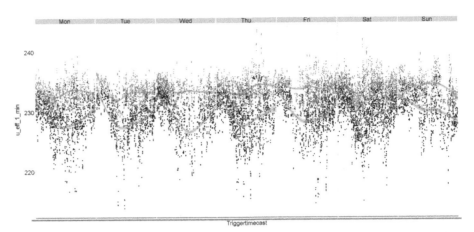

FIG. 18

Maximum and minimum voltages of phase 1 per weekday and the corresponding averaging.

and maximal value is compared to the actual—stored in a hash map—and updated if necessary. Aggregating and visualizing the maximum and minimum voltages of the snapshots per weekday (Fig. 18) shows beside the typical voltage characteristic extreme values, which—in case of unbalance—limit the available voltage band for in-feed. Continuous evaluation of the measurements can identify trends and potentials for improvements (unbalance counter measurements).

6 PERFORMANCE EVALUATIONS

This section investigates performance of MapReduce functions directly performed within the distributed database compared to a traditional relational single host database. The performance of two different custom MapReduce functions—implemented in Java—are investigated when executed as a MapReduce function and executed in an external application which is fetching the data from the database and inserting results back to it.

6.1 Comparison Requirements

The following criteria have been met in order to make the data processing comparable in terms of execution time.

6.1.1 Code

The program code for processing the data is effectively identical. While the MapReduce function implements specific interfaces for handling the input and output rows (output emitter), the non-MapReduce function implements standard "insert row" Java Data Base Connectivity (JDBC) statements (e.g., INSERT INTO table VALUES (...)). Table inserting operation time is separated from the execution time for the non-MapReduce application.

6.1.2 Data
The tests are conducted with the same raw data and results are validated to get the same outcomes. PostgreSQL vector data type for holding the three-phase values is not supported by Aster; therefore, it is transformed into three separate columns.

6.1.3 Database
In order to compare the execution time on the functions, processing power has to be comparable. PostgreSQL was installed on one of the virtual machine (VM) worker nodes to have identical processing power. While one database runs idle without work load during the test of the other, the influence is expected to be neglectable and the PostgreSQL server service has been stopped during tests.

6.1.4 System
The Aster database is processing in parallel for every node running in a VM on a server. Table 5 shows the system configuration of the VMs and the additional PC benchmark system. The server has 6/2 CPU/Cores per CPUs and 24 GB RAM and the PC 2/2 CPU/Cores per CPU and 16 GB RAM. The single-thread CPU Mark (second number) is comparable of the two systems. The Aster system was tested with 2 * 3 GB RAM for the workers and 2 GB for the queen and with double memory of 2 * 6 GB and 4 GB RAM. The second configuration matches the 16 GB of the PC system.

Table 5 Definition of Unsymmetry Events

Node	CPU Type	RAM	CPU/Cores	CPU Mark
Worker (VM)	Intel Xeon CPU W3690 3.47 GHz	3/6 GB	2/1	9729 (1576)[a]
Queen (VM)	Intel Xeon CPU W3690 3.47 GHz	2/4 GB	1/1	9729 (1576)[a]
PC	Intel CPU i5-4300U 1.90 GHz	16 GB	2/2	9729 (1607)[a]

[a]Single-thread CPU Mark.

6.2   Evaluation Setup

Three setups are used: Aster MapReduce in-database processing, Java JDBC connection to Aster database and processing external, and Java JDBC connection to PostgreSQL with external processing.

- Aster database cluster `beehive`: The used version of distributed database "Aster Express 5.10" comes in a preconfigured setup of one managing node (`queen`) and two worker nodes (`worker1` and `worker2`). It runs under SUSE Linux Enterprise 11. Due to redundancy reasons the workers host virtual workers (primaries and secondaries) distributed on different nodes. Keeping them synchronized implies some data exchange, but is not considered and investigated in detail for these tests.
- PostgreSQL `worker1`: The database was installed on `worker1` and has identical processing power. Concurrencies of database workloads have been eliminated (stopping postgres service) or avoided (no activity on Aster). Also the Java application code is executed locally (and remote for comparison). The performance is expected to be approximately half of the performance of two workers under the Aster system.
- PostgreSQL `localhost`: Additional performance measurements on a local PC.

6.3   MapReduce Function `CalcEventsLongFormat`

The MapReduce function was executed on approximately 2M rows for each of the three networks. According to the Aster management console, the processing time within the Aster database was about 2 seconds and the fetch time 3–4 s. This was comparable to both JDBC Java application runs, connecting to Aster and to PostgreSQL, which needed slightly above 2 s for processing.

6.4   MapReduce Function `MeterMinMax`

The size of the dataset was increased to 800M rows with about 100M voltage measurement rows. While the execution of the MapReduce function had no problems with the 3 GB per worker node setup, it was not possible to run the query over the JDBC connection. And as stated earlier for the PostgreSQL setup it has been reduced to 100M voltage measurements and sliced to 10M queries using SQL `LIMIT BY` and `OFFSET` statements.

Table 6 shows comparisons results of the different configurations.

Table 6 Benchmark Results

Benchmark	Fetch Time (min)	In-DB/Java (s)	Total (min)
Aster MapReduce 2 * 3 GB + 2 GB	14	2	14
Aster MapReduce 2 * 6 GB + 4 GB	9	2	9
Aster Java JDBC (local) 6 GB	–	179	–
PostgreSQL JDBC (worker) 3 GB	50	127	52
PostgreSQL JDBC (worker) 6 GB	36	63	40

7 CONCLUSION

This contribution shows the value and applicability to analyze large set of smart meter and sensor data for gaining deeper insight into interesting network states. The approach demonstrates application of open source-based data analytics in combination with commercial software packages. Visualization of complex data is clearly an advantage when comprehending, understanding, and interpreting data. The benefit of a distributed parallel processing database shows up when it comes to large datasets and complex functions for evaluating and processing data. Performance evaluations show an advantage over traditional database concepts, which can be further improved by intelligent data distribution to avoid costly data fetch processes. In 2016 the AIT Energy Data Analytics Lab was put into operation on a 24 nodes cluster with 12 CPUs/128 GB RAM per node including a Teradata Aster research installation containing 12 worker nodes. Current work includes analysis of smart meter and grid monitoring data from various networks.

References

[1] J. Wu, Y. He, N. Jenkins, A robust state estimator for medium voltage distribution networks, IEEE Trans. Power Syst. 28 (2) (2013) 1008–1016, https://doi.org/10.1109/TPWRS.2012.2215927.

[2] R. Silipo, P. Winters, Big Data, Smart Energy, and Predictive Analytics—time series prediction of Smart Energy Data, KNIME, 2013.

[3] P. Zhang, X. Wu, X. Wang, S. Bi, Short-term load forecasting based on Big Data technologies, CSEE J. Power Energy Syst. 1 (3) (2015) 59–67, https://doi.org/10.17775/CSEEJPES.2015.00036.

[4] D. De Silva, A data mining framework for electricity consumption analysis from meter data, IEEE Trans. Ind. Inform. 7 (3) (2011) 399–407.

[5] H. Maass, H.K. Cakmak, W. Suess, A. Quinte, W. Jakob, K.U. Stucky, U.G. Kuehnapfel, First evaluation results using the new electrical data recorder for power grid analysis, IEEE Trans. Instrum. Meas. 62 (9) (2013) 2384–2390, https://doi.org/10.1109/TIM.2013.2270923.

[6] V. Arya, R. Mitra, Voltage-based clustering to identify connectivity relationships in distribution networks, in: 2013 IEEE International Conference on Smart Grid Communications (SmartGridComm), 2013, pp. 7–12.

[7] K. Diwold, M. Stifter, P. Zehetbauer, Network and feeder assignment of smart meters based on communication and measurement data, in: 2015 International Symposium on Smart Electric Distribution Systems and Technologies (EDST)2015, , pp. 541–546.

[8] M. Yigit, V.C. Gungor, S. Baktir, Cloud computing for smart grid applications, Comput. Netw. 70 (2014) 312–329, https://doi.org/10.1016/j.comnet.2014.06.007.

[9] Y. Simmhan, S. Aman, A. Kumbhare, R. Liu, S. Stevens, Q. Zhou, V. Prasanna, Cloud-based software platform for Big Data analytics in Smart Grids, Comput. Sci. Eng. 15 (4) (2013) 38–47, https://doi.org/10.1109/MCSE.2013.39.

[10] M. Uslar, M. Specht, S. Rohjans, J. Trefke, J.M. González, The Common Information Model CIM-IEC 61968/61970 and 62325—A Practical Introduction to the CIM, Springer-Verlag, Berlin, Heidelberg, 2012. ISBN 978-3-642-25214-3.

[11] S. Zhang, J. Wang, B. Wang, Research on data integration of smart grid based on IEC61970 and cloud computing, in: D. Jin, S. Lin (Eds.), Advances in Electronic Engineering, Communication and Management, Vol. 1, Lecture Notes in Electrical Engineering, Springer, Berlin, Heidelberg, 2012, pp. 577–582. 139 ISBN 978-3-642-27286-8 978-3-642-27287-5.

[12] B. Bletterie, S. Kadam, R. Pitz, A. Abart, Optimisation of LV networks with high photovoltaic penetration—balancing the grid with smart meters, 2013 IEEE Grenoble Conference, 2013, pp. 1–6, https://doi.org/10.1109/PTC.2013.6652366.

[13] A. Abart, B. Bletterie, M. Stifter, H. Brunner, D. Burnier, A. Lugmaier, A. Schenk, Power Snap-Shot Analysis: a new method for analyzing low voltage grids using a smart metering system, 21st International Conference on Electricity Distribution, CIRED, Frankfurt, 2011.

[14] M. Stifter, B. Bletterie, D. Burnier, H. Brunner, A. Abart, Analysis environment for low voltage networks, 2011 IEEE First International Workshop on Smart Grid Modeling and Simulation (SGMS), 2011, pp. 61–66, https://doi.org/10.1109/SGMS.2011.6089199.

[15] PostgreSQL Global Development Group, PostgreSQL, 2015.

[16] J. Dean, S. Ghemawat, MapReduce: simplified data processing on large clusters. Commun. ACM 51 (1) (2008) 107–113, https://doi.org/10.1145/1327452.1327492.

[17] R. Core Team, R: A Language and Environment for Statistical Computing, R Foundation for Statistical Computing, Vienna, Austria, 2014.

[18] Teradata Aster, Teradata Aster—Aster Discovery Platform, 2016. http://www.teradata.com/products-and-services/Teradata-Aster/teradata-aster-database (accessed 04.10.17).

[19] Teradata Aster, Aster Analytics Foundation User Guide, 2016. Version 6.20.

CHAPTER 16

Predictive Analytics for Comprehensive Energy Systems State Estimation

Yingchen Zhang*, Rui Yang*, Jie Zhang†, Yang Weng‡, Bri-Mathias Hodge*

*National Renewable Energy Laboratory, Golden, CO, United States, †University of Texas at Dallas, Richardson, TX, United States, ‡Arizona State University, Tempe, AZ, United States

OVERVIEW CHAPTER

Energy sustainability is a subject of concern to many nations in the modern world. It is critical for electric power systems to diversify energy supply to include systems with different physical characteristics, such as wind energy, solar energy, electrochemical energy storage, thermal storage, bioenergy systems, geothermal, and ocean energy. Each system has its own range of control variables and targets. To be able to operate such a complex energy system, big-data analytics become critical to achieve the goal of predicting energy supplies and consumption patterns, assessing system operation conditions, and estimating system states—all providing situational awareness to power system operators. This chapter presents data analytics and machine learning-based approaches to enable predictive situational awareness of the power systems.

1 INTRODUCTION

Historically, the power system has been designed with dispatchable generation providing enough electricity to meet demand and additional reserves to meet contingencies. With the increase in wind and solar integration, better methods to forecast grid conditions will be needed to accommodate high penetrations of these variable and uncertain clean energy technologies. Another challenge with future grids is the ability to account for distributed energy resources (DER) that may not be under the operational control of the utility. Electricity end users increasingly actively participate in power system operations by providing energy or demand services, such as distributed photovoltaic (PV) or demand response (DR). System operators need to dramatically enhance their capability in monitoring, estimating, and predicting resource adequacy, demand fluctuations, as well as system health to be able to address the ever-changing grid challenges.

New sensing and monitoring capabilities provide power system operators with new opportunities to understand and predict the system conditions at all levels. At the transmission level, deployment of phasor measurement units (PMUs) has provided new capabilities in most major interconnections. In distribution

Big Data Application in Power Systems. https://doi.org/10.1016/B978-0-12-811968-6.00016-4

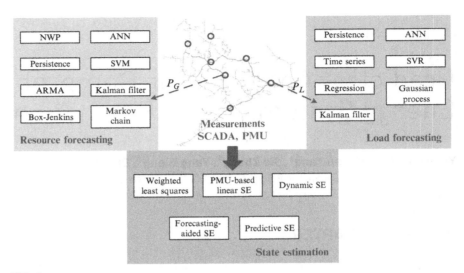

FIG. 1
Overview.

systems, smart meter-based advanced metering infrastructure has enabled two-way communication between system operators and customers. The adoption of solar and wind generators have created the need for power system operators to have better awareness of macro and microclimate conditions, such as irradiance and wind speeds. Smart devices such as inverters and electrical vehicle charging stations possess self-awareness and monitoring capabilities that can be used for centralized controls. When all of these new sensing technologies are adopted in large numbers, it creates an enormous amount of data at finer geographic and temporal scales than power system operators are used to handling. Therefore, big data analytics can play an increasingly important role in the modern power system operations in terms of estimation and prediction of comprehensive system conditions.

In this chapter, various methods to forecast the output of variable renewable energy resources, such as wind and solar are first reviewed. Then, different load forecasting methods are discussed. Finally, how to estimate and forecast grid states is explained. Fig. 1 gives an overview of the topics covered in this chapter.

2   RESOURCE FORECASTING

2.1   Renewable Forecasting

Renewable energy resources, particularly wind and solar energy, have become a primary focus in government policies, academic research, and the power industry. Among various renewables, wind energy is considered as one of the most promising alternatives [1]. However, the variable and uncertain nature of the renewable resources, such as wind and solar, may affect the economic and reliable operations of the power system [2], especially with increasing penetration levels of wind and solar power [3]. Therefore, it is important and desired to improve the accuracy of the wind and solar forecasting that is used in power system scheduling.

2.2 Wind Forecasting

2.2.1 Wind Forecasting Overview

Different wind forecasting models have been developed in the literature, and they can be generally classified into three groups [4]: (i) physical models that are usually based on numerical weather prediction (NWP) models; (ii) statistical methods, most of which are intelligent algorithms based on data-driven approaches; and (iii) hybrid physical and statistical models.

NWP models simulate the physics of the atmosphere utilizing physical laws and boundary conditions. There exist a variety of challenges when directly adopting NWP models for wind forecasting, such as the accuracy, spatial and temporal resolutions, domain, and hierarchical importance of the physical processes. Based on the domain coverage, the NWP models could be divided into limited area models (LAMs) and global models (GMs) [5]. Several GMs [6–8] have been developed to fulfill different forecasting needs, such as the Global Forecast System (GFS) and the Integrated Forecast Model. LAMs normally produce higher-resolution forecasts than GMs. Different LAMs have been developed for forecasting in different domains. Some of these are the High-Resolution Limited Area Model [9], ALADIN [10], the Fifth-Generation Mesoscale Model [11], and High-Resolution Rapid Refresh (HRRR) [12].

Statistical models are trained using historical data and usually outperform NWP models in very short-term forecasting (within 1-hour ahead) [13], partially due to the fact that NWP models normally take long time (e.g., hours) to run. Both linear and nonlinear methods have been widely applied to wind forecasting. Linear models, such as autoregressive moving average (ARMA) methods [14, 15], Box-Jenkins methods [16], Kalman filter [17], and Markov Chain models [18, 19], are most widely used in the literature. Artificial neural networks (ANN) and support vector machine (SVM) are the two most popular nonlinear methods for wind forecasting. Different ANN and SVM models have been compared in many studies, and they performed inconsistently under different conditions [20–22].

2.2.2 Big Data-Driven Wind Forecasting

Much of the historical attention on wind and solar power forecasting has focused on the day-ahead time frame due to the economic impacts of variable renewables interacting with electricity markets and the unit commitment process. Many of the new applications that require wind and solar power forecasting require more frequent updates and more granular forecasts. Big datadriven methods have been recently used to improve the accuracy of wind power forecasting at different temporal and spatial scales [23–25].

Hours- to Day-Ahead NWP-Based Wind Forecasting

The impact of atmospheric dynamics becomes more important for short-term horizons from a few hours to day ahead, and NWP models often produce more accurate forecasts on these timescales. For example, the big data-driven Wind Forecast Improvement Project (WFIP) was performed to improve short-term wind power forecasts and determine the value of these improvements to grid operators [23]. WFIP encompassed two study regions: the northern study region and the southern study region. The WFIP southern study region covers most of the Electric Reliability Council of Texas (ERCOT) service area, as shown in Fig. 2. The data from additional sensors deployed for the WFIP project, as well as tower data from a set of participating wind power plants within Texas, were assimilated into most of the ensemble members; however, the data from the project sensors were withheld from some ensemble members to gauge their impact on the forecasts [23]. The existing wind forecasting system at ERCOT used Mesoscale Atmospheric Simulations System model forecasts with initial

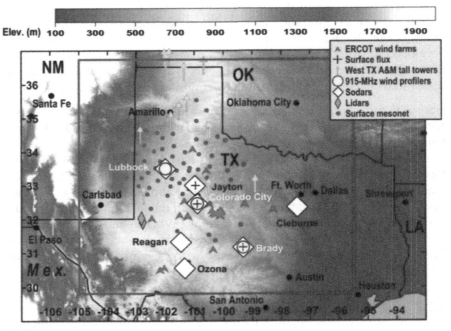

FIG. 2
WFIP southern study region in ERCOT [24].

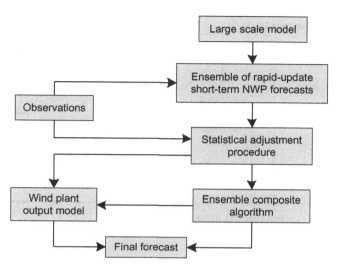

FIG. 3
Overall framework of the wind power forecasting system [24].

conditions and boundary conditions from the GFS and the North American Mesoscale Model. Ensemble methods have been shown to produce more accurate forecasts. The WFIP forecast system consists of an ensemble of high-resolution rapid-update NWP models. Each of these ensemble members incorporates a variety of model configurations, physics parameterizations, and data assimilation techniques. The purpose of integrating all of these ensemble members into one system is to construct an optimized composite forecast able to predict forecast uncertainty and assess the relative performance of different modeling approaches. Fig. 3 shows the overall framework of the wind power forecasting system. The WFIP ensemble members include [23, 24]:

(1) The National Oceanic and Atmospheric Administration's 3-km HRRR model, updated hourly.
(2) Nine NWP models updated every 2 hours on a 5-km grid:
    (a) Three configurations of the Advanced Regional Prediction System.
    (b) Three configurations of the Weather Research and Forecasting (WRF) model.
    (c) Three configurations of the Mesoscale Atmospheric Simulations System.
(3) An Advanced Regional Prediction System model updated every 6 hours on a 2-km grid.

Minutes- to 2-Hour-Ahead Machine-Learning-Based Wind Forecasting
Due to the nonlinear and nonstationary characteristics of wind speed, it is challenging to develop a generic model based on a single machine learning

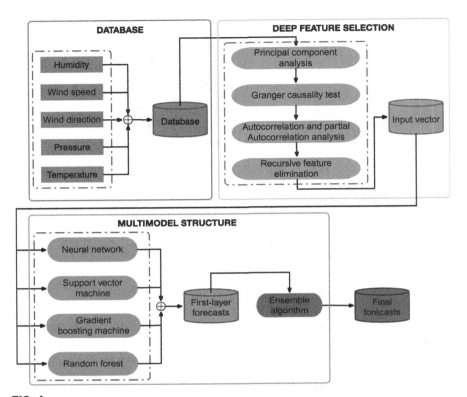

FIG. 4

The framework of the ensemble forecasting model [25].

algorithm that can produce the best forecasts at many different spatial and temporal scales. Big data-driven methodologies could potentially improve the accuracy of the wind forecasting. Fig. 4 illustrates a big data-driven multi-model wind forecasting methodology with deep feature selection methodology [25]. First, features extracted from the data variables are determined by a deep feature selection procedure and serve as inputs to the model. Four independent feature selection methods are included in the procedure and implemented sequentially. The first-layer machine learning models are built based on the selected feature combination. These models forecast wind speed or wind power as the output. A blending model is developed in the second layer to combine the forecasts produced by different algorithms from the first layer, and to generate both deterministic and probabilistic forecasts. Parameters of these models are optimally tuned by the grid search technique. Machine learning algorithms have distinctive advantages. For instance, ANN algorithms are adaptive by choosing different learning functions and loss functions, but have overfitting issues when the training dataset is not long enough. SVM is efficient to train and can provide relatively accurate results, but they are memory intensive and hard to tune. Tree ensemble algorithms like random forests and gradient boosting machines can avoid overfitting issues. The blending model is expected to integrate the advantages of different algorithms by canceling or smoothing the local forecasting errors.

2.2.3 Wind Forecasting Datasets

The Wind Integration National Dataset (WIND) Toolkit was created through the collaborative efforts of National Renewable Energy Laboratory and 3TIER and has been funded by the US Department of Energy, Office of Energy Efficiency and Renewable Energy, Wind and Water Power Technologies Office [26, 27]. The WIND Toolkit supports the next generation of wind integration studies. The WIND Toolkit includes meteorological conditions and turbine power for over 126,000 sites in the continental United States for the years 2007–13 (Fig. 5), which was produced with the WRF model version 3.4.1. The meteorological dataset has 2×2-km horizontal resolution, nine vertical levels, a temporal resolution of 5 minutes, and covers a 7-year period (2007–13). The simulations include a spin-up period of 48 hours. The model

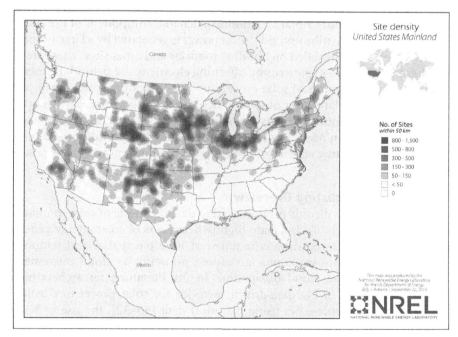

FIG. 5
Map showing the site density of the 126,692 wind sites [27].

was restarted every month, and we used scale selective grid nudging. There are three main datasets included in the WIND Toolkit [27]:

(1) The meteorological dataset includes basic information on the weather conditions in each 2 × 2-km grid cell, for example, wind profiles, atmospheric stability, and solar radiation data.
(2) A power dataset was created using the wind data and site-appropriate turbine power curves to estimate the power produced at each of the turbine sites.
(3) A forecast dataset includes forecasts for 1-, 4-, 6-, and 24-hour forecast horizons.

2.3 Solar Forecasting

Solar power penetration in the United States is growing rapidly, and the Sun-Shot Vision Study reported that solar power could provide as much as 14% of US electricity demand by 2030 and 27% by 2050 [28]. At these high levels of solar energy penetration, solar power forecasting will become very important for electricity system operations. Solar forecasting is a challenging task, and solar power generation presents different challenges for transmission and distribution networks [29]. On the transmission side, solar power takes the form of centralized solar power plants, a nondispatchable component of the generation pool. On the distribution side, solar power is generated by a large number of distributed arrays installed on building rooftops and other sites. These arrays can alter traditional load patterns by offsetting electricity use behind the meter. Integrating large amounts of solar power into the grid can magnify the impact of steep ramps in solar power output, which poses challenges to system operators' ability to account for solar variability. Forecast inaccuracies of solar power generation can result in substantial economic losses and power system reliability issues because electric grid operators must continuously balance supply and demand.

2.3.1 Solar Forecasting Overview

Solar power output is directly proportional to the magnitude of solar irradiance incident on PV panels. To integrate high penetrations of solar energy generation, accurate solar forecasting is required at multiple spatial and temporal scales. Solar irradiance variations are caused primarily by cloud movement, cloud formation, and cloud dissipation. In the literature, researchers have developed a variety of big data-driven methods for solar power forecasting, such as statistical approaches using historical data [30–32], the use of NWP models [33, 34], tracking cloud movements from satellite images [35], and tracking cloud movements from direct ground observations using sky cameras [35, 36]. NWP models are the most popular method for forecasting solar

Table 1 Solar Forecasting Methodologies [38]

	Methods	Description/Comment	Forecast Horizons
Physical approaches	NWP models	NWP models are the most popular method for forecasting solar irradiance more than 6 hours or days in advance	4 hours to days ahead
	Total Sky imagers (TSI)	TSIs are used to extract cloud features or to forecast short-term GHI	0–30 minutes ahead
Statistical approaches	Statistical methods	Statistical methods were developed based on autoregressive or artificial intelligence techniques for short-term forecasts	0–6 hours ahead
	Persistence forecasts	Persistence of cloudiness performs well for very short-term forecasts	<4 hours ahead

irradiance several hours or days in advance. Mathiesen and Kleissl [34] analyzed the global horizontal irradiance (GHI) in the continental United States forecasted by three popular NWP models: the North American Model, the GFS, and the European Centre for Medium-Range Weather Forecasts. Lorenz et al. [37] showed that cloud movement-based forecasts likely provide better results than NWP forecasts for forecast timescales of 4 hours or less; beyond that, NWP models tend to perform better. In summary, forecasting methods can be broadly characterized as physical or statistical. The physical approach uses NWP and PV models to generate solar power forecasts, whereas the statistical approach relies primarily on historical data to train models [38]. Recent solar forecasting studies [39, 40] integrated these two approaches by using both physical and historical data as inputs to train statistical models. These solar forecasting methods are summarized in Table 1.

2.3.2 Big Data-Driven Solar Forecasting

Hours- to Day-Ahead NWP-Based Solar Forecasting

As part of the project work performed under the SunShot Initiative's Improving the Accuracy of Solar Forecasting program, a system for improving solar forecast, Watt-sun, has been developed. Watt-sun uses big-data information processing technologies and applies machine-learnt, situation-dependent blending of multiple models to enhance system intelligence, adaptability, and scalability. The algorithm which provides the best accuracy for the last 2 days is selected for future solar power forecasts. Numerical results show 30% improvement in solar irradiance/power forecast accuracy compared with forecasts based on the best individual method, and 10% improvement compared with model forecasts processed by machine learning methods without situation categorization. Detailed information about the Watt-sun forecast method can be found in [40, 41].

Minutes- to 2-Hour-Ahead Sky Imaging-Based Solar Forecasting

Fig. 6 shows a short-term 1-hour-ahead GHI classification forecasting framework [42]. In addition to pattern recognition, the framework contains two other parts: the data preprocessing module and the GHI forecasting module. In the data preprocessing module, a three-step technique is applied to improve the pattern recognition and forecasting performance. The forecasting module is divided into three model sets: Model Set I (MSI), Model Set II (MSII), and Model Set III (MSIII). The first two sets of models forecast GHI for the first 4 daylight hours of each day. Then the GHI values in the remaining hours are forecasted by an optimal machine learning model determined based on a weather pattern classification model in the third model set. The weather pattern is determined by an SVM classifier.

2.4 Renewable Forecasting Performance Evaluation Metrics

A suite of generally applicable, value-based, and custom-designed metrics for renewable forecasting considering different time horizons, geographic locations, and applications was developed by Zhang et al. [29]. The developed renewable forecasting metrics can be broadly divided into four categories: (1) statistical metrics for different time and geographic scales, (2) uncertainty quantification and propagation metrics, (3) ramp characterization metrics, and (4) economic metrics. A brief description of the metrics is given in Table 2, and detailed information about each metric can be found in [29, 43]. A smaller value indicates a better forecast for most of the metrics, except for Pearson's correlation coefficient, skewness, kurtosis, distribution of forecast errors, and ramp characterization metrics.

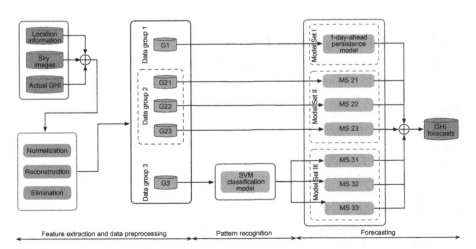

FIG. 6

Overall framework of the short-term GHI forecasting based on sky imaging and pattern recognition [42].

Table 2 Metrics for Renewable Energy Forecasting [29]

Type	Metric	Description/Comment
Statistical metrics	Distribution of forecast errors	Provides a visualization of the full range of forecast errors and variability of solar forecasts at multiple temporal and spatial scales
	Pearson's correlation coefficient	Linear correlation between forecasted and actual wind/solar power
	Root mean square error (RMSE) and normalized root mean square error (NRMSE)	Suitable for evaluating the overall accuracy of the forecasts while penalizing large forecast errors in a square order
	Maximum absolute error (MaxAE)	Suitable for evaluating the largest forecast error
	Mean absolute error (MAE) and mean absolute percentage error (MAPE)	Suitable for evaluating uniform forecast errors
	Mean bias error (MBE)	Suitable for assessing forecast bias
	Kolmogorov-Smirnov test integral (KSI) or KSIPer	Evaluates the statistical similarity between the forecasted and actual wind/solar power
	OVER or OVERPer	Characterizes the statistical similarity between the forecasted and actual wind/solar power on large forecast errors
	Skewness	Measures the asymmetry of the distribution of forecast errors; a positive (or negative) skewness leads to an over-forecasting (or under-forecasting) tail
	Excess kurtosis	Measures the magnitude of the peak of the distribution of forecast errors; a positive (or negative) kurtosis value indicates a peaked (or flat) distribution, greater than or less than that of the normal distribution
Uncertainty quantification metrics	Rényi entropy	Quantifies the uncertainty of a forecast; it can utilize all of the information present in the forecast error distributions
	Standard deviation	Quantifies the uncertainty of a forecast
Ramp characterization metrics	Swinging door algorithm	Extracts ramps in wind/solar power output by identifying the start and end points of each ramp
Economic metrics	95th percentile of forecast errors	Represents the amount of nonspinning reserves service held to compensate for wind/solar power forecast errors

3 USER ENERGY SYSTEM STATE ESTIMATION

3.1 Overview

Power system load forecasting has its use in planning as well as system operations. Better load forecasting can result in reduced operational costs and improved system reliability. Normally, system operators pay more attention to the aggregated load at a distribution substation, which could contain many distribution feeders because the load and generation are physically separated in the confidential hierarchy grid. With the increased DER penetration, the distribution level load forecasting has gained more attraction. The distribution level load forecasting faces more challenge because it contains more stochastically abrupt deviations. Further, the load itself can become resource when providing DR, the needs to estimate the DR capability at future time are critical for both operators and DR providers to better schedule the service. This section will introduce the data-driven load forecasting as well as DR forecasting using user energy system state estimation.

3.2 Load Forecasting

The load forecasting time horizon can be divided into short term, medium term, and long term. Short-term load forecasting (STLF) usually predicts the electricity demand from 1 hour to 1 week ahead and is used to facilitate the day-ahead and real-time resource scheduling in power systems [44]. The term "very short-term load forecasting" is also used, which focuses on demand forecasts less than 1 hour ahead [45, 46]. Medium-term load forecasting (MTLF) is usually from 1 week to 1 year ahead and is used for maintenance scheduling as well as contract negotiation in energy markets [47]. Long-term load forecasting (LTLF), which covers a time period of more than 1 year, is mainly used for expansion planning of power grids in order to accommodate future demand [47].

Many models and methods have been developed for load forecasting. The most commonly used methods for MTLF and LTLF are the end-user approach, which takes into account the end-user behaviors [48]; and the econometric approach, which studies the relationship between energy consumption and other factors, such as weather conditions and economic factors [47]. During the past decades, research efforts have been focused on STLF, since it plays an essential role in power system operation. Various approaches have been applied to STLF, including time-series methods [49, 50], regression-based methods [51, 52], and artificial intelligence-based methods [53–55].

An overview of the most commonly used STLF models and approaches is provided in the following. More comprehensive reviews of load forecasting methods, especially STLF methods, can be found in [47, 56, 57].

3.2.1 Conventional Methods

Conventional STLF approaches adopt statistical methods to model the relationship between demand and external factors that may influence demand. A variety of models have been used for load forecasting, including persistence method, time-series methods, regression analysis, and Kalman filtering-based methods.

Persistence model uses the previous day (or the corresponding day in the previous week) as a prediction. Such a method is sensitive to the rare events (e.g., the day in the last week is the Thanksgiving Eve day).

Time-series approaches model electricity demand as a time-varying random process and use historical data to predict future demand by exploiting the internal pattern of the time-series data. The most often used time-series methods include ARMA [58], autoregressive integrated moving average (ARIMA) [50], ARMA with exogenous variables (ARMAX) [59, 60], and autoregressive integrated moving average with exogenous variables (ARIMAX) [47]. These methods model future demand as a function of historical values of load consumption and other factors, such as time and weather.

Regression analysis is employed to model the relationship between electricity demand and external factors, such as weather, the hour of the day, and customer type. Multiple regression techniques have been applied to load forecasting, including linear regression [51], nonparametric regression [61], and robust regression [62].

State-space models have also been used to model the time-varying load, and Kalman filtering-based algorithms are adopted to provide a recursive update of the load in the near future [63, 64].

Since the actual relationship between the load and influencing factors is nonlinear and complex, the major challenge of using the aforementioned methods for load forecasting is how to develop an *accurate* model to represent electricity demand.

3.2.2 Artificial Intelligence-Based Methods

In recent years, artificial intelligence-based methods, such as ANNs and support vector regression (SVR), have been applied to better model the complex relationship between electricity demand and other factors and thereby better forecast future demand.

ANNs are composed of a number of neurons, each of which has an activation function converting its input to output. The neurons are interconnected to form a multilayer network, which is used to define the relationship between the input variables and output variables. In load forecasting, ANN models are trained to find the mapping between the load and influencing factors using historical data. The mapping function is then used to forecast the future load [53, 65]. An extensive review of neural network methods for STLF is given in [66].

The SVR method uses a nonlinear mapping function to transform the input data into a high-dimensional feature space, and then a linear regression is performed in this high-dimensional space [67].

Both the ANN and SVR methods have the capability to use nonlinear models to map the input and output variables and better capture the complex relationship between the load and other influencing factors. A detailed review of various computational intelligence techniques for STLF is provided in [68], including ANN, SVR, and other artificial intelligence-based methods.

Moreover, hybrid approaches are also common in load forecasting, which combines two or more aforementioned methods in order to achieve better forecasting accuracy [69–71].

3.2.3 Gaussian Process-Based Method

Most load forecasting methods predict the electricity demand at the aggregated level. For forecasting the demand of individual customers, different models will be needed for different customers since each may behave differently. Therefore, there is a need for accurate and personalized demand forecasting algorithms for individual customers.

In the following, a Gaussian process-based load forecasting algorithm is introduced. The Gaussian process-based method is highly flexible, since it can easily incorporate different existing regression-based methods. Therefore, it can provide a personalized demand forecasting algorithm for each individual customer by incorporating the suitable model for the customer's behaviors.

To visualize the distribution of load data, the Pacific Gas and Electric Company (PG&E) and OhmConnect data are analyzed for a proper statistical model. Fig. 7A and B show the distributions of two normalized loads at two different time stamps. The dashed lines are the fitted Gaussian distributions to the data. Fig. 7C shows the joint distribution of the same load in two time indices of a day. Together with these three figures, the data analysis shows that the loads at different time slots can be modeled as a multivariate Gaussian distribution. In order to check the Gaussianity rigorously, the K-squared test [72] and the Jarque-Bera test [73] can be conducted.

Here, the data before the demand forecasting period of a particular day are defined as a vector x of C elements, which may include features such as load, temperature [74], and day of the week. Without loss of generality, the case when the target demand forecasting is a scalar y is considered. An underlying function $y(x)$ is inferred from a training set $T = \{(x_i, y_i) | i = 1, ..., n\}$, where i is the historical day index and n is the total number of days in the training dataset. For compact notation, all input data are combined in a $C \times n$ matrix X, and the target variables in a vector y. Therefore, the training set can be written as $T = (X, y)$. Similarly, $\{(X^*, y^*)\}$ denotes the testing data, where X^* is known, and y^* is unknown.

The joint distribution of the training and testing data is:

$$\begin{pmatrix} y \\ y^* \end{pmatrix} = \left(\begin{bmatrix} m(X) \\ m(X^*) \end{bmatrix}, \begin{bmatrix} C(X,X), & C(X,X^*) \\ C(X^*,X), & C(X^*,X^*) \end{bmatrix} \right)$$

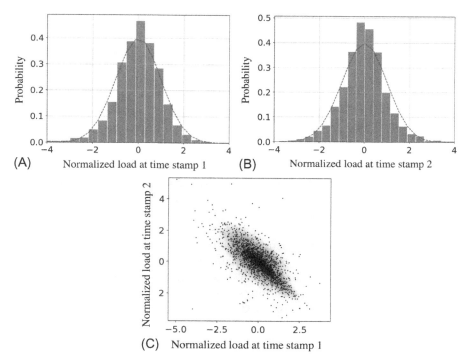

FIG. 7
Gaussianity of residential customers' loads at different time indices. (A) Histogram of normalized load distribution at one time index. (B) Histogram of normalized load distribution at another time index. (C) Joint load distribution of the two time indices.

Therefore, a Gaussian process is specified by a mean, $m(\cdot)$, and a covariance function, $C(\cdot, \cdot)$. The mean estimate under the Gaussian process framework is:

$$y^* = m(X^*) + C(X^*,X)C(X,X)^{-1}(y - m(X)) \tag{1}$$

And the covariance estimate is:

$$Cov(y^*) = K(X^*,X^*) - K(X^*,X) \cdot [C(X,X) + \lambda I]^{-1} C(X,X^*) \tag{2}$$

where λ is a hyperparameter.

After setting up the framework, data can be used to design the covariance functions.

- Embedding distance-based correlation: A future load equals the summation of a past load and the change between them. This leads to a stronger correlation of load values between two time slots closer to each other. As shown in Fig. 8, the correlation decreases from a 1-hour interval to an 11-hour interval for the PG&E dataset and OhmConnect dataset.

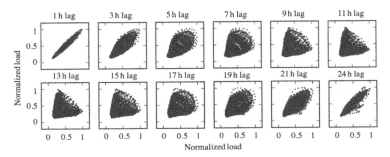

FIG. 8

Autocorrelation with different intervals. The x-coordinate represents load at time t. The y-coordinate represents load at time $t + t_0$, where t_0 is the time interval between loads at two indices.

Therefore, a squared exponential covariance is employed based on Euclidean distance.

- Embedding periodic pattern: From Fig. 8, the correlation increases again to a 24-hour interval, indicating a daily periodic pattern. As different residential customers may have different periodicities with different weights, periodic patterns are automatically detected and embedded into the covariance function.
- Embedding piecewise linear pattern in temperature: As loads are usually sensitive to temperature, the temperature is also used in the covariance function.

Fig. 9 shows the simulations of the Gaussian process-based method with the PG&E dataset with different training and testing lengths.

- Use 2 weeks of hourly load as a training set to forecast 1 week of hourly load.
- Use 4 weeks of hourly load as a training set to forecast 1 week of hourly load.
- Use 1 week of hourly load as a training set to forecast 4 weeks of hourly load.

As shown in Fig. 9, the mean estimate learns features of periodicity and temperature, and the 95% confidence zone covers true values for the week in the forecast. This happens even when the training set length is very short. It gives both highly confident mean estimates and trends of the uncertainty.

In addition to the load domain analysis in Fig. 9, the results for the error domain analysis of three different methods are shown in Fig. 10 for all users in the PG&E data. These methods include a moving average model, a regression

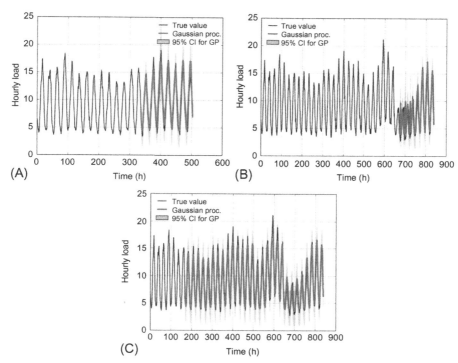

FIG. 9

Estimation comparisons with different training set lengths and different forecasting time horizons. (A) 2 weeks training, 1 week testing. (B) 4 weeks training, 1 week testing. (C) 1 week training, 4 weeks testing.

model taking into account the time of week and the temperature, and the Gaussian process-based method.

As shown in Fig. 10, the Gaussian process-based approach has the smallest error, no matter how long the forecasting is. Especially for very long-time horizons—up to 17 weeks—the Gaussian process-based method is much more robust than the other two methods.

Comparing to other load forecasting methods, the Gaussian process-based method has a mean estimate that is equivalent or better than the estimates generated by currently used baseline estimation methods. Also, Gaussian process-based method naturally provides the prediction of uncertainties inherent in the customer loads while deterministic method does not provide such functionality. Finally, the accuracy and the ability for providing confidence intervals are enhanced by Gaussian process-based method's flexibility of an adaptive component design according to customer behaviors.

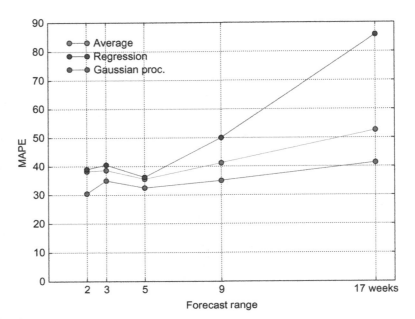

FIG. 10
Error domain analysis.

3.3 User Energy System State Estimation: Demand Response

Electricity consumers are becoming active resources in terms of providing DR. The Federal Energy Regulatory Commission defines DR as electric usage adjustments by the consumers from their normal consumption patterns [75, 76]. Such adjustments are in response to (1) changes in the price of electricity over time, or (2) incentive payments designed to induce lower electricity consumption at usage peaks or when system reliability is jeopardized [77]. Different than load forecasting, DR capability of certain user energy systems can be estimated through the baseline evaluation.

Traditional DR programs are usually designed for large commercial customers, as their usage is predictable. Therefore, the deterministic baseline evaluation can be used for the electricity consumption estimation based on the no-DR period [78, 79]. The difference between an estimated normal consumption and the actual usage is subsequently used to calculate the savings as shown in Fig. 11 [80–82].

Deterministic methods such as simple load average and temperature-based linear regression are used for commercial customers with satisfactory results [74, 83, 84]. For example, DR programs had great success with large power consumption users [85]. GreenTech Media reported in 2013 that 8.7 million in revenue had been generated within 7 months in the PJM Interconnection by conducting DR in system operation with mostly large customers, reducing the use of 80 traditional coal power plants in peak power supply [85].

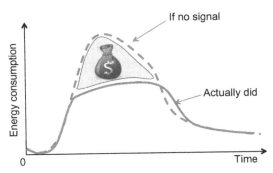

FIG. 11
Deterministic baseline estimation and its rewarding mechanism.

Although large customers currently create a significant portion of the revenue in the DR programs, the smaller residential consumers hold the key to potential growth in the DR customer number and the DR revenue. For example, 9.3 million customers had participated in DR programs by March 2016 in the United States, but more than 90% of them are in the residential sector [86]. In addition to making profits, DR at the residential level is becoming an attractive solution for the radically increased renewable generation to balance local power flow [87].

The drawback to traditional baseline estimation methods lies in their failure to utilize the historical data to capture the dynamics of complex user behaviors [84, 88], which is of particular importance for small to medium consumers with more variability [89]. One can also use the idea of control groups [90–92]; however, it may be hard to uniquely define the best control group that properly captures the uncertainty in user behavior of the treatment group (enrolled customers). As large utility companies have started to make historical data available to approved third parties (e.g., the Green Button initiative in California [93]) machine learning-based techniques can be used to capture the uncertainty at the consumer level. Similar to the content in load forecasting, Gaussian process regression can be used for DR as well. After which, automatic feature extractions can be employed to dynamically embed consumer behaviors into the mean and variance estimates.

4 POWER SYSTEM STATE ESTIMATION

4.1 Overview

Power system state estimation was developed decades ago and now forms the backbone of all control center applications. Operators collect thousands of measurements from meters and relays through supervisory control and data acquisition (SCADA) systems to solve for the system states, namely voltage magnitude and angle at all buses. With the fully solved system state variables, the power flow of the system can be calculated. The SCADA system can be seen as an early "big data" resource in power system. With the addition of PMUs, system state variables at many locations can be directly monitored. Advanced state estimation techniques have been developed and implemented for a better awareness of grid state conditions.

4.2   Conventional Nonlinear State Estimation

Power system state estimation aims to find the system state at a point in time (i.e., voltage magnitudes and phase angles) based on a set of measurements. Conventional static estimation uses the measurements, which are taken at one snapshot in time to estimate the corresponding system states of this time instance.

Let $x = [\theta_2, \theta_3, ..., \theta_n, |V_1|, |V_2|, ..., |V_n|]^T$ denote the vector of system states including the voltage magnitude and the phase angle at every bus except the phase angle at the reference bus 1 is set to be $\theta_1 = 0$. Let $z = [z_1, z_2, ..., z_m]^T$ represent the vector of collected measurements including the active and reactive power injections at buses, power flows in the network, and voltage magnitudes. The relationship between the collected measurements and the system states can be written as:

$$z = h(x) + \omega \tag{3}$$

where $h = [h_1(x), h_2(x), ..., h_m(x)]^T$ is the vector of nonlinear functions mapping the states to measurements and $\omega = [\omega_1, \omega_2, ..., \omega_m]^T$ is the measurement noise vector. Usually, the measurement noises are assumed to be independent Gaussian random variables with zero mean and variance σ_i^2 (i.e., $\omega \sim \mathcal{N}(0, \Sigma)$), where the covariance matrix Σ is a diagonal matrix with the ith diagonal element as σ_i^2. In practice, the measurement set is redundant, resulting in an over-determined nonlinear equation system (3).

The goal of state estimation is to find an estimate $\hat{x}$ of the true state x given the measurement set z. The most widely used approach for static state estimation is the weighted least squares (WLS) method. The best estimate of system states is found by:

$$\hat{x} = \arg \min_x J(x) = \arg \min_x (z - h(x))^T \Sigma^{-1} (z - h(x)) \tag{4}$$

Since the optimization problem is a nonlinear least squares problem, the Gauss-Newton method is used to find the best estimate in an iterative manner. At iteration k, the estimated state vector x^k is updated using [94]:

$$x^{k+1} = x^k + \Delta x^k \tag{5}$$

$$G(x^k)\Delta x^k = H^T(x^k)\Sigma^{-1}(z - h(x^k)) \tag{6}$$

$$G(x^k) = H^T(x^k)\Sigma^{-1}H(x^k) \tag{7}$$

where Δx^k is the update on the estimated system state at iteration k, and $H(x^k)$ is the Jacobian matrix of the vector function $h(x)$ evaluated at x^k as:

$$H(x^k) = \frac{\partial h(x)}{\partial x}\bigg|_{x=x^k} \tag{8}$$

The matrix $G(x)$ is the gain matrix, and Eq. (6) is referred to as the normal equation of the WLS algorithm.

The iterative procedure is terminated when the mismatch between the calculated measurement values using the system states and the actual measurement values is below a predetermined threshold.

4.3   PMU Data-Based Linear State Estimation and Dynamic State Estimation

With the broad development of PMUs in power systems to directly measure the system states, state estimation problems have naturally expanded to include PMU measurements. Several new techniques of state estimation have been developed to better understand grid conditions under all circumstances. Among them, linear state estimation [95] and dynamic state estimation [96] have seen broader development.

Because of the simultaneous collection of measurements across the system through PMUs, the linear model of the system can be constructed and solved [95]. The synchrophasor-based linear state estimation problem considers no correlation between adjacent states.

$$z = \begin{bmatrix} V \\ I_{\text{flow}} \end{bmatrix} = \begin{bmatrix} II \\ \gamma A + \gamma_s \end{bmatrix} x + e \tag{9}$$

where z is the measurement vector vertically concatenated voltage and current measurements and x is the complex state vector. The II matrix is an incidence matrix which identically relates the state vector to voltage measurements. The admittance matrix relates the state vector to current measurements.

Assume PMU measurement is available at two adjacent nodes, the linear equation can be solved as

$$\begin{bmatrix} V_i \\ V_j \\ I_{ij} \\ I_{ji} \end{bmatrix} = \begin{bmatrix} 1 & 0 \\ 0 & 1 \\ \gamma_{ij} + \gamma_{i0} & -\gamma_{ij} \\ -\gamma_{ij} & \gamma_{ij} + \gamma_{i0} \end{bmatrix} \cdot \begin{bmatrix} V_i \\ V_j \end{bmatrix} \tag{10}$$

Another important feature of the PMU measurements is that they can capture the electromechanical dynamics of the power system [96]. Therefore, dynamic state estimation was developed to use PMU measurements to derive the dynamic states of the power system, such as the generator speed. The system dynamic model can be generally represented by a set of differential and algebraic questions:

$$\frac{dx(t)}{dt} = f(x(t), y(t), t) \tag{11}$$

$$0 = g(x(t), y(t), t) \tag{12}$$

where x and y are the dynamic and algebraic states of the system. A set of PMU measurements z can be directly related to the system states and measurement error by:

$$z = h(x(t), y(t), t) + \eta \tag{13}$$

A standard least square estimation is then performed to minimize the measurement error compared with the system dynamic model outputs, deriving a dynamic state estimation.

It is also worth mentioning that a linear state estimator can be used for system dynamic state estimation. The system dynamic models are constructed the same way, but instead of least square solutions, robust linear models (least absolute value)-based solutions can be derived [97].

4.4 Predictive State Estimation

4.4.1 Forecasting-Aided State Estimation

Unlike conventional power system state estimation which estimates the system states at a certain time sample based on measurements taken at this snapshot, forecasting-aided state estimation provides a recursive update of the state estimate by taking into account the state transition over time. Mathematical models [98] describing the state trajectory have been employed to forecast the system states in the near future, resulting in a priori state estimate. With the newly received measurement data, the forecast system states are once again refined, rendering a more accurate state estimate. By incorporating short-term state forecasting, the performance of the state estimation can be improved, especially in dealing with missing measurements.

Most existing forecasting-aided state estimation approaches are based on the Kalman filter [98, 99]. Machine learning algorithms, such as ANNs, have also been used to provide state forecasting and thereby aid the state estimation [100–102]. A more comprehensive review of forecasting-aided state estimation is given in [103].

4.4.2 Predictive State Estimation

With the increasing penetration of renewable energy resources, the system states become more variable and unpredictable. Besides individual forecasting for variable resources and loads, state forecasting can greatly help operators proactively operate the systems and reduce the operation cost [104]. This section introduces a fast and high-accurate system state forecasting approach based on historical data and a machine learning algorithm.

The historical system states from measurements such as PMUs are the training data for the state forecaster. An extreme learning machine (ELM)-based algorithm is used to forecast the future system states for the distribution system [105]. In the ELM algorithm, the input weights and biases are randomly generated, and the output weights are computed. This means the computational burden of calculating input weights and biases can be eliminated.

As shown in Fig. 12, the ELM-based system state estimator is designed as follows. In the training part, M observations are collected, and the observation set $\{\alpha, \gamma\}$ can be built as follows:

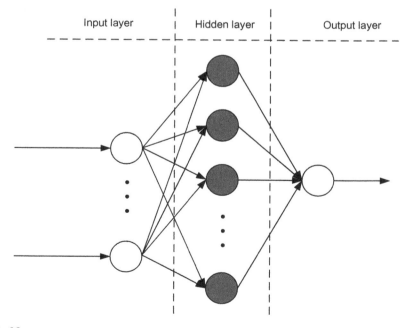

FIG. 12
Concept of the ELM method.

$$\alpha = \begin{pmatrix} s(1) & s(2) & \cdots & s(m) \\ s(2) & s(3) & \cdots & s(m+1) \\ \cdots & \cdots & \cdots & \cdots \\ s(M-m) & s(M-m+1) & \cdots & s(M-1) \end{pmatrix} \tag{14}$$

and

$$\gamma = \begin{pmatrix} s(m+1) \\ s(m+2) \\ \cdots \\ s(M) \end{pmatrix} \tag{15}$$

where $s(j)$ is the jth observation, α is the input matrix, and γ is the output vector. Consider an ELM algorithm with K neurons in the hidden layer. The activation function ψ used to model the data observation set $\{\alpha, \gamma\}$ can be written as:

$$\sum_{k=1}^{K} \psi_k(\xi_k, b_k, \alpha_i)\beta_k = \Psi_i \tag{16}$$

where $i = 1, 2, \ldots, M - m$, $k = 1, 2, 3, \ldots, K$. ξ_k is the input weight vector connecting the input α and the kth hidden neuron, β_k is the output weight connecting the kth hidden neuron and output Ψ, and b_k is the bias of the kth hidden neuron. The objective is to minimize the errors between the output of the activation function Ψ and γ. The objective function is as follows:

$$\min_{\beta} \mathcal{J} = \sum_{i=1}^{M-m} [\Psi_i - \gamma_i]^2 \tag{17}$$

where the optimization variable set is $\beta = \{\beta_1, \ldots, \beta_K\}$. $\xi = \{\xi_1, \ldots, \xi_K\}$ and $\mathbf{b} = \{b_1, \ldots, b_K\}$ are randomly generated at the first iteration. Subsequently, the ELM can be used to forecast future system states with the optimized parameters.

The IEEE 123-bus distribution system is shown as a test example to illustrate the behavior. A set of load data from a real utility SCADA system is used for evaluating the ELM-based state forecasting approach. As shown in Fig. 13, the sample rate is 1 Hz, and the time period is 12 days, with a total data length of 1,036,800 points. There are a lot of abrupt stochastic deviations in the load profile of the distribution system, as shown in Fig. 13, which can differ greatly from the transmission system. In the numerical results, bus 47, 49, 68, 76, and 83 are chosen with the load profile in Fig. 13. Then the system states are computed to evaluate.

The training data are five times the testing data. To evaluate the ELM-based forecasting method comprehensively, the sliding window test is used to traverse the whole system state data. For example, the sliding window test employed with 1-hour-ahead forecasting can be illustrated as follows.

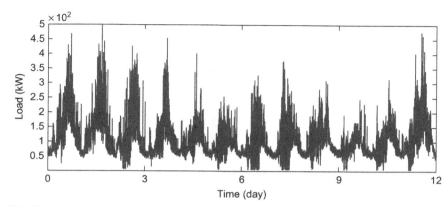

FIG. 13
Load profile of IEEE 123-bus distribution system.

1. First, the system state data section from the first to the fifth hours is taken as training data to determine the best parameters and build the forecasting model.
2. In the second step, the system state data section from the fifth to the sixth hours is taken as the test data to evaluate the performance of the forecasting model.
3. In the third step, for the next round forecasting, the training data section moves forward from the second to the sixth hours, and the test data section moves from the sixth to the seventh hours.
4. When the test data section moves to the end of the load data, the sliding window test with 1-hour forecasting is completed.

Short-term system state forecasting is studied at an hourly level in the literature [106, 107]. The performance of the ELM-based state forecasting method at different time scales is shown in Table 3. Using the ELM-based state forecasting method, the 1 hour-ahead has the best performance and the 16 hour-ahead forecasting has the largest forecasting errors. However, the MAPEs of voltage forecasting are below 2.00%, with an average of 1.370%. The MAPEs of angle forecasting are below 2.50%, with an average of 1.872%.

Table 3 Performance of the ELM approach

Forecasting Type	Voltage (%)	Angle (%)
1 hour ahead	1.131	1.578
2 hours ahead	1.182	1.628
4 hours ahead	1.289	1.752
8 hours ahead	1.474	1.975
16 hours ahead	1.775	2.431

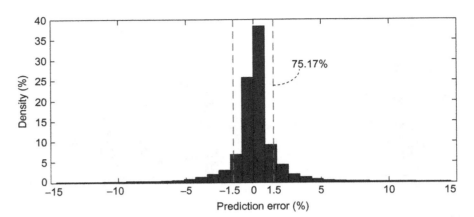

FIG. 14
Percent error of 8-hour-ahead forecasting using ELM: voltage magnitude.

As shown in Fig. 14, in the histogram of voltage forecasting error, it is noticed that more than 75% of the errors are accumulated between (−1.5%, 1.5%). Similarly, as shown in Fig. 15, in the histogram of voltage forecasting error, it is noticed that more than 75% of the errors are accumulated between (−2.3%, 2.3%). Hence, in most of the time, the ELM-based approach can forecast the system states with very high accuracy. In some rare cases, the forecasting errors for voltage magnitudes and angles are larger than ±10%. However, the cases with forecasting errors larger than ±10% are below 1% for both the voltage magnitudes and angles. The center of the voltage and angle forecasting error is 0, which indicates that the ELM-based approach performs accurately and is unbiased.

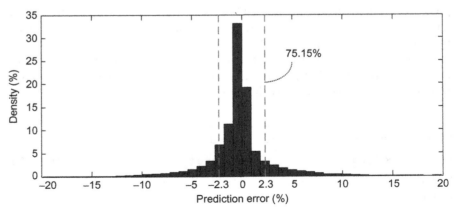

FIG. 15
Percent error of 8-hour-ahead forecasting using ELM: voltage angle.

Table 4 Comparison With Different Forecasting Approaches

	ANN	ARIMA	GA-SVR	Proposed Approach
MAPE (%)	1.702	9.972	3.146	1.725
Time (seconds)	897.2	25.7	767.3	17.1

To compare the performance of the ELM forecasters with the most common ones in [108–111], the 8-hour-ahead forecasting is chosen as an example. The MAPE and time consumption are the average of voltage and angle forecasting. As shown in Table 4, the ANN has the best accuracy performance in MAPE comparison; however, it takes more than 40 times longer than the ELM approach. Furthermore, the accuracy of ANN is very close to the ELM approach. Compared to the other approaches, the ELM approach has the best balance between forecasting accuracy and time consumption. Overall, machine learning methods demonstrate potential of accurately forecasting the power system states.

4.5 Distribution System State Estimation

The distribution grid has been undergoing a dramatic transformation in recent years. The penetration level of DERs such as PV systems keeps increasing and the electricity consumers have become active players because of their capabilities of performing demand side management. The distribution system becomes more complex while more resources are available to provide controllability and flexibility at the same time. These changes impose great challenges as well as opportunities on distribution system operations. Distribution system operators are in need of advanced energy monitoring and management systems in order to properly monitor the system states and determine the appropriate control actions of available resources. Therefore, distribution system state estimation plays an important role in the distribution system management as it provides full visibility into a distribution system by estimating the voltage magnitude and phase angle at each system bus, as well as the power flow on each distribution circuit branch, and thereby initial input to various control functions.

Even though most of the existing distribution state estimation algorithms inherit the paradigm of transmission-level state estimation [112, 113], distribution system state estimation has its unique features and challenges [114–116].

First, unlike the transmission system having redundant measurements, there exist a limited number of measurement devices in the distribution system, resulting in the distribution system not being fully observable. Pseudo-measurements, usually the load power consumption obtained from historical data, can be used for the distribution system state estimation. However, the accuracy of these pseudo-measurements may be limited.

Second, different models and algorithms are needed for the distribution system state estimation. Distribution systems are naturally unbalanced due to the different system configurations as well as unbalanced parameters and loads. Consequently, three-phase modeling of the distribution system state estimation problem is needed in order to accurately estimate the system states. Furthermore, the low reactance/resistance (X/R) ratios render the commonly used decoupled WLS algorithm for the transmission system state estimation not suitable for the distribution system.

Finally, as the distribution system is more often a radial or weakly meshed grid, branch currents can be used as possible primary state variables instead of nodal voltages, which may simplify the formulation of the distribution system state estimation problem.

With the deployment of more and more measurement devices in the distribution system, such as smart meters, synchrophasors, and distribution sensors, models and methods will be needed to facilitate the accurate estimation of distribution system states by utilizing the available measurement data, which is crucial for the implementation of distribution system state estimation.

5   CONCLUSIONS

Large volumes of heterogeneous data have been made available for power system coupled energy systems, from the variable generation resource measurements, smart homes and buildings measurements, to high-resolution and dynamic grid measurements. Data analytics are becoming more important and effective for power system operators, end users, generation operators, and aggregators, to manage the grid and its assets in real time and even in a look-ahead mode. This chapter presented data-driven applications for energy resource forecasting, energy consumption forecasting, as well as power system state estimation and forecasting. Together, they will provide system operators with advanced awareness of system future conditions. Operators can then proactively dispatch generators, schedule power flows and reactive power support, preconfigure system topology, or preprocure ancillary services. Eventually, the capability of data analytics can enable a predictive operation paradigm of the power systems in the future.

References

[1] B.K. Sahu, M. Hiloidhari, D.C. Baruah, Global trend in wind power with special focus on the top five wind power producing countries, Renew. Sust. Energ. Rev. 19 (2013) 348–359.

[2] J. Wang, A. Botterud, R. Bessa, H. Keko, L. Carvalho, D. Issicaba, J. Sumaili, V. Miranda, Wind power forecasting uncertainty and unit commitment, Appl. Energy 88 (11) (2011) 4014–4023.

[3] M. Cui, J. Zhang, A.R. Florita, B.-M. Hodge, D. Ke, Y. Sun, An optimized swinging door algorithm for identifying wind ramping events, IEEE Trans. Sust. Energy 7 (1) (2016) 150–162.

[4] J. Mendes, J. Sumaili, R. Bessa, H. Keko, V. Miranda, A. Botterud, Z. Zhou, Very short-term wind power forecasting: state-of-the-art, Tech. Rep., Argonne National Laboratory (ANL), 2014.

[5] S. Al-Yahyai, Y. Charabi, A. Gastli, Review of the use of Numerical Weather Prediction (NWP) models for wind energy assessment, Renew. Sust. Energ. Rev. 14 (9) (2010) 3192–3198.

[6] N.P. Wedi, P.K. Smolarkiewicz, A framework for testing global non-hydrostatic models, Q. J. R. Meteorol. Soc. 135 (639) (2009) 469–484.

[7] J.J. Traiteur, D.J. Callicutt, M. Smith, S.B. Roy, A short-term ensemble wind speed forecasting system for wind power applications, J. Appl. Meteorol. Climatol. 51 (10) (2012) 1763–1774.

[8] J. Côté, S. Gravel, A. Méthot, A. Patoine, M. Roch, A. Staniforth, The operational CMC-MRB global environmental multiscale (GEM) model. Part I: design considerations and formulation, Mon. Weather. Rev. 126 (6) (1998) 1373–1395.

[9] D.G. Jensen, C. Petersen, M.R. Rasmussen, Assimilation of radar-based nowcast into a HIRLAM NWP model, Meteorol. Appl. 22 (3) (2015) 485–494.

[10] C. Fischer, T. Montmerle, L. Berre, L. Auger, S.E. Ştefănescu, An overview of the variational assimilation in the ALADIN/France numerical weather-prediction system, Q. J. R. Meteorol. Soc. 131 (613) (2005) 3477–3492.

[11] E. Pichelli, R. Ferretti, D. Cimini, G. Panegrossi, D. Perissin, N. Pierdicca, F. Rocca, B. Rommen, InSAR water vapor data assimilation into mesoscale model MM5: technique and pilot study, IEEE J. Select. Top. Appl. Earth Observ. Remote Sens. 8 (8) (2015) 3859–3875.

[12] N.S. Wagenbrenner, J.M. Forthofer, B.K. Lamb, K.S. Shannon, B.W. Butler, Downscaling surface wind predictions from numerical weather prediction models in complex terrain with WindNinja, Atmos. Chem. Phys. 16 (8) (2016) 5229–5241.

[13] Q. Hu, P. Su, D. Yu, J. Liu, Pattern-based wind speed prediction based on generalized principal component analysis, IEEE Trans. Sust. Energy 5 (3) (2014) 866–874.

[14] E. Erdem, J. Shi, ARMA based approaches for forecasting the tuple of wind speed and direction, Appl. Energy 88 (4) (2011) 1405–1414.

[15] H. Liu, E. Erdem, J. Shi, Comprehensive evaluation of ARMA-GARCH (-M) approaches for modeling the mean and volatility of wind speed, Appl. Energy 88 (3) (2011) 724–732.

[16] H. Silaghi, C. Costea, Wind speed prediction using Box-Jenkins method, J. Comput. Sci. Control Syst. (1) (2008) 208.

[17] M. Poncela, P. Poncela, J.R. Perán, Automatic tuning of Kalman filters by maximum likelihood methods for wind energy forecasting, Appl. Energy 108 (2013) 349–362.

[18] A. Carpinone, R. Langella, A. Testa, M. Giorgio, Very short-term probabilistic wind power forecasting based on Markov chain models, in: IEEE 11th International Conference on Probabilistic Methods Applied to Power Systems (PMAPS), 2010, pp. 107–112.

[19] Z. Song, Y. Jiang, Z. Zhang, Short-term wind speed forecasting with Markov-switching model, Appl. Energy 130 (2014) 103–112.

[20] M.A. Ghorbani, R. Khatibi, M.H. FazeliFard, L. Naghipour, O. Makarynskyy, Short-term wind speed predictions with machine learning techniques, Meteorol. Atmos. Phys. 128 (1) (2016) 57–72.

[21] H. Chitsaz, N. Amjady, H. Zareipour, Wind power forecast using wavelet neural network trained by improved Clonal selection algorithm, Energy Convers. Manag. 89 (2015) 588–598.

[22] G. Li, J. Shi, On comparing three artificial neural networks for wind speed forecasting, Appl. Energy 87 (7) (2010) 2313–2320.

[23] J.M. Freedman, J. Manobianco, J. Schroeder, B. Ancell, K. Brewster, S. Basu, V. Banunarayanan, B.-M. Hodge, I. Flores, The Wind Forecast Improvement Project (WFIP): a public/private partnership for improving short term wind energy forecasts and quantifying the benefits of utility operations. The Southern Study Area, Final Report, Tech. Rep., AWS Truepower, LLC, Albany, NY, 2014.

[24] J. Zhang, M. Cui, B.-M. Hodge, A. Florita, J. Freedman, Ramp forecasting performance from improved short-term wind power forecasting over multiple spatial and temporal scales, Energy, 122 (2017) 528–541.

[25] C. Feng, M. Cui, B.-M. Hodge, J. Zhang, A data-driven multi-model methodology with deep feature selection for short-term wind forecasting, Appl. Energy 190 (2017) 1245–1257.

[26] National Renewable Energy Laboratory, Wind Integration National Dataset (WIND) Toolkit, Available from: https://www.nrel.gov/grid/wind-toolkit.html (accessed 10.10.17).

[27] C. Draxl, B. Hodge, A. Clifton, J. McCaa, Overview and meteorological validation of the Wind Integration National Dataset (WIND) toolkit, National Renewable Energy Laboratory, Tech. Rep., Golden, CO. NREL/TP-5000-61740, 2015.

[28] R. Margolis, C. Coggeshall, J. Zuboy, SunShot Vision Study, vol. 2, US Dept. of Energy, 2012. Available from: https://energy.gov/sites/prod/files/2014/01/f7/47927.pdf (accessed 10.10.17).

[29] J. Zhang, A. Florita, B.-M. Hodge, S. Lu, H.F. Hamann, V. Banunarayanan, A. M. Brockway, A suite of metrics for assessing the performance of solar power forecasting, Sol. Energy 111 (2015) 157–175.

[30] A. Hammer, D. Heinemann, E. Lorenz, B. Lückehe, Short-term forecasting of solar radiation: a statistical approach using satellite data, Sol. Energy 67 (1) (1999) 139–150.

[31] A. Sfetsos, A.H. Coonick, Univariate and multivariate forecasting of hourly solar radiation with artificial intelligence techniques, Sol. Energy 68 (2) (2000) 169–178.

[32] C. Paoli, C. Voyant, M. Muselli, M.-L. Nivet, Forecasting of preprocessed daily solar radiation time series using neural networks, Sol. Energy 84 (12) (2010) 2146–2160.

[33] R. Marquez, C.F.M. Coimbra, Forecasting of global and direct solar irradiance using stochastic learning methods, ground experiments and the NWS database, Sol. Energy 85 (5) (2011) 746–756.

[34] P. Mathiesen, J. Kleissl, Evaluation of numerical weather prediction for intra-day solar forecasting in the continental United States, Sol. Energy 85 (5) (2011) 967–977.

[35] R. Perez, K. Moore, S. Wilcox, D. Renné, A. Zelenka, Forecasting solar radiation—preliminary evaluation of an approach based upon the national forecast database, Sol. Energy 81 (6) (2007) 809–812.

[36] C.W. Chow, B. Urquhart, M. Lave, A. Dominguez, J. Kleissl, J. Shields, B. Washom, Intra-hour forecasting with a total sky imager at the UC San Diego solar energy testbed, Sol. Energy 85 (11) (2011) 2881–2893.

[37] E. Lorenz, D. Heinemann, H. Wickramarathne, H.G. Beyer, S. Bofinger, Forecast of ensemble power production by grid-connected PV systems, in: 20th European PV Conference, Milano, 2007, pp. 3–9.

[38] S. Pelland, J. Remund, J. Kleissl, T. Oozeki, K. De Brabandere, Photovoltaic and solar forecasting: state of the art, 2013, IEA PVPS, Task 14, Available from: http://www.iea-pvps.org/fileadmin/dam/public/report/technical/Photovoltaic_and_Solar_Forecasting_State_of_the_Art_REPORT_PVPS__T14_01_2013.pdf (accessed 10.10.17).

[39] Y. Chu, H.T.C. Pedro, L. Nonnenmacher, R.H. Inman, Z. Liao, C.F.M. Coimbra, A smart image-based cloud detection system for intrahour solar irradiance forecasts, J. Atmos. Ocean. Technol. 31 (9) (2014) 1995–2007.

[40] S. Lu, Y. Hwang, I. Khabibrakhmanov, F.J. Marianno, X. Shao, J. Zhang, B.-M. Hodge, H. F. Hamann, Machine learning based multi-physical-model blending for enhancing renewable energy forecast-improvement via situation dependent error correction, in: IEEE European Control Conference (ECC), 2015, pp. 283–290.

[41] IBM, Watt-Sun: a multi-scale, multi-model, machine-learning solar forecasting technology, Available from: https://energy.gov/eere/sunshot/watt-sun-multi-scale-multi-model-machine-learning-solar-forecasting-technology (accessed 10.10.17).

[42] C. Feng, M. Cui, M. Lee, J. Zhang, B.M. Hodge, S. Lu, H.F. Hamann, Short-term global horizontal irradiance forecasting based on sky imaging and pattern recognition, in: IEEE Power & Energy Society General Meeting, Chicago, IL, 2017.

[43] J. Zhang, B.-M. Hodge, S. Lu, H.F. Hamann, B. Lehman, J. Simmons, E. Campos, V. Banunarayanan, J. Black, J. Tedesco, Baseline and target values for regional and point PV power forecasts: toward improved solar forecasting, Sol. Energy 122 (2015) 804–819.

[44] G. Gross, F.D. Galiana, Short-term load forecasting, Proc. IEEE 75 (12) (1987) 1558–1573.

[45] K. Liu, S. Subbarayan, R.R. Shoults, M.T. Manry, C. Kwan, F.I. Lewis, J. Naccarino, Comparison of very short-term load forecasting techniques, IEEE Trans. Power Syst. 11 (2) (1996) 877–882.

[46] W. Charytoniuk, M.S. Chen, Very short-term load forecasting using artificial neural networks, IEEE Trans. Power Syst. 15 (1) (2000) 263–268.

[47] E.A. Feinberg, D. Genethliou, Load forecasting, in: J.H. Chow, F.F. Wu, J. Momoh (Eds.), Applied Mathematics for Restructured Electric Power Systems: Optimization, Control, and Computational Intelligence, chap. 12, Springer US, Boston, MA, ISBN 978-0-387-23471-7, 2005, pp. 269–285, https://doi.org/10.1007/0-387-23471-3_12.

[48] C.W. Gellings, Demand Forecasting for Electric Utilities, Fairmont Press, Inc., Lilburn, GA, 1992.

[49] M.T. Hagan, S.M. Behr, The time series approach to short term load forecasting, IEEE Trans. Power Syst. 2 (3) (1987) 785–791.

[50] N. Amjady, Short-term hourly load forecasting using time-series modeling with peak load estimation capability, IEEE Trans. Power Syst. 16 (4) (2001) 798–805.

[51] A.D. Papalexopoulos, T.C. Hesterberg, A regression-based approach to short-term system load forecasting, IEEE Trans. Power Syst. 5 (4) (1990) 1535–1547.

[52] T. Haida, S. Muto, Regression based peak load forecasting using a transformation technique, IEEE Trans. Power Syst. 9 (4) (1994) 1788–1794.

[53] K.Y. Lee, Y.T. Cha, J.H. Park, Short-term load forecasting using an artificial neural network, IEEE Trans. Power Syst. 7 (1) (1992) 124–132.

[54] A.G. Bakirtzis, J.B. Theocharis, S.J. Kiartzis, K.J. Satsios, Short term load forecasting using fuzzy neural networks, IEEE Trans. Power Syst. 10 (3) (1995) 1518–1524.

[55] B.-J. Chen, M.-W. Chang, et al., Load forecasting using support vector machines: a study on EUNITE competition 2001, IEEE Trans. Power Syst. 19 (4) (2004) 1821–1830.

[56] E. Kyriakides, M. Polycarpou, Short term electric load forecasting: a tutorial. in: K. Chen, L. Wang (Eds.), Trends in Neural Computation, chap. 16, Springer, Berlin, ISBN 978-3-540-36122-0, 2007, pp. 391–418, https://doi.org/10.1007/978-3-540-36122-0_16.

[57] H. Hahn, S. Meyer-Nieberg, S. Pickl, Electric load forecasting methods: tools for decision making, Eur. J. Oper. Res. 199 (3) (2009) 902–907.

[58] S.-J. Huang, K.-R. Shih, Short-term load forecasting via ARMA model identification including non-Gaussian process considerations, IEEE Trans. Power Syst. 18 (2) (2003) 673–679.

[59] H.-T. Yang, C.-M. Huang, C.-L. Huang, Identification of ARMAX model for short term load forecasting: an evolutionary programming approach, IEEE Trans. Power Syst. 11 (1) (1996) 403–408.

[60] C.-M. Huang, C.-J. Huang, M.-L. Wang, A particle swarm optimization to identifying the ARMAX model for short-term load forecasting, IEEE Trans. Power Syst. 20 (2) (2005) 1126–1133.

[61] W. Charytoniuk, M.S. Chen, P. Van Olinda, Nonparametric regression based short-term load forecasting, IEEE Trans. Power Syst. 13 (3) (1998) 725–730.

[62] L. Jin, Y.J. Lai, T.X. Long, Peak load forecasting based on robust regression model, in: International Conference on Probabilistic Methods Applied to Power Systems, IEEE, 2004, pp. 123–128.

[63] T. Zheng, A.A. Girgis, E.B. Makram, A hybrid wavelet-Kalman filter method for load forecasting, Electr. Power Syst. Res. 54 (1) (2000) 11–17.

[64] H.M. Al-Hamadi, S.A. Soliman, Short-term electric load forecasting based on Kalman filtering algorithm with moving window weather and load model, Electr. Power Syst. Res. 68 (1) (2004) 47–59.

[65] D.C. Park, M.A. El-Sharkawi, R.J. Marks, L.E. Atlas, M.J. Damborg, Electric load forecasting using an artificial neural network, IEEE Trans. Power Syst. 6 (2) (1991) 442–449.

[66] H.S. Hippert, C.E. Pedreira, R.C. Souza, Neural networks for short-term load forecasting: a review and evaluation, IEEE Trans. Power Syst. 16 (1) (2001) 44–55.

[67] E.E. Elattar, J. Goulermas, Q.H. Wu, Electric load forecasting based on locally weighted support vector regression, IEEE Trans. Syst. Man Cybern. Part C Appl. Rev. 40 (4) (2010) 438–447.

[68] S. Tzafestas, E. Tzafestas, Computational intelligence techniques for short-term electric load forecasting, J. Intell. Robot. Syst. 31 (1–3) (2001) 7–68.

[69] M. Hanmandlu, B.K. Chauhan, Load forecasting using hybrid models, IEEE Trans. Power Syst. 26 (1) (2011) 20–29.

[70] R.-A. Hooshmand, H. Amooshahi, M. Parastegari, A hybrid intelligent algorithm based short-term load forecasting approach, Int. J. Electr. Power Energy Syst. 45 (1) (2013) 313–324.

[71] A. Kavousi-Fard, H. Samet, F. Marzbani, A new hybrid modified firefly algorithm and support vector regression model for accurate short term load forecasting, Expert Syst. Appl. 41 (13) (2014) 6047–6056.

[72] R.B. D'agostino, A. Belanger, R.B. D'Agostino Jr., A suggestion for using powerful and informative tests of normality, Am. Stat. 44 (1990) 316–321.

[73] C.M. Jarque, A.K. Bera, A test for normality of observations and regression residuals, Int. Stat. Rev. 55 (1987) 163–172.

[74] K. Coughlin, M.A. Piette, C.A. Goldman, S. Kiliccote, Statistical analysis of baseline load models for non-residential buildings, Energy Build. 41 (2009) 374–381.

[75] FERC, Reports on Demand Response & Advanced Metering, Available from: http://www.ferc.gov/industries/electric/indus-act/demand-response/dem-res-adv-metering.asp (accessed 10.10.17).

[76] US Department of Energy, Benefits of demand response in electricity markets and recommendations for achieving them Tech. Rep. 2006.

[77] V.M. Balijepalli, V. Pradhan, S.A. Khaparde, R.M. Shereef, Review of demand response under smart grid paradigm, in: IEEE PES Innovative Smart Grid Technologies, India, 2011, pp. 236–243.

[78] H.P. Chao, Demand response in wholesale electricity markets: the choice of customer baseline, J. Regul. Econ. 39 (1) (2011) 68–88.

[79] R. Yin, P. Xu, M.A. Piette, S. Kiliccote, Study on Auto-DR and pre-cooling of commercial buildings with thermal mass in California, Energy Build. 42 (7) (2010) 967–975.

[80] A. Buege, M. Rufo, M. Ozog, D. Violette, S. McNicoll, Prepare for impact: measuring large C/I customer response to DR programs, in: ACEEE Summer Study on Energy Efficiency in Buildings, 2006.

[81] J. MacDonald, P. Cappers, D. Callaway, S. Kiliccote, Demand Response Providing Ancillary Services: A Comparison of Opportunities and Challenges in the US Wholesale Markets, Grid-Interop, 2012.

[82] H. Zhong, L. Xie, Q. Xia, Coupon incentive-based demand response: theory and case study, IEEE Trans. Power Syst. 28 (2) (2013) 1266–1276.

[83] N. Addy, S. Kiliccote, J. Mathieu, D.S. Callaway, Understanding the effect of baseline modeling implementation choices on analysis of demand response performance, in: ASME International Mechanical Engineering Congress and Exposition, 2012.

[84] K. Coughlin, M.A. Piette, C.A. Goldman, S. Kiliccote, Estimating demand response load impacts: evaluation of baseline load models for non-residential buildings in California, Lawrence Berkeley National Laboratory, 2008.

[85] K. Tweed, Demand Response Payments Increase Significantly in PJM, 2013, Available from: http://www.greentechmedia.com/articles/read/demand-response-payments-up-significantly-in-pjm (accessed 10.10.17).

[86] Federal Energy Regulatory Commission, State of the Markets Report, 2016, Available from: https://www.ferc.gov/market-oversight/reports-analyses/st-mkt-ovr/2015-som.pdf (accessed 10.10.17).

[87] R. Walton, EIA: FERC Order 745 to spark swift growth in demand response markets, 2016Available from: http://www.utilitydive.com/news/eia-ferc-order-745-to-spark-swift-growth-in-demand-response-markets/414997/ (accessed 10.10.17).

[88] Y. Wi, J. Kim, S. Joo, J. Park, J. Oh, Customer baseline load (CBL) calculation using exponential smoothing model with weather adjustment, in: Transmission and Distribution Conference and Exposition: Asia and Pacific, 2009, pp. 1–4.

[89] J.L. Mathieu, D.S. Callaway, S. Kiliccote, Examining uncertainty in demand response baseline models and variability in automated response to dynamic pricing, in: IEEE Conference on Decision and Control and European Control, 2011.

[90] G.R. Newsham, B.J. Birt, I.H. Rowlands, A comparison of four methods to evaluate the effect of a utility residential air-conditioner load control program on peak electricity use, Energy Policy 39 (2011) 6376–6389.

[91] J.L. Bode, M.J. Sullivan, D. Berghman, J.H. Eto, Incorporating residential AC load control into ancillary service markets: measurement and settlement, Energy Policy 56 (2013) 175–185.

[92] L. Hatton, P. Charpentier, E. Matzner-Lober, Statistical estimation of the residential baseline, IEEE Trans. Power Syst. 31 (2016) 1752–1759.

[93] D.S. Sayogo, A.P. Theresa, Understanding smart data disclosure policy success: the case of Green Button, in: ACM Proceedings of the 14th Annual International Conference on Digital Government Research, 2013.

[94] A. Abur, A.G. Exposito, Power System State Estimation: Theory and Implementation, CRC Press, Boca Raton, FL, 2004.

[95] K.D. Jones, J.S. Thorp, R.M. Gardner, Three-phase linear state estimation using phasor measurements, in: IEEE Power & Energy Society General Meeting, Vancouver, BC, 2013.

[96] E. Farantatos, G.K. Stefopoulos, G.J. Cokkinides, A.P. Meliopoulos, PMU-based dynamic state estimation for electric power systems, in: IEEE Power & Energy Society General Meeting, Calgary, AB, 2009.

[97] A. Abur, A. Rouhani, Linear phasor estimator assisted dynamic state estimation, IEEE Trans. Smart Grid, 2016.

[98] A.M.L. Da Silva, M.B. Do Coutto Filho, J.F. De Queiroz, State forecasting in electric power systems., in: IEE Proceedings C (Generation, Transmission and Distribution), vol. 130, 1983, pp. 237–244.

[99] G. Valverde, V. Terzija, Unscented Kalman filter for power system dynamic state estimation, IET Gener. Transm. Distrib. 5 (1) (2011) 29–37.

[100] A.P.A. da Silva, A.M.L. da Silva, J.C.S. de Souza, M.B. Do Coutto Filho, State forecasting based on artificial neural networks., in: Proc. 11th PSCC, 1993, pp. 461–467.

[101] J.C.S. Souza, A.M.L. Da Silva, A.P.A. Da Silva, Data visualisation and identification of anomalies in power system state estimation using artificial neural networks, IEE Proc. Gener. Transm. Distrib. 144 (5) (1997) 445–455.

[102] J.C.S. Souza, A.M.L. Da Silva, A.P.A. de Silva, Online topology determination and bad data suppression in power system operation using artificial neural networks, IEEE Trans. Power Syst. 13 (3) (1998) 796–803.

[103] M.B. Do Coutto Filho, J.C.S. de Souza, Forecasting-aided state estimation—Part I: panorama, IEEE Trans. Power Syst. 24 (4) (2009) 1667–1677.

[104] E. Sortomme, M.M. Hindi, S.D.J. MacPherson, S.S. Venkata, Coordinated charging of plug-in hybrid electric vehicles to minimize distribution system losses, IEEE Trans. Smart Grid 2 (1) (2011) 198–205.

[105] G.-B. Huang, Q.-Y. Zhu, C.-K. Siew, Extreme learning machine: theory and applications, Neurocomputing 70 (1) (2006) 489–501.

[106] J.M. Carrasco, L.G. Franquelo, J.T. Bialasiewicz, E. Galván, R.C.P. Guisado, M.A.M. Prats, J.I. León, N. Moreno-Alfonso, Power-electronic systems for the grid integration of renewable energy sources: a survey, IEEE Trans. Ind. Electron. 53 (4) (2006) 1002–1016.

[107] H. Jiang, Y. Zhang, J.J. Zhang, D.W. Gao, E. Muljadi, Synchrophasor-based auxiliary controller to enhance the voltage stability of a distribution system with high renewable energy penetration, IEEE Trans. Smart Grid 6 (2015) 2107–2115.

[108] D.M. Vinod Kumar, S.C. Srivastava, Power system state forecasting using artificial neural networks, Electr. Mach. Power Syst. 27 (6) (1999) 653–664.

[109] G. Zhang, B.E. Patuwo, M.Y. Hu, Forecasting with artificial neural networks: the state of the art, Int. J. Forecast. 14 (1) (1998) 35–62.

[110] P.-F. Pai, W.-C. Hong, Forecasting regional electricity load based on recurrent support vector machines with genetic algorithms, Electr. Pow. Syst. Res. 74 (3) (2005) 417–425.

[111] W.-C. Hong, Chaotic particle swarm optimization algorithm in a support vector regression electric load forecasting model, Energy Convers. Manag. 50 (1) (2009) 105–117.

[112] W.R. Cassel, Distribution management systems: functions and payback, IEEE Trans. Power Syst. 8 (3) (1993) 796–801.

[113] M.E. Baran, A.W. Kelley, State estimation for real-time monitoring of distribution systems, IEEE Trans. Power Syst. 9 (3) (1994) 1601–1609.

[114] Y.-F. Huang, S. Werner, J. Huang, N. Kashyap, V. Gupta, State estimation in electric power grids: meeting new challenges presented by the requirements of the future grid, IEEE Signal Process. Mag. 29 (5) (2012) 33–43.

[115] D. Della Giustina, M. Pau, P.A. Pegoraro, F. Ponci, S. Sulis, Electrical distribution system state estimation: measurement issues and challenges, IEEE Instrum. Meas. Mag. 17 (6) (2014) 36–42.

[116] A. Primadianto, C.-N. Lu, A review on distribution system state estimation, IEEE Trans. Power Syst. 32 (5) (2017) 3875–3883.

CHAPTER 17

Data Analytics for Energy Disaggregation: Methods and Applications

Behzad Najafi, Sadaf Moaveninejad, Fabio Rinaldi
Polytechnic University of Milan, Milan, Italy

CHAPTER OVERVIEW

Energy disaggregation, or nonintrusive load monitoring (NILM), aims at estimating the power demand of individual appliances from a household's aggregate electricity consumption. Due to the notable rise in the number of installed smart meters and owing to the numerous advantages of this approach over intrusive methods, NILM has received growing attention in the recent years. In this chapter, after reviewing different categories of household appliances, the state-of-the-art load signatures, including both macroscopic and microscopic features, are introduced. Next, commonly used supervised and unsupervised disaggregation algorithms, which are employed to classify the appliances based on the extracted features, are discussed. Publically accessible datasets and open-source tools, which have been released in the recent years to assist the NILM research and to facilitate the comparison of disaggregation algorithms, are then reviewed. Finally, main applications of energy disaggregation, including providing itemized energy bills, enabling more accurate demand prediction, identifying mal-functioning appliances, and assisting occupancy monitoring, are presented.

1 INTRODUCTION

During recent decades, concerns over energy crisis and global warming conse-quences have been significantly rising and many countries are introducing public policies in order to deal with these issues [1]. The increased volumes of green-house gases (GHGs), the major source of which is the combustion of fossil fuels, are believed to be the main cause of the global warming [2]. According to the third article of Climate Change General Law (CCGL), mitigation of progressive rise in GHG emissions, not only necessitates the substitution of conventional power generation units with more environment-friendly technologies, but also requires using electrical power in a more efficient manner [1]. Hence, reducing the energy demand, through improved energy consumption management, is a crucial measure to deal with the escalating energy crisis and global warming issues. A large portion of the global energy demand is due to energy consumption of buildings [3,4]. According to Buildings Energy Data Book [5], almost 40% of US primary energy consumption and 70% of the US electricity consumption are utilized in the buildings sector [4]; hence, any attempt to reduce the correspond-ing consumption can result in a notable benefit.

Big Data Application in Power Systems. https://doi.org/10.1016/B978-0-12-811968-6.00017-6

A proper measure for wiser energy management in this sector is providing real-time information about the electrical consumption of individual buildings [6]. For the case of the residential sector, such a detailed information about the instantaneous electrical load of individual apartments can be provided by smart meters. This information facilitates the implementation of smart grids [7] which in turn enable information exchange between energy providers and consumers through large-scale monitoring and control [8,9]. Furthermore, the data reported by smart meters enables the energy providers to estimate the aggregate consumption profile and formulate usage policies such as the hourly change in rate [7].

In addition to the information about the total energy usage in residential buildings, more precise information including itemized energy consumption profiles can lead to further benefits for both end users and grid managers. Such detailed information aids the utility companies and grid operators to improve their predictions of residential energy demand, facilitates the demand side management, and can even allow them to have more precise user segmentation. Besides, users typically have a poor estimation about the share of each appliance in their household's total energy consumption and are usually mistaken about the effective measures for saving energy [10–12]. For this reason, a report including itemized (appliance-by-appliance) energy consumption data gives the users a more accurate idea about the devices with the highest consumption and helps them reduce their monthly bill by taking more effective measures [7,13]. Kim et al. [14] demonstrated that 9%–20% of energy conservation can be obtained by implementing an energy consumption strategy based on appliance-by-appliance utilization information [1]. Furthermore, providing customers with detailed energy bills improves their energy literacy [15]. Moreover, a precise log of devices usage is beneficial for checking appliance status and detecting malfunctioning devices (MFDs).

The above-mentioned benefits of providing decomposed energy consumption information motivated the development of appliance load monitoring (ALM) techniques. ALM methods can be divided into two main categories: intrusive load monitoring (ILM) and nonintrusive load monitoring (NILM). ILM is a more traditional way in which each device of interest is attached to a set of sensors to record its energy consumptions [16]. Though, this methodology requires a large number of sensors and smart meters and is consequently an expensive and cumbersome approach [7]. Hence, differentiating energy consumption of individual equipment in a cost-effective way remains an open issue [14].

As one of the practical alternatives to deal with the mentioned issues, NILM or nonintrusive appliance load monitoring (NIALM) was invented by Hart et al. [17] from Massachusetts Institute of Technology (MIT). NILM requires simpler data gathering hardware but more complex software, for processing and analyzing signals, in comparison with ILM [18]. In NILM methodology, the aggregate electrical load of household is first measured and, through analyzing the obtained overall signal, the consumption profile of each individual appliance is then identified [19]. Several NILM methods have been introduced in the literature; though, all of them include three main steps of data acquisition, extracting features from data, and classifying appliances [16]. The data acquisition in the NILM context only requires a single set of sensors attached to service entry of the house in order to capture the total residential load [16]. Using these sensors, voltage and current signals are sampled at a proper rate [17] in order to detect the device pattern more precisely [16]. After conducting the data acquisition, particular appliance features or signatures are extracted from the obtained overall signal. Finally, the last step is dedicated to determining the contribution of individual appliances to the aggregate loads utilizing the extracted features. NILM can be defined as a machine learning (ML) problem, since appliance classification is conducted by running a mathematical algorithm and detecting the appliance signatures in the entire signal [9].

The objective of this chapter is to provide a comprehensive overview of energy disaggregation methodology and its main applications. Since the choice of monitoring approaches, extracted features, and event detection methods depends on the operational states and load types of individual appliances, the corresponding device categories are first explained. Next, state-of-the-art device signatures including the macroscopic and microscopic features, which have commonly been employed in the literature, are reviewed. Different classification algorithms, containing both supervised and unsupervised methods, are then reviewed and frequently employed accuracy metrics, which are utilized to evaluate the performance of NILM algorithms, are discussed. Publically available datasets and open-source tools, which have been provided by the NILM community in order to assist the energy disaggregation research and facilitating the comparison of algorithms, are subsequently presented. Finally, the main use cases of energy disaggregation, such as providing the users with itemized energy bills, detecting mal-functioning devices, occupancy monitory, facilitating more accurate energy consumption prediction, and enhancing the effectiveness of demand response, are reviewed.

2 APPLIANCE CATEGORIES

The choice of the approach employed to monitor the electrical devices and the suitable signature to identify their possible contribution to the aggregate load depends on their corresponding operation principle and load properties. Accordingly, one common criterion to categorize devices, in the NILM context, is based on their operational states [20]. Besides, the appliances can also be classified according to the type and linearity of their corresponding loads [21,22]. Information regarding the appliance categories provides the reader with a better understanding of the underlying reason behind the choice of the sampling hardware, employed signatures, and utilized event detection methods within the NILM framework. Hence, this section is dedicated to introducing different categories of devices based on these two above-mentioned criteria.

2.1 Device Classifications Based on the Operational States

Hart [18] classified the devices based on the possible variations in their steady-state measured values of real and reactive power and categorized them in three groups of On/Off, Finite State Machine (FSM), and continuously variable. He had also declared that nonintrusive ALM prototype, proposed in his first article [23], can just be valid for On/Off devices, and might lead to significant errors if applied to multistate appliances. In a report provided by Electric Power Research Institute (EPRI) it was similarly indicated that the NILM methodology has determined to be an effective approach to monitor the devices with two operation states [24]. Baranski and Voss [25] defined Permanent Consumer as the fourth class of devices [4,20]. The operation characteristics of the above-mentioned classes of devices along with the corresponding examples are provided below:

2.1.1 On/Off

This class of appliances consists of devices with merely two operation states (On/Off). Several Household appliances, including table lamp and toaster, belong to this category. Though, this category does not include devices with multiple discrete "On" states including lamps with three modes or washing machines with different functionalities [18].

2.1.2 Finite State Machines

This category includes devices with a finite number of operating states. The switching pattern of these devices has a periodically repeating nature, the fact that facilitates the identification of their operation through disaggregation. This class can properly describe several appliances including stove burners, dishwashers, washing machines, and cloth dryers [20]. It is noteworthy that employing an "On/Off" model for an FSM device can result in recognizing it as several individual appliances and even some operational states of the device might not be recognized at all [20].

The model that simulates the operation principles of FSM devices is represented by circles and arcs where the circles indicate the states (i.e., name and an operating power level) and the arcs denote switching between different states. While modeling the FSM appliances, zero loop-sum constraint (ZLSC) must be met.

According to ZLSC criterion, in each cycle of state transitions sum of the variations in power must be equal to zero [18].

2.1.3 Continuous Variable Device

The third category of devices, which is driven from the FSM model, is the general form of FSM with infinite number of states. Such appliances have a continuous range of power draw and have neither repeated cycle for state transitions nor specific step-change features. Examples of such appliances are sewing machines, light dimmers, power drills with variable speed, and other power electronic controlled loads [18]. Distinguishing this type of electrical devices in the aggregated load is a challenging task. The original NILM approach and the following extended method, developed by Hart [18], were suitable for "On/Off" and FSM classes and were not applicable for identifying this type of devices. Later, other researchers investigated different features of appliances with variable power loads in order to find a promising NILM methodology for identifying these devices. The study conducted by Wichakool et al. [26] is one of the latest studies focused on proposing NILM methods which are suitable for appliances with various power loads.

2.1.4 Permanent Consumer

The last category of devices are the ones that remain on the "on" mode continuously such as telephone sets, TV receivers, and smoke detectors. The appliances belonging to this category are called "permanent consumer devices" since their rate of energy consumption remains approximately constant [20].

2.2 Device Classifications Based on the Corresponding Load Characteristics

Each type of residential appliance results in a specific signature in the aggregated load; hence, in addition to the above-mentioned categorization according to the operational states, devices can also be classified based on their load type [21,22]. According to Dong et al. [27], residential electrical loads can be broadly categorized as: resistive, inductive, capacitive, and other groups. The last class, which includes the devices not included in the first three categories, can itself be divided into two groups of switch-mode power supply (SMPS) and composite loads. Different types of electrical devices along with their load characteristics are summarized in Table 1.

3 NILM METHODOLOGY

The original NILM methodology was introduced by Hart et al. from MIT [17] which was followed by his proposed extended methods [18,28]. His approach was extended in the studies conducted by Bouloutas and Schwartz [29],

Table 1 Appliance Categories Based on the Corresponding Features (Load Types) [21,22]

Load Types	Examples of Household Appliances	Load Characteristics
Resistive	Appliances with heating elements: – Resistive kettle – Toasters, ovens – Space heaters – Coffee makers – Incandescent lighting	– Turning on: higher power, no transient – In use: decay to relatively flat power level – Zero reactive power – No harmonics of the current
Inductive	Appliances with AC motors: – Compressors (in fridges and – air conditioners) – Dishwasher – Washing-machine drain pump – Fans – Various types of mixers – Vacuum cleaners	– Turning on: initial spike in power, long transient – In use: growth or decay to a flat power level – Large reactive power – Odd-numbered harmonic current
Capacitive	– No significant capacitive loads in buildings (due to the fact that although many loads have capacitive elements, their overall behavior is dominated by inductive and resistive characteristics)	– A purely capacitive load is the one that draws current through a sinusoidal shape that peaks in advance compared to the voltage sine wave
Including SMPS (nonlinear)	– TVs – Personal computers – Video recorders	– Turning on: short but very high amplitude transient – Large amount of harmonic contents – Notable power fluctuations limited by ceilings and floors in power level – Draw current in nonsinusoidal form
Composite loads (nonlinear) (no pure resistive, inductive and capacitive)	– Air conditioner (including compressor, fan, duct dampeners, and central humidifiers) – Refrigerator (consisting of compressor (inductive load), door lights, ice maker, water dispenser), electric dryers, washing machines, dishwashers (consisting of a motor and a heating element)	– Operate in repetitive cycles that utilize each of their constituent loads differently – Demonstrate specific behavior in different parts of their operation cycle determined by the load type of the component of the appliance that is in use

Leeb et al. [30,31], Cole and Albicki [32,33], Baranski and Voss [25,34,35], and several other studies, which will be discussed in this section. Although the proposed NILM methodologies follow different approaches, most of them include the common steps of data acquisition, feature extraction, and appliance disaggregation [36]. Data acquisition refers to measuring the aggregate load of a household, which can be conducted with a high- or low-frequency sampling rate. The next step is dedicated to extracting specific features from the acquired overall load signal. These features can include macroscopic or microscopic signatures depending on the utilized sampling rate. Since each electrical appliance is characterized by a specific set of features, this acquired information allows identifying the operation of devices. Therefore, the last step is dedicated to disaggregating the overall load into the consumption of individual appliance by employing the extracted features and the available characteristic signatures of devices. Accordingly, in the present section, commonly used device signatures, including the macroscopic and microscopic features along with their corresponding required data acquisition methodology, are first described. Next, different supervised and unsupervised disaggregation algorithms are discussed in detail. Finally, state-of-the art accuracy metrics, which are employed in order to compare the accuracy of different energy disaggregation, are presented.

3.1 Device Signatures

Each operating electrical appliance is characterized by a specific set of measurable features, also termed as "signatures," which reveal information about its consumption pattern, nature, and operation [18]. Investigations on device signatures were initiated by General Electric and Oak Ridge laboratories in the late 1980s, in which current was studies as a signature of motor-driven appliances [37]. In the early 1990s, researchers of MIT, EPRI in the United States [18], and Electricite de France (EDF) [22] proposed active and reactive power draw (P and Q) as signatures to track electrical equipment. Later, more parameters such as transient waves [31], harmonics [38], voltage and current [39], electromagnetic interference (EMI) spectrum [40], electrical current startup [41], and electrical noise on the voltage, created by the abrupt switching of electrical devices [42], were addressed by other researchers.

Distinct signatures are captured while electrical devices are operating in steady or transient states. In case of steady-state operation, the extracted load features have no variations, i.e., their variations with respect to a specific tolerance are negligible. Steady operation states of devices can be differentiated by corresponding signatures which include power changes, root mean square (RMS) voltage, RMS current, power factor, harmonics, and V-I trajectory [20,36]. Identifying electrical devices based on their steady-state signatures permits tracking their behavior in a continuous manner, which is easier to implement in comparison with conducting instantaneous measurements. Furthermore, these signatures satisfy the above-mentioned ZLSC criteria for FSM devices [6].

The transient state instead refers to the operation period in which an electrical device is between the off and the steady states. Some of the parameters which represent a transient event are their shape, size, duration, and time constant [18]. Transient behavior of the load is caused by abrupt variations in circuit condition such as turning a device on or off [36]. Electrical appliances typically consist of several components which could introduce such changes and create a specific signature [38]. Detailed explanations regarding the transient behavior of different household appliances were given in Section 2. In comparison with the steady-state signatures, signatures corresponding to transient condition provide less information while they commonly require sampling with high frequency [43]. In addition, network geometry and position of the device influences transient pulses [44]. Nevertheless, different electrical equipment with similar steady-state feature could be distinguished in total load by analyzing their transient signatures. The mentioned notable benefit motivated researchers to investigate several transient signatures [31,45].

In another categorization, signatures can be divided into two main groups of macroscopic and microscopic features [4]. In this classification, macroscopic features refer to the real and reactive power changes, which are obtained using low frequency sampling. On the other hand, harmonics, noise, acquired by high frequency sampling, are categorized as microscopic signatures [4]. Details of different microscopic and microscopic signatures that have been frequently employed in the literature are discussed below:

3.1.1 Macroscopic Signatures
Signatures extracted from the total load using low frequency sampling (around 1 Hz) would result in obtaining macro level (macroscopic) features [46]. Real and reactive power variations are the most commonly used macroscopic signatures. Real power refers to the energy consumed by electrical appliance during operation; while the reactive power is not delivered to the load and is dissipated in the electrical cables as heat [21]. However, the reactive power, which is generated by capacitive and inductive components [20], provides an additional information that can facilitate the procedure of identifying electrical devices. These macro level signatures were investigated in the primary works carried out by EPRI and MIT [18,38,47]. In these studies, the magnitude and sign variations of real (P) and reactive power (Q) were tracked over time and the corresponding positive and negative changes were then matched in order to identify the event of turning a device on or off. MIT researchers later extended their initial approach in order to apply it to the total load of an industrial building [31]. Their study demonstrated that after filtering the sudden peaks, the obtained filtered electrical loads would have a small reactive power and a long transient (i.e., Startup time). Hence, they concluded that load detection based on changes in real and reactive power has some limitations which can be compensated by employing transient events as additional signatures [4]. In the same context, Albicki and Cole [32,33] proposed utilization of slopes and edges in the power draw as additional features alongside the steady-state signatures. In their extended approach, which is a promising method for identifying motor-driven appliances, the edges and slopes are defined as upward power spike at the first moment of operation and gradual (with slower pace) variations during the period in which the device is turning on [4,27].

3.1.2 Microscopic Signatures

Features that are acquired using measurement devices with a high sampling frequency are referred to as microlevel (microscopic) signatures [48]. Harmonics and Fourier transform, Fast Fourier Transform (FFT) of signal noise, unprocessed waveforms, and some features beyond FFT are examples of these high frequency signatures [46]. In order to extract current waveform and noise features in high frequency, more complex hardware with high sampling rate are required [46].

In a study conducted by Laughman et al. [38], harmonics were considered as the third dimension added to ΔP-ΔQ plane in order to enhance load identification. The latter provides complementary features in the situations in which there is an ambiguous overlapping in ΔP-ΔQ plane [44,49]. According to the Nyquist criteria, the minimum sampling rate to obtain the highest harmonic must be twice that frequency. Hence, since the 11th harmonic is often the highest employed harmonic, the sampling frequency of 1.2–2 kHz will be needed to obtain all of the required features. It is worth mentioning that high frequency sampling of the waveform is limited by the storage and transmission capacities [4]. Unique information provided by harmonics of the current is specifically useful for identification of nonlinear appliances with nonsinusoidal current. Small electrical devices, which draw similar amount of real and reactive power, could be identified by harmonic features of their currents. Moreover, nonlinear devices, such as motor-driven appliances, produce substantial low-order odd harmonics due to the triangular wave form of their current signal [18].

A study conducted by Leeb et al. [30] employed some of the first coefficients of the short time FFT, known as spectral envelope, as an extension to harmonics in order to facilitate the detection of variable-load appliances. Patel et al. [42] employed the spectrum of electrical noise which is created on the voltage signal due to the abrupt switching of electrical devices. This approach permits the identification of appliances in the frequency domain when their transient features are overlapped in time domain [24]. However, extracting and examining the mentioned transient noise in the signal is computationally expensive and it is also necessary to train the system based on noise FFT of each household appliance and their combinations [4].

Apart from the harmonics and FFT, other signatures such as wavelet transform, geometrical shape of the *I-V* waveform, and transient energy have been utilized in the literature [4]. Wavelet transform represents load's physical behavior in the transient state [20]. It provides information about concurrent localization versus time and frequency, an information that cannot be obtained using FFT [4]. The geometrical properties of *I-V* curve, independent of time, were employed by Lee et al. [50] and Lam et al. [51] as a distinct appliance signature [4]. It is worth mentioning that several distinct signatures can be used simultaneously in order to increase the device identification accuracy [48].

3.1.3 Nontraditional Signatures

In addition to previously mentioned features, which are obtained from the aggregate load's signal, some nontraditional features have also been employed in the recent NILM studies. These signatures are utilized in order to provide additional information about the appliance operation, which are not contained in the ordinary features. Examples of such features are time-related parameters including the time of the day, appliance run times along with temperature and light sensing [27,36].

3.2 Disaggregation Algorithms

Disaggregation algorithms are the ones aiming at recognizing the appliances which have contributed to the aggregate consumption load. Several disaggregation algorithms have been proposed in the literature and, in a categorization based on the system training, they can be divided into two main classes of supervised and unsupervised methods. Supervised approaches are the algorithms that need their classifiers to be trained with labeled dataset, which includes the signatures of different appliances.

In the unsupervised methods neither an event nor a-priori information and labeled data are needed. In this class of algorithms, probabilistic models such as factorial hidden Markov models (FHMM) [14] can be employed in order to simulate the behavior of the appliances. A recent review article on NILM methodologies has shown increased interest toward unsupervised techniques [20]. Supervised approaches require system training, which makes them to be less scalable for disaggregating large number of household devices. Though, unsupervised algorithms are less difficult to be implemented but provide less information [24]. More detailed information regarding the above-mentioned categories of disaggregation algorithms are provided in the next parts of this section:

3.2.1 Supervised Algorithms

Supervised methods require labeled dataset including the features of different appliances in order to train the classifiers. Such system training can be conducted both through on-line and off-line approaches [20]. In the on-line approach, data is labeled based on real-time event detection and is used to train the system concurrently. In the off-line training method, appliances are monitored in a specific environment and during a certain period and their corresponding signatures are labeled.

Installing a measurement unit for each individual appliance, in order to obtain the required labeled data, is an expensive and time-consuming process. In an alternative approach, proposed by Hart et al. [17], appliances can be switched on sequentially in order to detect them individually from the aggregate load [9]. This method was extended by labeling the operation of each device through a smart phone [52]. Several datasets, which include labeled data obtained through investigating the signatures of several household appliances, are now publically available. These open datasets, which are introduced in the next section, permit the researchers to train their disaggregation algorithms without conducting the mentioned cumbersome measurement procedure.

Supervised disaggregation algorithms can be divided into two main categories of pattern recognition and optimization methods. The former relies on the events to identify the appliances while the latter does not depend on the events [20].

Pattern Recognition (Event-Based) Methods

The original NILM algorithm, proposed by Hart et al. [17], was a pattern recognition or event-based method. The pattern recognition methods commonly include the following three steps: event detection, feature extraction, and pattern matching.

The event detection step in Hart's method was conducted through identifying variations in steady-state power levels by an edge detector. In the extended methods, proposed by other researchers, additional criteria including spiking, ramping and small oscillating behavior [33], large oscillation [53], and power fluctuations [54] were proposed in order to detect the events. Apart from the mentioned criteria, several other steady-state and transient features, which were introduced in the previous section, have been employed to detect events. Once events are detected, labeled, and time stamped, a set of signature is captured from the measured samples around each event in order to characterize it. Hence, event detection significantly reduces the mass of data to those just related to the events [55,56]. In order to carry out the last step, several pattern matching algorithms have been proposed in the literature, which are briefly reviewed in this section:

- Original MIT algorithm

Hart's approach [23] was based on simple clustering of real and reactive power changes. Accordingly, neglecting the time stamps, recognized steady-state changes of real and reactive power are first mapped to a scatter plot in P-Q (real power vs reactive power) space. In the next step, in the P-Q plane, detected events, which have powers with similar magnitude but different signs, are paired. The latter step is conducted on the grounds of the fact that frequently observed power changes with negative and positive signs could be referred to switching appliance on and off, respectively [23]. Whereas, irregular changes may occur due to noise, measurement error, or simultaneous operation of appliances. The last step is focused on addressing the events which could not be matched using the existing clusters. In order to do so, maximum likelihood estimation is employed to match new feature vector with available clusters and to find the class of appliances that has most likely caused the event. Other approaches have also been proposed in the literature in order to group multi-dimensional scatter plots to clusters [23,57,58]. The most challenging issue in this step is to automatically specify the number of different clusters to look for. Number of defined clusters represents how many devices constitute the aggregated load [23].

On/Off and FSM models are next developed based on the clusters of step changes. In order to develop the FSM model, ZLSC and uniqueness constraint (UC) are taken into account [18]. As was previously mentioned, the former indicates that the sum of state transition sequences in any loop is zero, while the later says that in each cycle there could be only a single off state with power level equal to zero [18]. As soon as the system learns an FSM, the corresponding events are removed from the data and the procedure is continued by learning other FSM from the remaining data. It is noteworthy that in the mentioned procedures, some tolerance must be considered to deal with possible errors due to the noise, load variations or concurrent operation of several small appliances. An optimal decoding technique known as Viterbi algorithm amends errors for the cases in which one symbol is corrupted into another [18].

Once the above-mentioned procedures are conducted, each appliance forms a unique cluster in P-Q plane. In general, household appliances could be identified by matching each event with a database of known appliances features [48]. This information can be collected either through a training process or using historical data [4].

It is worth mentioning that some electrical devices change their resistance after turning on. This can cause mismatch in power changes and apply power drift around 10% [4]. Furthermore, variable loads and multistate household appliances could not be recognized by this method. Moreover, low power electrical devices were usually grouped into clusters near the origin. Hence P-Q plane does not provide sufficient information to distinguish this kind of electrical devices [48].

- Extensions to MIT method

Other event-based disaggregation algorithms, employing additional features, were later developed to extend original MIT approach. These studies employed transient features [59] and a hybrid system which utilized both transient and steady-state signatures [31,60]. NorFord and Leeb [31] also developed a method to recognize overlapping transients which is a challenging issue in event-based algorithms.

- Bayesian approach

In an approach proposed in Marchiori et al. [61] and Liang et al. [48], for each appliance, a naïve Bayes classifier can be trained based on power level and features corresponding to the state change. Hence, a set of trained classifiers will be able to identify appliance-specific states. This method is based on the assumption that states of household devices are uncorrelated which is not always a correct hypothesis [20].

- Heuristic method

In this methodology, real and reactive power is clustered using the histogram thinning approach. Comparing this method with Bayesian approach reveals that for situations when appliance draw stable power, Bayesian classifiers have a higher performance [20].

- Supervised ML

In order to handle the cases in which a large number of electrical devices must be distinguished with a high accuracy, supervised ML algorithms can be employed. These methods enable conducting the training procedure using different features including the state transitions and temporal information [49]. Artificial neural networks (ANN) [62] and Hidden Markov models (HMMs) [63–66] are two common ML algorithms which have been utilized in the literature. Performance of ML classifiers depends on the set of signatures, type, and number of devices which have been employed for training [20]. ML methods generally require a notable memory and are computationally intensive. In addition, with an increase in the number of input parameters, their required training and classification time escalates [43]. Due to the complexity issues, applicability of HMM is limited and such complexity grows exponentially with the number of devices. Besides, installing new household appliances requires repeating the learning procedure [67]. Both ANN and HMM need massive amount of data for training and building the model for each individual appliance, a problem that clearly becomes more critical while dealing with a large number of devices. However, input feedback can cause ANN to be more adjustable and have a better performance [7,39].

K-nearest neighbors algorithm is utilized to deal with the situations in which there are several unlabeled items and K-labeled nearest neighbors can be accordingly used to train the classifier [7,40,43,68].

Other research activities presented in Srinivasan et al. [49], Kato et al. [69], Lin et al. [70], Figueiredo et al. [68], Figueiredo et al. [71], and Kolter et al. [72] have employed support vector machines (SVMs) in order to accomplish the disaggregation task. Harmonic signatures and low frequency features could be used to train classifiers in SVM. Moreover, SVM accompanied by Gaussian Mixture Model Classifier (GMM) can create a hybrid model [73]. In hybrid SVM/GMM method, GMM demonstrates how current waveforms are distributed while extracted power features are classified by SVM [20]. To enhance the obtained accuracy, different algorithms can be combined and a hybrid model, called committee decision mechanisms, can be developed [46,48,74].

Optimization (Eventless) Methods

The second category of supervised algorithm consists of the optimization methods, which are also called eventless algorithms. An example of supervised eventless disaggregation algorithm is the optimization method proposed by Suzuki et al. [67]. This technique attempts to find the most optimized matching between the unknown measured aggregate load and the known loads available in database [20]. The mentioned matching is conducted employing features extracted from load measurements, which could belong to a single or a composite load [4,48,75]. Different optimization methods including integer programming and genetic algorithm have been utilized in the literature to accomplish the matching task [25,67,76].

The main challenge in implementing optimization methods is disaggregating a combination of devices concurrently which may cause elevated complexity or lack of recognition of appliances due to signature overlapping [46]. The complexity problem can become more critical when the aggregated signal contains unknown loads which are not already available in appliance feature database [20].

3.2.2   Unsupervised Algorithms

Unsupervised load disaggregation methods, which do not rely on labeled data, have recently attracted researchers' attention [14]. In these methodologies, there is no need to have a prior knowledge about the appliances or training data, the fact that results in minimizing the initial setting costs and human interaction. This characteristic makes unsupervised algorithms promising alternatives for being adopted in inexpensive large-scale load disaggregation systems. In contrary to most supervised ML disaggregation, unsupervised ML algorithms are eventless [20]. Appliance features are learned automatically from aggregated load over a specific period such as several hours, days, or months [14,77] and these features are then assigned to a certain class [9]. It is required to assign a label to each class which can be done either manually or trough recent proposed approach known as Bayesian inference framework [78]. FHMM [14,79] and Additive Factorial HMMs [80] are some of the unsupervised algorithms utilized in the literature for disaggregation purposes.

Moreover, several FHMM could be combined to estimate sequence of hidden states and build unsupervised algorithms which can decompose aggregated load into its contributing appliances [43]. FHMM and three main extensions of it, called conditional FHMM, factorial hidden semi-Markov model, and conditional factorial hidden semi-Markov model, were utilized by Makonin [43] for accomplishing residential load disaggregation.

Besides supervised and unsupervised disaggregation method, an intermediated semi-supervised technique was introduced by Parson et al. [9] which is composed of both supervised and unsupervised modules. User interaction dependency of supervised methods could be reduced through semisupervised algorithms. Supervised module trains its classifiers through off-line learning using available dataset of labeled appliances and provides general appliance models to be used in unsupervised module.

3.3 Accuracy Metrics

Rapid expansion of the NILM sector and the recent development of numerous NILM methods have made providing a standard method for evaluating their corresponding performance an essential task [20,48]. According to Liang et al. the performance of NILM algorithms can be evaluated by using three accuracy meters such as disaggregation accuracy, detection accuracy, and overall accuracy. Several accuracy metrics have been proposed in the recent studies [4,20,43]. Some of the common metrics employed in the literature are given below:

- True/false positive rate

To evaluate performance accuracy of an event detector, corresponding true positives rate (TPR) and false positives rate (FPR) are compared. The TPR and FPR rate can be defined in terms of true positive (TP), true negative (TN), false positive (FP), and false negative (FN) samples as:

$$TPR = \frac{TP}{TP + FN} \tag{1}$$

$$FPR = \frac{FP}{FP + TN} \tag{2}$$

These rates can be visualized in a receiver operating characteristics (ROC) curve. ROC is a well-known technique to compare performance of detection and classification methods which are mainly used in pattern recognition [4,20]. In ROC curve, the best detector must be located as close as possible to the point corresponds to the ideal detector which is $TPR = 1$ and $FPR = 0$ [14]. In NIALM framework, event detection is the starting point for classification, disaggregation, and energy tracking. Hence, the more accurate the detection the less error propagates to the following steps [24].

- Precision and recall

Similar to TPR and FPR, Precision and Recall are two accuracy metrics which depend on TP, TN, FP, and FN and are defined as follows:

$$\text{Precision} = \frac{\text{TP}}{\text{TP} + \text{FP}} \qquad (3)$$

$$\text{Recall} = \frac{\text{TP}}{\text{TP} + \text{FN}} \qquad (4)$$

- F-score

F-score which was defined by Kim et al. [14] as harmonic mean of Precision recall can be defined in the general form [81] as:

$$F_\beta = \frac{\left(\beta^2 + 1\right) \cdot \text{precision} \cdot \text{recall}}{\beta^2 \cdot \text{precision} + \text{recall}} \qquad (5)$$

- Modified F-score

A modified version of F-score was employed in Makonin [43] as a more suitable approach to evaluate accuracy of NILM. Previously, only accuracy of state classifications was measured for an appliance. In addition to previous measurements, this new method also measures how accurate appliance consumptions were predicted. In modified F-score, TP is replaced with Accurate TP + Inaccurate TP which affects other metrics such as precision, recall, and F-score.

- Confusion matrix

Each element of this matrix represents how many times there is confusion between each state of an electrical device and other states or classified correctly without confusion [82].

- Total power change

In previous metrics, it was assumed that all events are of same importance despite the fact that some events are related to devices which draw high power while other events relate to small appliances. Therefore, considering a weight for each event could be beneficial. To address this issue, in Anderson [24] sum of the power changes due to missed events and false positive events were investigated. Similarly, the same author proposed another metric termed as average power change.

- Hamming loss

In NIALM appliances the whole information lost due to incorrect classification of appliance could be defined by a metrics known as Hamming loss [82].

- Energy disaggregation accuracy metrics

In addition to detection metrics, energy disaggregation metrics are provided to satisfy accuracy evaluation requirements [82]. These metrics include error in total energy assigned, fraction of total energy assigned correctly, normalized error in assigned power, and RMS error in assigned power, the details of which can be found in Batra et al. [82].

4 AVAILABLE OPEN DATASETS

In order to assess the accuracy of disaggregation algorithms, it is essential to employ datasets, which contain both the total power demand of households and the corresponding submetered (appliance by appliance) power data. The following list includes the characteristics of the publicly available datasets, which can be utilized in order to evaluate the disaggregation algorithms:

- REDD

The Reference Energy Disaggregation dataset (REDD) [83], which includes both total and itemized (submetered) power data obtained from six households, was released in 2011. REDD, which was the first dataset particularly provided for NILM purposes, has become the most commonly used dataset for assessing disaggregation algorithms [84].

- BLUED

The Building-Level fully-labeled dataset for Electricity Disaggregation (BLUED) was introduced in 2012 [85]. BLUED dataset contains voltage and current measurements, with a sampling rate of 12 kHz, obtained from a single household for a week. Although the data set does not contain submetered data, it includes labeled and time-stamped records regarding the transition of each appliance and can thus be a suitable ground truth for assessing the event-based algorithms.

- Smart*

The Smart* project, released in 2012, consists of measurement data from three households located in Massachusetts (United States). Although the submetered data is just provided for one of the houses, the data set contains the data obtained from many additional sensors including aggregate electricity usage, with a sampling rate of 1 s, temperature and humidity data in indoor rooms and out-door weather data.

- Household Electricity Survey

The Household Electricity Survey dataset [86], also introduced in 2012, includes submetered appliance level data from 251 houses out of which the aggregate data is also collected for 14 households.

- Tracebase

Tracebase dataset [87] provides recorded power consumption traces of several devices (both residential and office appliances) which were acquired using Plugwise (a commercial measurement tool) units.

- AMPds

The original Almanac of Minutely Power dataset (AMPds) [88], released in 2013, consisted of submetered and total power data obtained from a single house for 1 year. The dataset was next extended in the following release (AMPds 2), to contain 2 years of measurement data.

- *iAWE*

Also introduced in 2013, the Indian data for Ambient Water and Electricity Sensing (iAWE) [89] was also released in 2013, which contains both aggregate and submetered power data captured from a single house located in New Delhi for 73 days.

- BERDS

BERDS—BERkeley EneRgy Disaggregation Dataset [90] includes a set of power-related measurement data (real, reactive, and apparent power) from various devices operating in the UC Berkeley campus. The monitored devices include lighting units, Hot Water Pump and heating, ventilating, and air conditioning (HVAC) Fan loads, and some additional information including indoor and outdoor temperatures and HVAC system airflow have also been provided.

- ACS-F1

ACS-F1 (Appliance Consumption Signatures-Fribourg 1) dataset [91] includes the electrical consumption measurement data (real power (W), reactive power (var), RMS current (A), and phase of voltage relative to current (ϕ)) obtained through two acquisition sessions of 1 h on about 100 home appliances. The measurements have been conducted using plug-based sensors at low frequency (typically every 10 s) and the monitored devices include 10 categories of mobile phones, coffee machines, computer stations, fridges and freezers, CD players, lamps, laptops, microwave oven, printers, and televisions.

- UK-DALE

UK-DALE (the UK Domestic Appliance-Level Electricity) dataset [92] was released in 2014. This data set includes the measured aggregate power data (with the sampling rate of 16 kHz) and appliance-level power consumption (with the sampling rate of 1/6 Hz) from five households for duration of 655 days.

- ECO

ECO (Electricity Consumption and Occupancy) dataset [93], including power data collected from 6 Swiss households, was also released in 2014. The ECO data set provides 1 Hz aggregate consumption data (current, voltage, and phase shift for each of the three phases in the household) and 1 Hz submetered data obtained from selected appliances. This dataset also includes occupancy information, obtained through manual labeling and passive infrared sensors.

- GREEND

The GREEND dataset [94], introduced in 2014, includes appliance level power measurement data with the sampling rate of 1 Hz from 9 households in Austrian region of Carinthia and the Italian region of Friuli-Venezia Giulia.

- SustData

SustData dataset [95] includes numerous energy consumption measurements (real, reactive, and apparent power), captured at 1 min time intervals, from 50 homes (6 individual houses and 44 apartments).

- COMBED

The Commercial Building Energy Dataset (COMBED) dataset [96] includes recorded power-related measurements, at 30 s intervals, obtained from 200 smart meters installed in an academic campus in India.

- PLAID

Plug-Level Appliance Identification Dataset (PLAID) [97] contains current and voltage measurements captured, with a sampling rate of 30 kHz, from 11 different appliance types present in 55 households in Pennsylvania, United States.

- DRED

DRED (Dutch Residential Energy Dataset) [98], released in 2015, provides appliance level electricity measurements along with occupancy information and ambient parameters obtained from a household in the Netherlands.

- Dataport (Pecan Street)

The Dataport database, owned by Pecan Street Inc., is the world's largest source of disaggregated customer energy data [84] and is available free of charge for academic use. The database contains electricity data captured from 722 households (including 501 single-family homes, 183 apartments, 35 town homes, and 3 mobile homes) located in the American states of Texas, Colorado and California. Both aggregate load and individual appliance consumptions were monitored, at 1-min intervals, in most of the houses.

- COOLL

The Controlled On/Off Loads Library (COOLL) [99], provided by PRISME laboratory of the University of Orléans, France, is a dataset of electrical current and voltage measurements obtained from 42 appliances of 12 types with 100 kHz of sampling frequency.

- WHITED

Worldwide Household and Industry Transient Energy (WHITED) Dataset [100] includes the start-up transient (first 5 s) records of 110 different appliances (including 47 different device types) in 6 different regions (4 regions in Germany, 1 in Austria, and 2 in Indonesia). The measurements have been carried out using a low-cost custom sound card meter with a sampling rate of 44 kHz.

- REFIT

The REFIT dataset [101] includes raw electrical consumption data (including aggregate load and submetered measurements) obtained from 20 households sampled at 8 s intervals.

5   AVAILABLE ENERGY DISAGGREGATION OPEN-SOURCE TOOLS

Despite the recent rapid expansion of the energy disaggregation field, comparing the proposed disaggregation algorithms has been a very challenging task. This has been due to the utilization of different data sets, the lack of benchmark implemented algorithms, and various accuracy metrics which have been employed in different studies to access the accuracy of their proposed algorithms [82]. In order to tackle these issues, a publically available metadata (NILM metadata) and an open source toolkit (Nonintrusive load monitoring toolkit, NILMTK) have been recently developed, by researchers of the NILM community, in order to provide a means for comparing the energy disaggregation algorithms.

- Metadata for energy disaggregation

As was explained in the previous section, several energy disaggregation datasets have been released over the last few years, though the lack of accompanying standard metadata makes processing these datasets a cumbersome and time-consuming procedure. In order to tackle these issues, Kelly et al. [102] proposed a hierarchical metadata schema for energy disaggregation, which models appliances (including corresponding prior knowledge and models of appliances), meters, dwellings, and datasets. Their schema, which has been presented as an open-source project, has successfully been used to capture metadata for many of the above-mentioned datasets.

- NILMTK

Motivated by the lack of benchmark implemented disaggregation algorithms and standard accuracy metrics, Batra et al. [82] proposed an open-source toolkit, called nonintrusive load monitoring toolkit (NILMTK), which enables the comparison of energy disaggregation algorithms in a reproducible manner. NILMTK provides parser for several existing datasets, a set of preprocessing methods, and a collection of statistics in order to describe the datasets. It also includes two reference benchmark disaggregation algorithms (combinatorial optimization and HMM) and a set of accuracy metrics, which enable comparing disaggregation approaches through a common set of accuracy measures. However, the toolkit was designed to handle relatively small datasets [103]. To address this issue, the second release of this tool (NILMTK v0.2) was next developed by Kelly et al. [103] which is able to handle arbitrarily large datasets. In contrary to the first version of the tool, in which the entire data set is loaded into the memory, in NILMTK v0.2 the available data is loaded in chunks and the result of the disaggregation algorithm is saved to disk chunk by chunk. Hence, the tool is able to apply disaggregation on large datasets including Data-Port [84]. In addition, this extended version of the tool provides much richer metadata support by being integrating with the above-mentioned NILM Metadata [102]. Through this integration, NILMTK v0.2 includes dataset converters for many of the aforementioned datasets including REDD [83], iAWE [89], DataPort [84], UK-DALE [92], COMBED [96], and GreenD [94].

6 MAIN USE CASES OF ENERGY DISAGGREGATION

In this section, the main applications of energy disaggregation are reviewed. The benefits that this method provides the users and grid operators with are as follows:

- Providing itemized energy bills and personalized energy savings recommendations

Providing energy bills including the consumption of individual appliance, in order to help people reduce their energy, might be the most commonly cited application of NILM [104]. A study carried out by Kempton and Montgomery [105] demonstrated that the estimation of most of the residential consumers about the share of home devices in the total consumption is relatively poor. As an instance, although the residential lighting constitutes a small portion of the overall residential energy consumption, consumers often indicate lighting and even most commonly mention it first while expressing their idea about the main energy consumption causes. Whereas, hot water heating, which typically consumes about seven times the corresponding consumption of lighting, was pointed out less frequently. Hence, consumer's failure to accurately estimate the share of each device in the overall energy consumption leads to higher total consumption or makes their attempts to save energy ineffective. Accordingly, if the residents are provided with an accurate feedback about the devices with the highest energy consumption, they can modify their consumption behavior in order to save energy and thus reduce their monthly energy bills [104]. Furthermore, apart from the possible saving, it has also been demonstrated that consumers that are given appliance level energy consumption information in their households have higher energy literacy [15].

Energy disaggregation can also enable providing the customers with more accurate energy saving recommendations. Fischer et al. [106] presented a system which delivers energy-related recommendations based on household usage profiling. Their proposed system provides comparisons of the user's current energy tariffs to the tariffs available on the market, and provides advices on how much the user can save by shifting detected deferrable loads (e.g., washing machine or tumble dryer) to off-peak times. In order to do so, their proposed system makes use of appliance-level load disaggregation along with real-time energy tariff API, an energy data store, and a set of algorithms for usage prediction. The evaluation based on task-driven walkthroughs with 10 users with 3 months of monitored consumption data showed that system found cheaper tariffs for most of them (9/10).

- Detecting MFDs

Disaggregated electrical consumption can also be employed in order to identify faulty appliances. As an instance, the frosting cycle of a fridge with a damaged seal is more frequent than a normal one, the fact that can be identified using itemized fridge consumption data. Furthermore, appliance level data can facilitate individuating the appliances (e.g., printers) that do not shift to stand-by mode within an acceptable interval. A study by Martin and Poll [107] determined that by changing the time-to-sleep setting of the determined MFD, the corresponding annual energy consumption can be reduced by 39%. One of their recommendations to tackle the problem of devices failing to go to sleep mode was applying automated analysis in order to detect problems with device performance, a task which can effectively be accomplished through online energy disaggregation [104].

- Occupancy monitoring

Another possible use case of NILM is estimating the occupancy state in the building. Such a possibility can be employed in order to remotely monitor the health of the occupants and specifically elderly people. Belley et al. [108–110] proposed an NILM-based methodology for activity recognition and applied their approach on a smart home prototype by simulating daily scenarios taken from clinical trials that were previously performed with Alzheimer patients. They demonstrated that by utilizing this method, with a minimal investment and the exploitation of relatively limited data, activities of daily living could be efficiently recognized. In their latest study [110], they also presented a new NILM-based algorithmic approach capable of recognizing erratic behaviors related to cognitive deficits and of providing hints/prompts and reminders to guide a cognitively impaired person in the completion of daily activities. Kalogridis and Dave [111] implemented an NILM-based behavior anemology detection within a healthcare context and their results suggested that analyzing the TV and microwave consumption profiles, obtained through disaggregation, can be a promising approach for identifying behavior irregularities. Alcalá et al. [112] proposed a similar nonintrusive health monitoring method which detects activities of daily living through disaggregation of smart meter power consumption profile. In their approach, unique daily routines are learned automatically from the obtained appliance usage data via a log Gaussian Cox process. By applying their method on two real-world data sets, they showed that their method is able to identify over 80% of the kettle usages. Furthermore, they demonstrated that their approach permits earlier interventions in houses with a consistent routine, in comparison with a fixed-time intervention benchmark.

The NILM-based activity recognition can also facilitate implementing a smart heating strategy [104,113]. Spiegel and Albayrak [114] employed energy disaggregation techniques to detect the appliance usages and in turn the occupancy state of the household. Next, they employed the inferred occupancy states to optimize the heating schedules.

- Helping utility companies with customer segmentation

Knowledge about the customer's household characteristics (including the number of occupants per household, their employment status, and the properties of their buildings) allows the utilities to personalize their energy efficiency campaigns. Such personalization enhances the participation rates, results in larger energy savings, and increases the customer retention. However, such information is commonly gathered through surveys which are an expensive and cumbersome process [104]. Disaggregation of smart meter data can be employed as an alternative nonintrusive and cheaper approach for obtaining the mentioned information [115–118]. Beckel et al. [119] employed supervised ML methodologies to automatically estimate the characteristics of a household from its electricity consumption profile. They evaluated their analysis by analyzing smart meter data collected from 4232 households in Ireland and demonstrated that an accuracy of more than 70% over all households can be obtained. In their next study [116], they employed the smart meter data together with the corresponding ambient parameters and developed a method to determine the sensitivity of a household to outdoor temperature and the times of sunset/sunrise. This information was next utilized to improve their household classification system and in turn the corresponding customer segmentation.

- Improved demand side response

Demand side response provides the possibility of shifting demand away from the peak and thus decreasing the corresponding cost of energy, although the average daily consumption is not altered [104]. Furthermore, it provides helps modifying the demand based on the intermittent power generation of renewable energy sources; hence decreasing the need for expensive energy storage systems [120]. Pipattanasomporn et al. [121] provided the load profiles of selected major household appliances in the United States (clothes washers, clothes dryers, air conditioners, electric ovens, dishwashers, electric water heaters, and refrigerators) and discussed the demand response opportunities provided by these appliances. Energy disaggregation permits the utility companies to identify a device with a high consumption rate at a peak hour in a household and send a message to the corresponding user asking them to postpone their usage in order to smooth out the current peak in the demand [104]. In this context, Kong et al. [122] proposed an upgraded architecture for existing smart meter infrastructure in which it is upgraded by embedding an energy disaggregation algorithm. They claimed that their proposed architecture enhances both the interactivity between utility companies and customers and the demand side management.

- Enhancing the accuracy of energy demand prediction

Disaggregation of households' power consumption allows grid operators to improve their predictions of corresponding energy demand. Accordingly, Basu et al. [123] conducted a study in which the consumption profile of appliances is first identified through disaggregation and the obtained appliance-by-appliance load profiles along with meteorological information are next employed to predict the future usage. Rao et al. [124] presented a novel methodology for determining active devices and predicting future usage to aggregate power usage profile together with demographic data. By applying different models on their dataset, they demonstrated that SVM with Edge Analysis is the most suitable model for device identification while autoregressive moving average model is the most promising method for predicting future usage. Using the mentioned models, they obtained device identification and future consumption prediction accuracies of 75% and 90% respectively.

7 CONCLUSION

Our review revealed that although numerous appliance features have been investigated and several disaggregation algorithms have been proposed in the literature, no set of signatures or algorithms have been determined to be appropriate for all types of household appliances. Hence, developing a comprehensive methodology for detecting all appliance types with a high accuracy is still an open problem.

Considering the fact that the measurement procedure, which is required in order to obtain the training data for the supervised disaggregation algorithms, is both expensive and cumbersome, the recently developed unsupervised algorithms will receive more attention.

Furthermore, several publically accessible datasets and open-source tools, which have been released in the recent years, have notably facilitated the performance comparison of different NILM methodologies. Besides, several possible benefits of energy disaggregation for the users, energy companies, and grid operators have been demonstrated in the recently conducted studies. Hence, the authors believe that the potential economical profit for the customers and utility companies along with the recent facilitated public access to the disaggregation data and tools will result in further expansion of this field in the upcoming years with even a faster pace.

References

[1] J.A. Hoyo-Montano, C.A. Pereyda-Pierre, J.M. Tarin-Fontes, J.N. Leon-Ortega, in: Overview of non-intrusive load monitoring: a way to energy wise consumption, International Power Electronics Congress—CIEP, 2016, pp. 221–226.

[2] N.N. Oreskes, The scientific consensus on climate change: how do we know we're not wrong? in: J.F. DiMento, P. Doughman (Eds.), Climate Change: What It Means for Us, Our Children, and Our Grandchildren, MIT Press, 2007, p. 65.

[3] K.X. Perez, W.J. Cole, J.D. Rhodes, A. Ondeck, M. Webber, M. Baldea, T. F. Edgar, Nonintrusive disaggregation of residential air-conditioning loads from sub-hourly smart meter data, Energy Build. 81 (2014) 316–325.

[4] M. Zeifman, K. Roth, Nonintrusive appliance load monitoring: review and outlook, IEEE Trans. Consum. Electron. 57 (2011) 76–84.

[5] Energy, U. S., Department of Energy, Buildings Energy Data Book, Department of Energy, (2009).

[6] N.F. Esa, M.P. Abdullah, M.Y. Hassan, A review disaggregation method in non-intrusive appliance load monitoring, Renew. Sust. Energ. Rev. 66 (2016) 163–173.

[7] M.S. Tsai, Y.H. Lin, Modern development of an adaptive non-intrusive appliance load monitoring system in electricity energy conservation, Appl. Energy 96 (2012) 55–73.

[8] J.S. Donnal, J. Paris, S.B. Leeb, Energy applications for an energy box, IEEE Internet Things J. 3 (2016) 787–795.

[9] O. Parson, S. Ghosh, M. Weal, A. Rogers, An unsupervised training method for non-intrusive appliance load monitoring, Artif. Intell. 217 (2014) 1–19.

[10] S. Attari, M. Dekay, C. Davidson, W. de Bruin, Public perceptions of energy consumption and savings, Proc. Natl. Acad. Sci. 107 (2010) 16054–16059.

[11] S. Geman, D. Geman, Stochastic relaxation, Gibbs distributions, and the Bayesian restoration of images, IEEE Trans. Pattern Anal. Mach. Intell. PAMI-6 (1984) 721–741.

[12] B. Ritchie, G. McDougall, J. Claxton, Complexities of household energy consumption and conservation, J. Consum. Res. 8 (1981) 233–242.

[13] W. Abrahamse, L. Steg, C. Vlek, T. Rothengatter, A review of intervention studies aimed at household energy conservation, J. Environ. Psychol. 25 (2005) 273–291.

[14] H. Kim, M. Marwah, M. Arlitt, G. Lyon, J. Han, in: Unsupervised disaggregation of low frequency power measurements, Proceedings of the 11th SIAM International Conference on Data Mining, SDM 2011, 2011, pp. 747–758.

[15] T. Schwartz, S. Denef, G. Stevens, L. Ramirez, V. Wulf, in: Cultivating energy literacy: results from a longitudinal living lab study of a home energy management system, Proceedings of the SIGCHI Conference on Human Factors in Computing Systems, ACM, Paris, 2013.

[16] Abubakar, I., Khalid, S. N., Mustafa, M. W., Shareef, H. & Mustapha, M. An overview of non-intrusive load monitoring methodologies, 2015 IEEE Conference on Energy Conversion CENCON, 2015.54–59.

[17] Hart, G.W., Kern, E. C. & Schweppe, F. C. 1989. Non-intrusive appliance monitor apparatus. Google Patents.

[18] G.W. Hart, Nonintrusive appliance load monitoring, Proc. IEEE 80 (1992) 1870–1891.

[19] S.R. Shaw, S.B. Leeb, L.K. Norford, R.W. Cox, Nonintrusive load monitoring and diagnostics in power systems, IEEE Trans. Instrum. Meas. 57 (2008) 1445–1454.

[20] A. Zoha, A. Gluhak, M.A. Imran, S. Rajasegarar, Non-intrusive load monitoring approaches for disaggregated energy sensing: a survey, Sensors 12 (2012) 16838–16866.

[21] S. Barker, S. Kalra, D. Irwin, P. Shenoy, in: Empirical characterization and modeling of electrical loads in smart homes, 2013 International Green Computing Conference Proceedings, 27–29 June, 2013, pp. 1–10.

[22] F. Sultanem, Using appliance signatures for monitoring residential loads at meter panel level, IEEE Trans. Power Delivery 6 (1991) 1380–1385.

[23] G.W. Hart, E.C. Kern Jr., F.C. Schweppe, Non-Intrusive Appliance Monitor Apparatus, U.S. Patent 4,858,141, Massachusetts Institute of Technology and Electric Power Research Institute, Inc., 1989.

[24] K.D. Anderson, Non-Intrusive Load Monitoring: Disaggregation of Energy by Unsupervised Power Consumption Clustering (Ph.D. dissertation), Carnegie Mellon University, 2014.

[25] Baranski, M. & Voss, J. 2003. Non-intrusive appliance load monitoring based on an optical sensor. Proceedings of IEEE Power Tech Conference, 8–16.

[26] W. Wichakool, Z. Remscrim, U.A. Orji, S.B. Leeb, Smart metering of variable power loads, IEEE Trans. Smart Grid 6 (2015) 189–198.

[27] M. Dong, P.C.M. Meira, W. Xu, C.Y. Chung, Non-intrusive signature extraction for major residential loads, IEEE Trans. Smart Grid 4 (2013) 1421–1430.

[28] G.W. Hart, Correcting dependent errors in sequences generated by finite-state processes, IEEE Trans. Inf. Theory 39 (1993) 1249–1260.

[29] A. Bouloutas, M. Schwartz, Two extensions of the Viterbi algorithm, IEEE Trans. Inf. Theory 37 (1991) 430–436.

[30] S.B. Leeb, S.R. Shaw, J.L. Kirtley, Transient event detection in spectral envelope estimates for nonintrusive load monitoring, IEEE Trans. Power Delivery 10 (1995) 1200–1210.

[31] L.K. Norford, S.B. Leeb, Non-intrusive electrical load monitoring in commercial buildings based on steady-state and transient load-detection algorithms, Energy Build. 24 (1996) 51–64.

[32] A.I. Cole, A. Albicki, in: Algorithm for non-intrusive identification of residential appliances, Proceedings—IEEE International Symposium on Circuits and Systems, 1998, pp. 338–341.

[33] A.I. Cole, A. Albicki, in: Data extraction for effective non-intrusive identification of residential power loads, Conference Record—IEEE Instrumentation and Measurement Technology Conference, 1998, pp. 812–815.

[34] M. Baranski, J. Voss, in: Detecting patterns of appliances from total load data using a dynamic programming approach fourth, IEEE International Conference on Data Mining (ICDM'04), 2004.

[35] M. Baranski, J. Voss, in: Genetic algorithm for pattern detection in NIALM systems, Conference Proceedings—IEEE International Conference on Systems, Man and Cybernetics, 2004, pp. 3462–3468.

[36] I. Abubakar, S.N. Khalid, M.W. Mustafa, H. Shareef, M. Mustapha, Application of load monitoring in appliances' energy management—a review, Renew. Sust. Energ. Rev. 67 (2017) 235–245.

[37] J.W.M. Cheng, G. Kendall, J.S.K. Leung, in: Electric-load intelligence (E-LI): concept and applications, TENCON 2006—2006 IEEE Region 10 Conference, 14–17 November, 2006, pp. 1–4.

[38] C. Laughman, K. Lee, R. Cox, S. Shaw, S. Leeb, L. Norford, P. Armstrong, Power signature analysis, IEEE Power Energ. Mag. 99 (2) (2003) 56–63.

[39] H.H. Chang, C.L. Lin, J.K. Lee, in: Load identification in nonintrusive load monitoring using steady-state and turn-on transient energy algorithms, Proceedings of the 2010 14th International Conference on Computer Supported Cooperative Work in Design, CSCWD, 2010, pp. 27–32.

[40] S. Gupta, M.S. Reynolds, S.N. Patel, in: ElectriSense: single-point sensing using EMI for electrical event detection and classification in the home, UbiComp'10—Proceedings of the 2010 ACM Conference on Ubiquitous Computing, 2010, pp. 139–148.

[41] M. Berenguer, M. Giordani, F. Giraud-By, N. Noury, in: Automatic detection of activities of daily living from detecting and classifying electrical events on the residential power line, 2008 10th IEEE Intl. Conf. on e-Health Networking, Applications and Service, HEALTHCOM, 2008, pp. 29–32.

[42] Patel, S. N., Robertson, T., Kientz, J. A., Reynolds, M. S., Abowd, G. D. 2007. At the flick of a switch: detecting and classifying unique electrical events on the residential power line, UbiComp, 271–288.

[43] Makonin, S. 2012. Approaches to non-intrusive load monitoring (NILM) in the home. SFU Computing Science PhD Depth Exam.

[44] H. Najmeddine, K. El Khamlichi Drissi, C. Pasquier, C. Faure, K. Kerroum, A. Diop, T. Jouannet, M. Michou, in: State of art on load monitoring methods, PECon 2008—2008 IEEE 2nd International Power and Energy Conference, 2008, pp. 1256–1258.

[45] H.H. Chang, H.T. Yang, C.L. Lin, Load identification in neural networks for a non-intrusive monitoring of industrial electrical loads, in: International Conference on Computer Supported Cooperative Work in Design, Springer Berlin Heidelberg, 2007, pp. 664–674.

[46] M. Zeifman, C. Akers, K. Roth, in: Nonintrusive monitoring of miscellaneous and electronic loads, 2015 IEEE International Conference on Consumer Electronics, ICCE, 2015, pp. 305–308.

[47] S. Drenker, A. Kader, Nonintrusive monitoring of electric loads, IEEE Comput. Appl. Power 12 (1999) 47–51.

[48] J. Liang, S.K.K. Ng, G. Kendall, J.W.M. Cheng, Load signature study part I: basic concept, structure, and methodology, IEEE Trans. Power Delivery 25 (2010) 551–560.

[49] D. Srinivasan, W.S. Ng, A.C. Liew, Neural-network-based signature recognition for harmonic source identification, IEEE Trans. Power Delivery 21 (2006) 398–405.

[50] W.K. Lee, G.S.K. Fung, H.Y. Lam, F.H.Y. Chan, M. Lucente, in: Exploration on load signatures, International Conference on Electrical Engineering (ICEE), 2004, pp. 1–5.

[51] H.Y. Lam, G.S.K. Fung, W.K. Lee, A novel method to construct taxonomy electrical appliances based on load signatures, IEEE Trans. Consum. Electron. 53 (2007) 653–660.

[52] M. Weiss, A. Helfenstein, F. Mattern, T. Staake, in: Leveraging smart meter data to recognize home appliances, IEEE International Conference on Pervasive Computing and Communications, 19–23 March, 2012, 2012, pp. 190–197.

[53] S.B. Leeb, A Conjoint Pattern Recognition Approach to Nonintrusive Load Monitoring (Ph.D. dissertation), Massachusetts Institute of Technology, 1993.

[54] L. Farinaccio, R. Zmeureanu, Using a pattern recognition approach to disaggregate the total electricity consumption in a house into the major end-uses, Energy Build. 30 (1999) 245–259.

[55] M. Berges, E. Goldman, H.S. Matthews, L. Soibelman, in: Learning systems for electric consumption of buildings, Proceedings of the 2009 ASCE International Workshop on Computing in Civil Engineering, 2009, pp. 1–10.

[56] M.E. Berges, E. Goldman, H.S. Matthews, L. Soibelman, Enhancing electricity audits in residential buildings with nonintrusive load monitoring, J. Ind. Ecol. 14 (2010) 844–858.

[57] M.R. Anderberg, Cluster Analysis for Applications: Probability and Mathematical Statistics: A Series of Monographs and Textbooks, Academic Press, New York, 2014.

[58] Hartigan, J. A. 1975. Cluster algorithms. IRF Scientific Report, vol. 214, John Wiley & Sons, 1993.

[59] Leeb, S. B. & Kirtley Jr, J. L. 1996. Transient event detector for use in nonintrusive load monitoring systems. Google Patents.

[60] K.D. Lee, Electric Load Information System Based on Non-Intrusive Power Monitoring (Ph.D. dissertation), Massachusetts Institute of Technology, 2003.

[61] A. Marchiori, D. Hakkarinen, Q. Han, L. Earle, Circuit-level load monitoring for household energy management, IEEE Pervasive Comput 10 (2011) 40–48.

[62] A.G. Ruzzelli, C. Nicolas, A. Schoofs, G.M.P. O'Hare, in: Real-time recognition and profiling of appliances through a single electricity sensor, IEEE SECON, 2010, pp. 1–9.

[63] Z. Ghahramani, An introduction to hidden Markov models and Bayesian networks, Int. J. Pattern Recognit. Artif. Intell. 15 (2001) 9–42.

[64] S. Marsland, Machine Learning: An Algorithmic Perspective, CRC Press, Boca Raton, FL, 2015.

[65] L. Rabiner, B. Juang, An introduction to hidden Markov models, IEEE ASSP Mag. 3 (1986) 4–16.

[66] T. Zia, D. Bruckner, A. Zaidi, in: A hidden Markov model based procedure for identifying household electric loads, IECON Proceedings (Industrial Electronics Conference), 2011, pp. 3218–3223.

[67] K. Suzuki, S. Inagaki, T. Suzuki, H. Nakamura, K. Ito, in: Nonintrusive appliance load monitoring based on integer programming, Proceedings of the SICE Annual Conference, 2008, pp. 2742–2747.

[68] Figueiredo, M. B., de Almeida, A. & Ribeiro, B. 2011. An experimental study on electrical signature identification of nonintrusive load monitoring (NILM) systems. Adaptive and Natural Computing Algorithms, 31–40.

[69] T. Kato, H.S. Cho, D. Lee, T. Toyomura, T. Yamazaki, in: Appliance recognition from electric current signals for information-energy integrated network in home environments, 7th

International Conference on Smart Homes and Health Telematics, ICOST, 5597, 2009, pp. 150–157.

[70] G.Y. Lin, S.C. Lee, J.Y.J. Hsu, W.R. Jih, in: Applying power meters for appliance recognition on the electric panel, Proceedings of the 2010 5th IEEE Conference on Industrial Electronics and Applications, ICIEA, 2010, pp. 2254–2259.

[71] M. Figueiredo, A. de Almeida, B. Ribeiro, Home electrical signal disaggregation for non-intrusive load monitoring (NILM) systems, Neurocomputing 96 (2012) 66–73.

[72] J.Z. Kolter, S. Batra, A.Y. Ng, Energy disaggregation via discriminative sparse coding, Adv. Neural Inf. Proces. Syst. 23 (2010) 1153–1161.

[73] Y.X. Lai, C.F. Lai, Y.M. Huang, H.C. Chao, Multi-appliance recognition system with hybrid SVM/GMM classifier in ubiquitous smart home, Inform. Sci. 230 (2012) 39–55.

[74] J. Liang, S.K.K. Ng, G. Kendall, J.W.M. Cheng, Load signature study part II: disaggregation framework, simulation, and applications, IEEE Trans. Power Delivery 25 (2010) 561–569.

[75] Y. Du, L. Du, B. Lu, R. Harley, T. Habetler, in: A review of identification and monitoring methods for electric loads in commercial and residential buildings, 2010 IEEE Energy Conversion Congress and Exposition, ECCE 2010—Proceedings, 2010, pp. 4527–4533.

[76] A. Schoofs, A. Guerrieri, D.T. Delaney, G. O'Hare, A.G. Ruzzelli, in: ANNOT: automated electricity data annotation using wireless sensor networks, Sensor Mesh and Ad Hoc Communications and Networks (SECON), 2010 7th Annual IEEE Communications Society Conference on, 2010, pp. 1–9.

[77] H. Goncalves, A. Ocneanu, M. Bergés, R.H. Fan, in: Unsupervised disaggregation of appliances using aggregated consumption data, Proc. KDD Workshop Data Mining Appl. Sustainability, 2011, pp. 21–24.

[78] M.J. Johnson, A.S. Willsky, Bayesian nonparametric hidden semi-Markov models, J. Mach. Learn. Res. 14 (2012) 673–701.

[79] Z. Ghahramani, M.I. Jordan, Factorial hidden Markov models, Mach. Learn. 29 (1997) 245–273.

[80] J.Z. Kolter, T. Jaakkola, Approximate inference in additive factorial HMMs with application to energy disaggregation, J. Mach. Learn. Res. 22 (2012) 1472–1482.

[81] Sokolova, M., Japkowicz, N. & Szpakowicz, S. 2006. Beyond accuracy, F-score and ROC: a family of discriminant measures for performance evaluation. In: Sattar, A. & Kang, B.-H. (eds.) AI 2006: Advances in Artificial Intelligence: 19th Australian Joint Conference on Artificial Intelligence, Hobart, Australia, December 4–8, 2006. Proceedings. Berlin, Heidelberg: Springer.

[82] N. Batra, J. Kelly, O. Parson, H. Dutta, W. Knottenbelt, A. Rogers, A. Singh, M. Srivastava, in: NILMTK: an open source toolkit for non-intrusive load monitoring, Fifth International Conference on Future Energy Systems (ACM E-Energy), 2014.

[83] J.Z. Kolter, M.J. Johnson, in: REDD: a public data set for energy disaggregation research, Proceedings of the SustKDD Workshop on Data Mining Applications in Sustainability, 2011, pp. 1–6.

[84] O. Parson, G. Fisher, A. Hersey, N. Batra, J. Kelly, A. Singh, W. Knottenbelt, A. Rogers, in: Dataport and NILMTK: a building data set designed for non-intrusive load monitoring, 2015 IEEE Global Conference on Signal and Information Processing, GlobalSIP, 2015, pp. 210–214.

[85] Anderson, K., Ocneanu, A., Benitez, D., Carlson, D., Rowe, A. & Berges, M. 2012. BLUED: A fully labeled public dataset for event-based non-intrusive load monitoring research.Proceedings of the 2nd KDD Workshop on Data Mining Applications in Sustainability (SustKDD)1-5.

[86] J.P. Zimmermann, M. Evans, J. Griggs, N. King, L. Harding, P. Roberts, C. Evans, Household electricity survey a study of domestic electrical product usage, Intertek Report R6614, 2012.

[87] Reinhardt, A., Baumann, P., Burgstahler, D., Hollick, M., Chonov, H., Werner, M. & Steinmetz, R. On the accuracy of appliance identification based on distributed load metering data. 2012 Sustainable Internet and ICT for Sustainability, SustainIT 2012, 2012.

[88] S. Makonin, F. Popowich, L. Bartram, B. Gill, I.V. Bajic, in: AMPds: a public dataset for load disaggregation and eco-feedback research, Electrical Power and Energy Conference (EPEC), 2013 IEEE, 2013, pp. 1–6.

[89] N. Batra, M. Gulati, A. Singh, M.B. Srivastava, in: It's different: Insights into home energy consumption in India, Proceedings of the Fifth ACM Workshop on Embedded Sensing Systems for Energy-Efficiency in Buildings (ACM BuildSys), 2013.

[90] Maasoumy, M., Sanandaji, B., Poolla, K. & Vincentelli, A. S. BERDS-Berkeley energy disaggregation data set. Proceedings of the Workshop on Big Learning at the Conference on Neural Information Processing Systems (NIPS), 2013.

[91] C. Gisler, A. Ridi, D. Zufferey, O.A. Khaled, J. Hennebert, in: Appliance consumption signature database and recognition test protocols, 2013 8th International Workshop on Systems, Signal Processing and Their Applications (WoSSPA), 12–15 May, 2013, pp. 336–341.

[92] J. Kelly, W. Knottenbelt, The UK-DALE dataset, domestic appliance-level electricity demand and whole-house demand from five UK homes, Sci. Data 2 (2015).

[93] C. Beckel, W. Kleiminger, R. Cicchetti, T. Staake, S. Santini, in: The ECO data set and the performance of non-intrusive load monitoring algorithms, BuildSys 2014—Proceedings of the 1st ACM Conference on Embedded Systems for Energy-Efficient Buildings, 2014, pp. 80–89.

[94] A. Monacchi, D. Egarter, W. Elmenreich, S. D'Alessandro, A.M. Tonello, in: GREEND: an energy consumption dataset of households in Italy and Austria, The 5th IEEE International Conference on Smart Grid Communications (SmartGridComm), 2014.

[95] L. Pereira, F. Quintal, R. Gonçalves, N.J. Nunes, in: SustData: a public dataset for ICT4S electric energy research, ICT for Sustainability 2014, ICT4S, 2014, pp. 359–368.

[96] Batra, N., Parson, O., Berges, M., Singh, A. & Rogers, A. 2014. A comparison of non-intrusive load monitoring methods for commercial and residential buildings arXiv preprint arXiv:1408.6595.

[97] J. Gao, S. Giri, E.C. Kara, M. Berg, in: PLAID: a public dataset of high-resolution electrical appliance measurements for load identification research: demo abstract, Proceedings of the 1st ACM Conference on Embedded Systems for Energy-Efficient Buildings, Memphis, TN, ACM, New York, NY, 2014.

[98] U. Akshay, S.N. Nambi, A.R. Lua, R.V. Prasad, Loced: location-aware energy disaggregation framework, Proceedings of the 2nd ACM International Conference on Embedded Systems for Energy-Efficient Built Environments, ACM, 2015, pp. 45–54.

[99] Picon, T., Meziane, M. N., Ravier, P., Lamarque, G., Novello, C., Bunetel, J.-C. L. & Raingeaud, Y. 2016. COOLL: controlled on/off loads library, a public dataset of high-sampled electrical signals for appliance identification. arXiv preprint arXiv:1611.05803.

[100] M. Kahl, A.U. Haq, T. Kriechbaumer, H.-A. Jacobsen, in: Whited—a worldwide household and industry transient energy data set, Workshop on Non-Intrusive Load Monitoring (NILM), 2016 Proceedings of the 3rd International, 2016.

[101] D. Murray, L. Stankovic, V. Stankovic, An electrical load measurements dataset of United Kingdom households from a two-year longitudinal study, Sci. Data 4 (2017).

[102] J. Kelly, W. Knottenbelt, in: Metadata for energy disaggregation, The 2nd IEEE International Workshop on Consumer Devices and Systems (CDS 2014), 2014.

[103] J. Kelly, N. Batra, O. Parson, H. Dutta, W. Knottenbelt, A. Rogers, A. Singh, M. Srivastava, in: NILMTK v0.2: A non-intrusive load monitoring toolkit for large scale data sets, BuildSys

2014—Proceedings of the 1st ACM Conference on Embedded Systems for Energy-Efficient Buildings, 2014, pp. 182–183.

[104] Kelly, J, Disaggregation of Domestic Smart Meter Energy Data (PhD), Imperial College London, 2016.

[105] W. Kempton, L. Montgomery, Folk quantification of energy, Energy 7 (1982) 817–827.

[106] J.E. Fischer, S.D. Ramchurn, M.A. Osborne, O. Parson, T.D. Huynh, M. Alam, N. Pantidi, S. Moran, K. Bachour, S. Reece, E. Costanza, T. Rodden, N.R. Jennings, in: Recommending energy tariffs and load shifting based on smart household usage profiling, International Conference on Intelligent User Interfaces, Proceedings IUI, 2013, pp. 383–394.

[107] R. Martin, S. Poll, Energy analysis of multi-function devices in an office environment, ASHRAE Trans. (2014) 120.1.

[108] C. Belley, S. Gaboury, B. Bouchard, A. Bouzouane, in: Activity recognition in smart homes based on electrical devices identification, ACM International Conference Proceeding Series, 2013.

[109] C. Belley, S. Gaboury, B. Bouchard, A. Bouzouane, An efficient and inexpensive method for activity recognition within a smart home based on load signatures of appliances, Pervasive Mob. Comput. 12 (2014) 58–78.

[110] C. Belley, S. Gaboury, B. Bouchard, A. Bouzouane, in: A new system for assistance and guidance in smart homes based on electrical devices identification, ACM International Conference Proceeding Series, 2014.

[111] G. Kalogridis, S. Dave, in: Privacy and eHealth-enabled smart meter informatics, 2014 IEEE 16th International Conference on e-Health Networking, Applications and Services, Healthcom 2014, 2015, pp. 116–121.

[112] J. Alcalá, O. Parson, A. Rogers, in: Detecting anomalies in activities of daily living of elderly residents via energy disaggregation and Cox processes, BuildSys 2015—Proceedings of the 2nd ACM International Conference on Embedded Systems for Energy-Efficient Built, 2015, pp. 225–234.

[113] S. Spiegel, Optimization of in-house energy demand, in: Smart Information Systems, Springer International Publishing, 2015, pp. 271–289.

[114] S. Spiegel, S. Albayrak, in: Energy disaggregation meets heating control, Proceedings of the ACM Symposium on Applied Computing, 2014, pp. 559–566.

[115] A. Albert, R. Rajagopal, Smart meter driven segmentation: what your consumption says about you, IEEE Trans. Power Syst. 28 (2013) 4019–4030.

[116] C. Beckel, L. Sadamori, S. Santini, T. Staake, in: Automated customer segmentation based on smart meter data with temperature and daylight sensitivity, 2015 IEEE International Conference on Smart Grid Communications, SmartGridComm, 2015, pp. 653–658.

[117] A. Kavousian, R. Rajagopal, M. Fischer, Determinants of residential electricity consumption: using smart meter data to examine the effect of climate, building characteristics, appliance stock, and occupants' behavior, Energy 55 (2013) 184–194.

[118] J. Kwac, C.W. Tan, N. Sintov, J. Flora, R. Rajagopal, in: Utility customer segmentation based on smart meter data: empirical study, 2013 IEEE International Conference on Smart Grid Communications, SmartGridComm, 2013, pp. 720–725.

[119] C. Beckel, L. Sadamori, T. Staake, S. Santini, Revealing household characteristics from smart meter data, Energy 78 (2014) 397–410.

[120] U.S. Department of Energy, Benefits of Demand Response in Electricity Markets and Recommendations for Achieving Them, A report to the United States congress pursuant to Section 1252 of the Energy Policy Act of 2005, 2006.

[121] M. Pipattanasomporn, M. Kuzlu, S. Rahman, Y. Teklu, Load profiles of selected major house-hold appliances and their demand response opportunities, IEEE Trans. Smart Grid 5 (2014) 742–750.

[122] W. Kong, Y. Xu, Z.Y. Dong, D.J. Hill, J. Ma, C. Lu, in: An extended prototypical smart meter architecture for demand side management, Proceeding—2015 IEEE International Conference on Industrial Informatics, INDIN, 2015, pp. 1008–1013.

[123] K. Basu, V. Debusschere, S. Bacha, in: Residential appliance identification and future usage prediction from smart meter, IECON Proceedings (Industrial Electronics Conference), 2013, pp. 4994–4999.

[124] Rao, K. M., Ravichandran, D. & Mahesh, K. Non-intrusive load monitoring and analytics for device prediction, Lecture Notes in Engineering and Computer Science, 2016, 132–136.

CHAPTER 18

Energy Disaggregation and the Utility-Privacy Tradeoff

Roy Dong*, Lillian J. Ratliff†

*University of California, Berkeley, Berkeley, CA, United States, †University of Washington, Seattle, WA, United States

CHAPTER OVERVIEW

The problem of energy disaggregation is the estimation of individual device usage patterns from available aggregate energy consumption measurements. In this work, we consider the fundamental limits of the energy disaggregation problem, and use these limits to quantify the tradeoff between the utilization of data for smart grid operations and the privacy provided to energy consumers. First, our fundamental limits build on a statistical testing framework to provide a theoretical bound to the accuracy of energy disaggregation that can be achieved by any algorithm. Then, we present a framework for understanding how variations in system design can affect the operational benefits of collecting data, as well as the privacy of users. We instantiate this framework in a direct load control example where we use thermostatically controlled loads and vary the frequency with which a centralized controller receives sensor measurements. Our work formalizes the process of incorporating privacy considerations into the design of modern energy systems.

1 INTRODUCTION

Energy disaggregation and nonintrusive load monitoring (NILM) are general terms, which refer to methods that estimate the energy consumption of individual devices, or statistics of the energy consumption signal, without installing individual sensors at the plug level.[1] The goals of different energy disaggregation algorithms include event detection (i.e., determine when certain devices switch states) and energy disaggregation (i.e., recover the power consumption signals of each device in its entirety from the aggregate signal).

In many cases, we would like to have the latter for many households, but installing sensors on every plug in each house is prohibitively expensive and intrusive. For example, studies have shown that merely providing users feedback on their energy consumption patterns is sufficient to improve their consumption behaviors [1–3]. Forecasts predict that 20% savings in residential buildings are attainable with the use of personalized recommendations based on

---

[1]Throughout this chapter, we will be using energy disaggregation and NILM interchangeably.

Big Data Application in Power Systems. https://doi.org/10.1016/B978-0-12-811968-6.00018-8

disaggregated data [3]. In addition, these savings are sustainable over long time periods, and are not transient effects of introducing new interfaces to users. These device-level measurements can further be used for strategic marketing of energy-saving programs and rebates, both improving efficacy of the programs and reducing costs.

Conversely, the availability of smart meter data presents privacy risks as well. Previous studies have shown that monitoring energy consumption at high granularity can allow the inference of detailed information about consumers' lives such as the times they eat, when they watch TV, and when they take a shower [4]. Such information is highly valuable and will be sought by many parties, including advertising companies [5], law enforcement [6], and criminals [7].

NILM algorithms can help guide regulation for privacy policies in advanced metering infrastructures (AMIs) [8]. Analyzing NILM algorithms is a way to determine how much device-level information is contained in an aggregate signal. This information is critical to understanding the privacy concerns in AMIs and which parties should have access to aggregate power consumption data. Governments, researchers, and organizations are working on privacy standards and policies to guide AMI deployments.

Previous work on this front has proved very fruitful; researchers have considered the issue of data privacy in smart grid infrastructures, and have proposed novel mechanisms for protecting the collected data (encryption, access control, and cryptographic commitments) [9, 10], by anonymization and aggregation [11, 12], and by preventing inferences and reidentification from databases that allow queries from untrusted third parties (via differential privacy) [13]. In the work presented in this chapter, we add to this literature by analyzing methods to minimize the quality and quantity of data transmitted while still achieving certain operational objectives.

To successfully understand the utility-privacy tradeoff in these smart grid operations, we must quantify two things. First, we must model the tradeoff between the quality of collected data and performance of smart grid operations. Second, we must understand how data quality affects an adversary's ability to infer a consumer's private information.

As a proof of concept, we consider the utility-privacy tradeoff in a direct load control (DLC) example. To analyze the utility of data, we consider how the performance of proposed DLC mechanisms change as fewer and fewer measurements are received by the controller. This allows us to quantify how much data is needed for smart grid operations.

The underlying philosophy of our work is that these data transmission policies often unintentionally transmit information about private parameters unrelated to the original control goal: we separate operational parameters from parameters users may consider "private." Furthermore, the operational goals of a systems operator are different from the inferential goals of a privacy-breaching adversary. Thus, different types of analyses are needed to understand the trade-off between data collection and smart grid performance versus the tradeoff between data collection and user privacy.

In this work, we consider the energy consumption patterns inside the home private. Thus, the fundamental limits of NILM algorithms can provide a good benchmark for defining privacy risk; the state-of-the-art NILM algorithm may be a reasonably conservative model for an adversary. For example, if we use the framework defined in [14, 15], we can analyze the accuracy of an adversary's inference when using energy disaggregation algorithms with a prior on device usage patterns and models for individual devices. An understanding of the fundamental limits can provide a theoretical guarantee of privacy, if we conclude that disaggregation is impossible in a certain scenario. It can be used in the design of AMIs, by determining a minimum sampling rate, sensor accuracy, and network capacity to achieve a desired goal. Further, it may allow us to determine how many measurements actually need to be stored and transmitted.

To quantify the privacy risk in these mechanisms, we use recent results in NILM to give theoretical guarantees on when NILM algorithms will fail: adversaries will not be able to infer the device usage of a consumer from observing the aggregate power consumption of a building. In addition, we model the private parameters of a consumer, and the inferences that can be made about private parameters from device usage patterns.

This chapter can be divided into two main contributions.

First, we study the fundamental limits of NILM algorithms. We consider a building containing a number of devices. Given the aggregate power consumption of these devices, we would like to distinguish between two scenarios (e.g., whether or not a light turns on, or whether it was a toaster or kettle that turned on). In particular, provided an arbitrary NILM algorithm, we seek bounds on the probability of distinguishing two scenarios given an aggregate power consumption signal. In addition, once we have this theory developed for two scenarios, we generalize to find an upper bound on the probability of distinguishing between a finite number of scenarios. With this theory of the fundamental limits of NILM in hand, we address questions about the possibility of NILM in the context of AMIs. Further, using high-frequency, high-resolution measurements of power consumption signals of common household devices as the ground truth, we analyze the probability of successfully identifying common scenarios in a household. We also analyze the tradeoff between successful NILM and sensor/model accuracy, as well as sampling rate.

Second, we provide a general framework for considering the utility-privacy tradeoff in the smart grid. This framework formalizes the tension between high-resolution, high-frequency data providing better control of our energy systems with the fact that this allows the inference of very personal information about individual lifestyles. We instantiate this on a DLC example. We define an operational objective and choose a privacy metric that represents the ability of an adversary to infer private facts about energy consumers. Then, we consider the operational performance and the privacy of consumers in a tradeoff analysis in this context.

The rest of this chapter is organized as follows. In Section 2, we review the relevant literature for both the fundamental limits of NILM and the utility-privacy tradeoff in the smart grid. In Section 3, we discuss the fundamental limits of energy disaggregation and how to calculate privacy bounds. In Section 4, we introduce a general framework for analyzing the utility-privacy tradeoff in energy systems, as well as explore an example implementing the fundamental limits of Section 3. We provide closing remarks in Section 5.

2   BACKGROUND

2.1   Energy Disaggregation Background

The problem of NILM is essentially a single-channel source separation problem: determine the power consumption of individual devices given their aggregated power consumption. The source separation problem has a long history in information theory and signal processing and well-known methods include the infomax principle [16], which tries to maximize some output entropy, the maximum likelihood principle [17], which uses a contrast function on some distribution on the source signals, and a time-coherence principle [18], which assumes time-coherence of the underlying source signals. These often lead to formulations, which use some variation of a principle component analysis or independent component analysis.

The most common applications of the source separation theory is to audio signals and biomedical signals. For these applications, it is often assumed that source signals are i.i.d. stationary processes. We note that power consumption signals are very different from these types of signals. The power consumption of a device has strong temporal correlations and is not stationary, for example, whether or not a device is on at a given time is correlated with whether or not it was on an instant ago, and the mean power consumption signal changes with the state of the device. The algorithmic and theoretical development in source separation have therefore not been successfully applied to NILM and most methods for NILM are rather different to those developed for classical source separation.

The field of NILM is much younger than source separation and most development has focused on algorithms. We briefly outline a few approaches here. One approach has focused on the design of hardware to best detect the signatures of distinct devices [19–21], but algorithms to handle the hardware's measurements are still an open problem. Another approach which has been taken by much of the machine learning community is to use hidden Markov models (HMMs), or some variation, to model individual devices [22–24]; energy disaggregation can be done with an expectation maximization algorithm. In recent publications [14, 15], we model individual devices as dynamical systems and use adaptive filtering. These are a few examples of concrete algorithms for NILM. For a more comprehensive review, we refer the reader to [3].

The discussion presented here focus on the theoretical limitations of an arbitrary NILM algorithm. To the best of our knowledge, there has not been any previous work attempting to model the NILM problem in its full generality and derive theoretical bounds. The work is inspired by recent work in differential privacy [25–28]. The underlying goal of differential privacy is to model privacy in a fashion that encapsulates arbitrary prior information on the part of the adversary and an arbitrary definition of what constitutes a privacy breach. The theory of differential privacy can be extended to give similar, but weaker, bounds to those derived in this paper.

2.2 Utility-Privacy Tradeoff Background

Some of the earliest literature in applied privacy was ensuring that surveys could be conducted in a privacy-preserving fashion. These methods were called *randomized response* methods [29, 30]. These researchers noticed that there was structural bias when surveys requested sensitive information, such as whether or not a subject was HIV-positive. The key component for guaranteeing privacy was to given individual subjects *deniability*: a positive answer could either be a true response or due to the randomness in the survey procedure.

The next advances in the applied privacy literature was in statistical databases. In [31], the author argues that any definition of privacy should satisfy the following desideratum: nothing can be learned about a user with the database that could not be learned without the database. One attempt to satisfy this desideratum was k-anonymity [32], which provides methods to ensure that for any one user, there are at least $k - 1$ users who appear indistinguishable from said user.

More recently, the advent of big data has introduced many databases with potentially sensitive data that could be utilized by an adversary as side information to infer private facts. For example, in [33], the authors are able to take anonymized Netflix data and, using publicly available information from IMDB, recover the identities of individual users. These results pushed researchers to no longer consider privacy of a database in isolation, but in the larger context of widely available side information.

Arguably the most popular privacy metric, *differential privacy* was introduced in [25]. Differential privacy requires an exogenous adjacency relationship, which specifies pairs of potential values for private parameters that we hope to keep indistinguishable. With this adjacency relationship, differential privacy is a bound on the change in the distribution of the observables between any two adjacent private parameters.

Differential privacy is attack-agnostic, in the sense that as a metric it does not suppose the adversary launches a particular type of inference attack. Furthermore, differential privacy is also agnostic to the amount of side information an adversary has, since it simply captures how much the distribution of the observable changes for small perturbations to private parameters.

An alternative definition uses an information-theoretic metric to quantify privacy loss. In particular, *mutual information* between a private parameter and a public observable has recently become a popular metric [34, 35]. One interpretation of the mutual information is the difference between the entropy of the prior distribution and the entropy of the posterior distribution [36]; from that perspective, this metric has an intuitive interpretation as quantifying the reduction in the uncertainty of an adversary due to a public observable. This metric is attack-agnostic, since it simply quantifies a statistical relationship between private and public variables. However, this requires a specification of the available side information to an adversary, as reflected in the prior distribution.

It is our belief that, similar to previous technological changes, new smart grid technologies will motivate a sea change in how privacy is perceived, defined, quantified, and treated. All the previously mentioned references consider privacy in the context of databases, but a nascent area of research is the investigation of how privacy can be understood in the context of systems with dynamics.

Recent work in this regard includes the extension of differential privacy to Kalman filtering [27], constrained optimization [37] and convex optimization [38], distributed control [39], and online learning [40]. In similar, there have been efforts to consider information-theoretic metrics in the context of dynamic systems such as the smart grid [41, 42].

Our framework is equipped to handle any privacy metric, but in our DLC example, we use our fundamental limits to provide a measure of the *inferential privacy* of users.

3 FUNDAMENTAL LIMITS OF NILM

3.1 Problem Statement

As mentioned in Section 1, NILM has a variety of end uses. For each of these potential applications, the statistics of interest may be different. Thus, when we state the problem of NILM, we remain as general as possible to accommodate all these applications.

We are given an aggregate power consumption signal for a building. Let $y[t] \in \mathbb{R}$ denote the value of the aggregate power consumption signal at time t for $t = 0$, ..., $N - 1$, and let $y \in \mathbb{R}^N$ refer to the entire signal. This signal is the aggregate of the power consumption signal of several individual devices:

$$y[t] = \sum_{i=1}^{D} y_i[t] \quad \text{for } t = 0, ..., N \tag{1}$$

where D is the number of devices in the building and $y_i[t]$ is the power consumption of device i at time t.

There are many possible goals of NILM. For example, the energy disaggregation problem is to recover y_i for $i = 1, 2, ..., D$ from y. Another goal commonly studied is to recover information about the y_i from y, such as when lights turn on or the power consumption of the fridge over a week.

In general, we will refer to the phenomena we wish to distinguish as scenarios throughout this paper.

3.2 Model of Energy Disaggregation Algorithms

We outline a general framework for analyzing the problem outlined in Section 3.1. At a high level, the framework can be summarized as follows. First, any NILM method must choose some representation for individual devices; these can be seen as functions from some input space to $\mathbb{R}^N$. Depending on the purpose of the NILM algorithm, the input space will vary; essentially, scenarios we wish to distinguish should correspond to different inputs in the input space. Then, we describe NILM algorithms as functions on the observed aggregate signal. The definition is meant to be general and hold across both generative and discriminative techniques.

3.2.1 Aggregate Device Model

Formally, let $(\Omega, \mathcal{F}, P)$ denote our probability space. As in Section 3.1, D denotes the number of devices and N denotes the length of our observed power signal.

Let Θ_i denote the input space for the ith device. Inputs represent scenarios we wish to distinguish. The output space, representing the power consumption signal of an individual device, is $\mathbb{R}^N$ for every device. Then, the model associated with the ith device can be denoted as $G_i : \Theta_i \times \Omega \to \mathbb{R}^N$. Here, we have the condition that, for any $u_i \in \Theta_i$, $G_i(\theta_i, \cdot)$ is a random variable. Finally, let $\Theta = \Theta_1 \times \Theta_2 \times \cdots \times \Theta_D$, and let $G : \Theta \times \Omega \to \mathbb{R}^N$ be defined as $G((\theta_1, \theta_2, ..., \theta_D), \omega) = \sum_{i=1}^{D} G_i(\theta_i, \omega)$. Here, G denotes our aggregated system (i.e., the model of our building).

Definition 1. Given that the input is $\theta \in \Theta$, the distribution of the power consumption is $G(\theta, \cdot)$.

We emphasize the generality of this framework. Many state-of-the-art methods can be formulated in this framework. For example, factorial HMM methods [22–24] can be thought of as single-input, single-output systems where the input is the state of the underlying Markov chains. The Markov transition probabilities become a prior on the input signal. In previous work [14, 15], we formulated the models as dynamical systems whose inputs are real-valued and correspond to the device usage. Thus, we now have a general way of expressing different models of devices in an NILM problem.

3.2.2   NILM Algorithms

An algorithm for NILM will be a function of our observed aggregate power consumption signal. Its result will depend on the goal of the algorithm, and the end use of the algorithm output. For example, it could be the set of possible estimated disaggregated energy signals, $\{\hat{y}_i\}_{i=1}^{D}$, or the set of possible discrete event-labels on our time-series data, or a set of statistics on the disaggregated data.

More formally, let S represent some NILM algorithm and $\mathcal{Z}$ represent its output space, discussed earlier. Then, the algorithm could be thought of as a function $S: \mathbb{R}^T \to \mathcal{Z}$. We will analyze a general S in the following section.

3.3   Fundamental Limits of Energy Disaggregation

In this section, we derive an upper bound on the probability of successfully distinguishing two scenarios with any NILM algorithm. Then, we extend these results to handle the case where we wish to upper bound the probabilities of distinguishing a finite set of scenarios, as well as two collections of scenarios. Note that in our framework, scenarios correspond to inputs to our device models, and we will use the two terms interchangeably.

3.3.1   Distinguishing Two Scenarios

First, fix any two inputs $v_0, v_1 \in \Theta$, which we wish to distinguish. For example, we may pick v_0 and v_1 so that they differ only in the usage of one device. In that case, we are analyzing the difference in observed output caused by whether or not, say, a microwave turns on in the morning. Alternatively, we may choose inputs that correspond to more dissimilar scenarios, such as whether or not a household uses an air conditioner at all. The choice of v_0, v_1 depends on which scenarios we wish to distinguish in our NILM algorithm.

As mentioned previously, let $S: \mathbb{R}^N \to \mathcal{Z}$ denote any NILM algorithm. Then, let $I: \mathcal{Z} \to \{0, 1\}$ be an indicator for whether or not an algorithm output satisfies some condition. For example, I could output 1 if a particular discrete phenomena (e.g., a light turning on) is detected in the algorithm output, and 0 otherwise. Or, I could output 1 if the estimated power consumption signals of individual devices lie in a certain set.

Suppose this indicator captures whether our algorithm believes the input is v_0 or v_1. That is, $(I \circ S)$ should output 1 if the NILM algorithm believes the input is v_1 and 0 if it believes the input is v_0. For this reason, from this point forward we will refer to I as our discriminator.

Definition 2. $(I \circ S)$ is measurable, that is, $(I \circ S)^{-1}(\{1\})$ is a measurable set in $\mathbb{R}^N$, with respect to the Borel field on $\mathbb{R}^N$.
We note that this is a reasonable assumption, as most, if not all, NILM algorithms in practice will be a finite composition of measurable functions.

In addition, we note that this is a very conservative understanding of an NILM algorithm. In general, these algorithms are not be designed simply to distinguish between v_0 and v_1, and are likely not to be optimal in this regard. Thus, by analyzing an optimal $(I \circ S)$, we have a conservative upper bound on the probability of distinguishing v_0 and v_1. In particular, the scenarios v_0 and v_1 may contain additional information, so our optimal separator is allowed to use side information, such as the switching times of devices, when doing inference, making our bound more conservative.

Furthermore, we can contrast our contribution with existing work in differential privacy. Whereas differential privacy would consider any v_0 and v_1 that are adjacent, and bound the change in distributions for a fixed mechanism, here we fix a particular v_0 and v_1 and consider a bound on the performance of any mechanism.

Thus, we can formulate this in classical hypothesis testing frameworks seen in the statistics literature [36]. Our main contribution is the abstraction of the task of NILM that allows us to use well-known results in detection theory.

Let y denote our observed signal. Suppose that $G(v_0, \cdot)$ has a probability density function (pdf) f_0 and similarly $G(v_1, \cdot)$ with f_1. Let our likelihood ratio be defined as:

$$L(y) = \frac{f_1(y)}{f_0(y)} \qquad (2)$$

The maximum likelihood estimator (MLE) finds the input that maximizes the likelihood of our observations. The MLE is given by:

$$\hat{\theta}_{\mathrm{MLE}}(y) = \begin{cases} v_1 & \text{if } L(y) \geq 1 \\ v_0 & \text{otherwise} \end{cases} \qquad (3)$$

If we have a prior p on the probability of v_0 or v_1 as inputs, we can find the maximum a posteriori (MAP) estimate. This finds the input that is most likely given our observations and prior. The MAP is:

$$\hat{\theta}_{MAP}(\gamma) = \begin{cases} v_1 & \text{if } L(\gamma) \geq \dfrac{p(v_0)}{p(v_1)} \\ v_0 & \text{otherwise} \end{cases} \tag{4}$$

Note that this prior can be a discrete distribution or a density. However, for simplicity, we will treat the prior as a discrete distribution throughout this paper; small notational changes are required for the prior to be a density.

Now, suppose we have a maximum acceptable probability of mislabeling the input v_1; let this parameter be denoted $\beta > 0$. Also, let u denote the true input. The optimal estimator with this constraint is:

$$\begin{aligned} \min_{\hat{\theta}} \quad & P(\hat{\theta} = v_1 | \theta = v_0) \\ \text{subject to} \quad & P(\hat{\theta} = v_0 | \theta = v_1) \leq \beta \end{aligned} \tag{5}$$

By the Neyman-Pearson lemma, the non-Bayesian detection problem in Eq. (5) has the following solution:

$$\hat{\theta}_{NB}(\gamma) = \begin{cases} v_1 & \text{if } L(\gamma) \geq \lambda \\ v_0 & \text{otherwise} \end{cases} \tag{6}$$

where λ is chosen such that $P(\hat{\theta}_{NB} = v_0 | \theta = v_1) = \beta$.

Throughout the rest of this paper, we will consider the MAP, but these can be extended to the other two cases. The probability of interest is the probability of successful NILM:

Definition 3. For the two-input case, the *probability of successful NILM* for an estimator $\hat{\theta}$ is:

$$\sum_{i=0}^{1} P(\hat{\theta}(\gamma) = v_i | \theta = v_i) p(\theta = v_i) \tag{7}$$

This can be explicitly calculated given the densities and the prior. In addition, any algorithm and discriminator $(I \circ S)$ will perform worse than $\hat{\theta}_{MAP}$, so the MAP estimate provides an upper bound on any algorithm's probability of successful NILM.

Proposition 1. *Any estimator $\hat{\theta}$ will have a probability of successful NILM bounded by:*

$$\sum_{i=0}^{1} P(\hat{\theta}_{MAP}(\gamma) = v_i | \theta = v_i) p(\theta = v_i) \tag{8}$$

3.3.2 Distinguishing a Finite Number of Scenarios

This easily extends to distinguishing between a finite number of scenarios. Let V denote a finite set of inputs. Then

Definition 4. For the N-input case, the *probability of successful NILM* for an estimator $\hat{\theta}$ is:

$$\sum_{i=1}^{N} P(\hat{\theta}(\gamma) = v_i | \theta = v_i) p(\theta = v_i) \tag{9}$$

The MAP is given by:

$$\hat{\theta}_{\text{MAP}}(\gamma) = \arg\max_{v \in \Theta} P(G(\theta, \cdot) = \gamma | \theta = v) p(\theta = v) \tag{10}$$

Proposition 2. *There is an upper bound to the probability of successful NILM provided by the MAP:*

$$\sum_{i=1}^{N} P(\hat{\theta}_{\text{MAP}}(\gamma) = v_i | \theta = v_i) p(\theta = v_i) \tag{11}$$

3.3.3 Distinguishing Two Collections of Scenarios

This philosophy of deriving an upper bound extends nicely to whenever we wish to distinguish two collections of scenarios. This corresponds to distinguishing two sets of inputs.

Now, suppose we have two sets of inputs: V_0 and V_1. We can still define the probability of successful NILM in this context:

Definition 5. For the case where we wish to distinguish two sets of inputs, the *probability of successful NILM* for an estimator $\hat{\theta}$ is:

$$\sum_{i=0}^{1} P(\hat{\theta}(\gamma) \in V_i | \theta \in V_i) p(\theta \in V_i) \tag{12}$$

Depending on the context, this quantity may be calculable. In other cases, it may be possible to find good approximations or upper bounds. We will see this arise in Section 3.4.

3.4 Gaussian Case

In this section, we instantiate our theory on the special case where our model is a deterministic function with additive Gaussian noise.

3.4.1 Two Scenarios

Suppose our system takes the following form:

$$G(\theta, \omega) = h(\theta) + w(\omega) \tag{13}$$

where $h: \Theta \to \mathbb{R}^N$ is a deterministic function and w is a random variable. Furthermore, fix any two inputs v_0, v_1 which we wish to distinguish, and suppose that w is a zero-mean Gaussian random variable with covariance Σ. Furthermore, suppose our prior is $p(\theta = v_0) = p(\theta = v_1) = 0.5$.

This can encapsulate the case where the uncertainty arises from measurement noise and model error. Referring to our motivating example, suppose that the only difference between v_0 and v_1 is the presence of a toaster turning on once in v_1. The question we are asking is: Can we detect the toaster turning on?

Then, let f_0 denote the Normal pdf with mean $h(v_0)$ and covariance Σ, and similarly let f_1 be the Normal pdf with mean $h(v_1)$ and the same covariance Σ. For shorthand, let $\mu_0 = h(v_0)$ and $\mu_1 = h(v_1)$.

Since the covariance matrix Σ is the same for both random variables, $\hat{\theta}_{\text{MAP}}$ is determined by a hyperplane. Let $a^T = (\mu_0 - \mu_1)^T \Sigma^{-1}$ and $b = \frac{1}{2}\left(\mu_1^T \Sigma^{-1} \mu_1 - \mu_0^T \Sigma^{-1} \mu_0\right)$. Then

$$\hat{\theta}_{\text{MAP}}(y) = \begin{cases} v_1 & \text{if } a^T y + b \leq 0 \\ v_0 & \text{otherwise} \end{cases} \tag{14}$$

Now, suppose the input is actually v_0. That is, y is distributed according to f_0. Then, the signed distance from y to the boundary of the hyperplane is given by $\frac{1}{\|a\|_2}(a^T y + b)$. This is a linear function of Gaussian random variable, and is thus also a Gaussian random variable. Furthermore, the mean of this random variable will be $\frac{1}{\|a\|_2}(a^T \mu_0 + b)$, and the variance will be:

$$\sigma^2 = \frac{1}{\|a\|_2^2} a^T \Sigma a = \frac{(\mu_0 - \mu_1)^T \Sigma^{-1}(\mu_0 - \mu_1)}{(\mu_0 - \mu_1)^T \Sigma^{-2}(\mu_0 - \mu_1)} \tag{15}$$

Thus, given that the input is actually v_0, the probability that $\hat{\theta}_{\text{MAP}}(y) = v_0$ is:

$$\begin{aligned} P(\hat{\theta}_{\text{MAP}}(y) &= v_0 | \theta = v_0) \\ &= \frac{1}{2}\left(1 - \text{erf}\left(\frac{-\frac{1}{\|a\|_2}(a^T \mu_0 + b)}{\sqrt{2\sigma^2}}\right)\right) \end{aligned} \tag{16}$$

where erf is the Gauss error function and Eq. (16) is simply the 1 minus the cumulative distribution function (cdf) of the distance to the hyperplane evaluated at 0, that is, the probability that the signed distance is positive.

The computations are exactly the same for the case where the input is v_1. Thus:

Proposition 3. *By Eq. (8), the probability of successfully distinguishing* v_0 *and* v_1 *with the MAP is given by:*

$$\frac{1}{2}\left(1 - \operatorname{erf}\left(\frac{-\frac{1}{\|a\|_2}(a^T\mu_0 + b)}{\sqrt{2\sigma^2}}\right)\right) \tag{17}$$

Note that, in general, disaggregation algorithms would not be designed simply to distinguish between v_0 and v_1, and are likely not to be optimal in this regard. That is, Eq. (17) provides a theoretical upper bound on how good any possible disaggregation algorithm could perform in distinguishing v_0 and v_1. Also, note that $\frac{1}{\|a\|_2}(a^T\mu_0 + b)$ will be positive if $\mu_0 \neq \mu_1$. It follows that the upper bound is always greater than 0.5 if $\mu_0 \neq \mu_1$, and the MAP achieves this upper bound. Thus, if the inputs cause different outputs from the system, there will always exist an algorithm that improves the discrimination between v_0 and v_1 over blind guessing.

3.4.2 K Scenarios

In this section, we build on the development in Section 3.4.1 to handle the case where we wish to distinguish several inputs.

Suppose now that we have a finite set of inputs that we wish to distinguish. Consider the set $\{v_i\}_{i=1}^K$, where $v_i \in \Theta$ for each i. Again, suppose all these inputs are equally likely, that is, $p(\theta = v_i) = \frac{1}{K}$ for all i. We wish to find the MAP. We carry over the assumption of Gaussian noise with variance Σ. The MAP will partition $\mathbb{R}^N$ with hyperplanes of the form given in Section 3.4.1.

So, suppose the actual input is v_i. We wish to ask: What is the probability the MAP will accurately identify v_i from the other $K - 1$ inputs? Let $\mu_i = h(v_i)$ for $i = 1, \ldots, N$. Then, let $a_i^T = (\mu_1 - \mu_i)^T\Sigma^{-1}$ and $b_i = \frac{1}{2}(\mu_i^T\Sigma^{-1}\mu_i - \mu_1\Sigma^{-1}\mu_1)$. Given our observation $y \in \mathbb{R}^T$, we wish to ask the probability that $\frac{1}{\|a_i\|_2}(a_i^Ty + b_i) > 0$ for $i = 2, \ldots, K$ (i.e., that the input u_1 is more likely than any of the other inputs). More succinctly, define

$$A = \begin{bmatrix} a_2^T/\|a_2\|_2 \\ a_3^T/\|a_3\|_2 \\ \vdots \\ a_K^T/\|a_K\|_2 \end{bmatrix} \quad b = \begin{bmatrix} b_2/\|a_2\|_2 \\ b_3/\|a_3\|_2 \\ \vdots \\ b_K/\|a_K\|_2 \end{bmatrix} \tag{18}$$

We wish to ask the probability that $Ay + b$ is in the positive orthant of $\mathbb{R}^N$. Recall that y is distributed according to mean μ_1 and covariance Σ. Thus, the random variable $Ay + b$ has mean $A\mu_1 + b$ with covariance $A\Sigma A^T$. The probability that this random variable is in the positive orthant cannot be analytically calculated, but can be approximated with high accuracy.

This can be done for $i = 2, \ldots, K$ as well, and provide an upper bound on the probability of successful NILM.

3.4.3   Linear Systems

In this section, we specialize the previous theory to the case where all our devices are linear systems. Suppose that the dynamics of our household are of the form $y = A\theta + e$, and our noise e has covariance $\hat{\sigma}^2 I$. Note that σ^2 as defined in Eq. (15) is equal to $\hat{\sigma}^2$.

Now, suppose the sets that we wish to distinguish are $V_0 = \{0\}$ and $V_1 = \{v: L \leq \|v\|_2 \leq U\}$, for some constants $0 < L \leq U$. That is, can we detect an input with magnitude in the range $[L, U]$? By Eq. (12), we have the probability of successful NILM for an estimator $\hat{u}$ is:

$$P(\hat{\theta}(y) = 0|\theta = 0)p(\theta = 0) + P(\hat{\theta}(y) \in V_1|\theta \in V_1)p(\theta \in V_1) \tag{19}$$

First, consider a fixed input $v \in V_1$. If we suppose that $\theta = v$, then the probability of an estimator $\hat{\theta}$ distinguishing v from 0 is bounded by:

$$P(\hat{\theta}(y) \neq 0|\theta = v) \leq \frac{1}{2}\left(1 + \text{erf}\left(\frac{\|Av\|_2}{2\sqrt{2\sigma^2}}\right)\right) \tag{20}$$

This can be seen by noting that, after a projection into one dimension, the separating hyperplane is the point $\pm\|Av\|_2/2$. Without loss of generality, let us suppose the separating point is $\|Av\|_2/2$.

Note that this equation is an increasing function of $\|Av\|_2$. This gives us:

$$\frac{1}{2}\left(1 + \text{erf}\left(\frac{\|Av\|_2}{2\sqrt{2\sigma^2}}\right)\right) \leq \frac{1}{2}\left(1 + \text{erf}\left(\frac{\sigma\max(A)U}{2\sqrt{2\sigma^2}}\right)\right) \tag{21}$$

where $\sigma\max(A)$ is the largest singular value of A. This held for any $v \in V_1$, so measure-theoretic properties give us:

$$P(\hat{\theta}(y) \in V_1|u \in V_1) \leq \frac{1}{2}\left(1 + \text{erf}\left(\frac{\sigma\max(A)U}{2\sqrt{2\sigma^2}}\right)\right) \tag{22}$$

Proposition 4. *In the linear system case, the probability of successful NILM is bounded above by:*

$$p(\theta = 0) + \frac{1}{2}\left(1 + \text{erf}\left(\frac{\sigma\max(A)U}{2\sqrt{2\sigma^2}}\right)\right)p(\theta \in V_1) \tag{23}$$

These are bounds which do not depend explicitly on a model, but rather only on the sensitivity of the model. Thus, even with just knowledge of the variance of the noise and the sensitivity of our linear systems, we can still find an upper bound on the probability of successful NILM.

4 UTILITY-PRIVACY TRADEOFF

In this section, we consider the utility-privacy tradeoff in the smart grid. We use the fundamental limits proved above as a privacy metric in this section. In other words, we focus on interpreting θ as the device usage patterns inside the household, rather than the "inputs" given to devices. In this section, u will correspond to the input given to a system using collected smart meter data for operational benefit. To emphasize the point: u, which is considered the "input" in this section, is distinct from θ, which is the private parameters of users. We have tried to ensure that the notation refers to the same objects between sections, but these objects might have different roles, for example, θ is the input to devices when proving fundamental limits, but θ is the private information in this section.

4.1 Framework

In this section, we introduce a framework for quantifying the tradeoff between the operational utility of data and the privacy levels of consumers.

Privacy-preserving mechanisms can be divided into two categories: mechanisms which control *access* to data, or mechanisms which vary the *quality* of data.

Access control methods have been researched primarily by the cryptography community, with very strong results [43]. The former can provide strong guarantees of privacy against outside adversaries, but does not protect users from privacy breaches by those who have access to the data. For example, your utility company should have access to your energy consumption, but they may be able to infer aspects of your lifestyle from these patterns [4, 15].

By contrast, quality-based methods have been researched by several communities. For example, most differential privacy mechanisms add noise to the data [25, 44]: as the noise levels increase, the quality of the data decreases, and privacy levels increase as well. As another example, systems can sample real-time data less frequently to increase the privacy levels of consumers; these mechanisms are considered in [8, 45]. By modifying the quality of the data prior to its transmission, these methods guarantee privacy against both outside adversaries and insiders. However, the modifications to the data's quality must be carefully designed to not erode its original utility; if the data are no longer useful for its intended purposes, then the efficiency and comfort benefits of these novel technologies will be lost.

In this chapter, we will focus on privacy-preserving mechanisms that vary the quality of data to achieve different levels of privacy.

4.1.1 The Utility of Data

The utility of a particular set of data comes from the improvement in the performance of some service due to said data. To model these systems, we follow a control theoretic framework.

We are interested in the performance of our system at some set of times $T \subset \mathbb{R}_+$. This includes the discrete time cases $T = \{0, 1, ..., N-1\}$ for some $N \in \mathbb{N}$ or $T = \{0, 1, ...\}$, as well as the continuous time cases $T = [0, T_f]$ for some $T_f \in \mathbb{R}$ or $T = [0, \infty)$. For simplicity, we will assume operation of the system begins at time $t = 0$ and $0 \in T$.

Our system has some state space $\mathcal{X}$, which represents all possible configurations of the system at one point in time. Usually, we will take $\mathcal{X} = \mathbb{R}^n$ for some $n \in \mathbb{N}$. We will denote the state at time $t \in T$ as $x(t)$. Similarly, the control actions we can take upon the system live in some input space, $\mathcal{U}$, with the input at time t denoted $u(t)$. The dynamics of the system are captured in a function $\phi : \mathcal{X} \times \mathcal{U}^T \rightarrow \mathcal{X}^T$ which takes an initial condition and an input signal across all T and specifies which trajectory in $\mathcal{X}^T$ the system will follow. For example, in the context of linear time-invariant systems, if $\phi(x_0, u) = x$, then x is the unique solution to differential equation $\dot{x}(t) = Ax(t) + Bu(t)$ with initial condition $x(0) = x_0$.

The performance of the system is evaluated with respect to a cost function $J : \mathcal{X}^T \times \mathcal{U}^T \rightarrow \mathbb{R}$. The system has an initial condition $x_0 \in \mathcal{X}$ and obeys the system dynamics ϕ. The system operator wants to pick a $u \in \mathcal{U}^T$ such that $J(\phi(x_0, u), u)$ is kept low. Ideally, the optimal control problem would be solved: $\min_u J(\phi(x_0, u), u)$. However, this often is difficult and, in practice, we will use controllers that will approximate the optimal control strategy subject to information and tractability constraints.

To attempt to minimize this cost, the system operators will design a controller. This controller will determine the input $u \in \mathcal{U}^T$ that will be given to the system. However, this controller will have a limited amount of data about the system. In our framework, we will consider how variations in the quality of the data affect the system operator's control decisions, and therefore affect the realized cost of the system. For some quality level q and time $t \in T$, we will let $Y(q, t)$ denote the data available to the controller at time t.[2] With this data, the controller will pick a control input $u(t) \in \mathcal{U}$. We let this process be denoted $u_c(Y(q,t), t) \in \mathcal{U}$.

---

[2] For generality, we have not included details of what space these objects q and $Y(q, t)$ live in. Formally, q can live in a general space, but we will often think of $q \in \mathbb{R}$. For example, q can denote the sampling period of our system, as we will explore in Section 4.2. Similarly, $Y(q, t)$ can live in some arbitrary space for each q and t. In the example in Section 4.2, $Y(q, t)$ will be a collection of random variables that the controller can observe at time t.

With this controller specified, we can consider the mapping from quality level q to realized cost J. That is, for a particular quality q, the controller will use the controller and issue control command $u_c(Y(q,t),t) \in \mathcal{U}$ at each time $t \in T$. This will cause the realized cost to be $J(\phi(x_0, u_q), u_q)$ where $u_q \in \mathcal{U}^T$ is defined as $u_q(t) = u_c(Y(q,t),t) \in \mathcal{U}$ for every $t \in T$.

Abstractly, this allows us to quantify the utility of data by showing how the control performance of the cyber-physical system erodes for different quality levels of data. As previously mentioned, we will instantiate this in a concrete example in Section 4.2.

4.1.2   The Privacy of Data

Data are collected from consumers with the intent of improving Internet of Things (IoT) operations. However, these data also allow the inference of private information about consumers, unrelated to IoT operations. This section quantifies how much information about the private lives of consumers is contained in data.

In the previous section, we fixed a set of time indices $T \subset \mathbb{R}_+$, and defined a data mechanism $Y(q, t)$ for each quality level q and time t. This data mechanism defined what information is collected and transmitted, and we quantified how a controller's performance changes as the quality level q is varied. In this section, we will consider how variations in q affect the privacy levels of consumers in the data mechanism $Y(q, t)$. We take a statistical perspective on privacy: What is the inferential power of these new observations relative to some private parameter? Our model is as follows.

Users have a private parameter $\theta \in \Theta$, which they wish to protect. These private parameters θ live in a space Θ with some particular structure, which depends on the privacy metric in use. In differential privacy, the private parameter space Θ is equipped with an "adjacency" relationship specifies which pairs $(\theta, \theta') \in \Theta \times \Theta$ which should be indistinguishable. For information theoretic metrics and the inferential privacy metric used in Section 4.1.5, θ is seen as a random variable taking finitely many values, that is, Θ has finitely many elements and there exist a prior distribution P_θ for the random variable θ.

These privacy metrics should be general enough in definition to allow evaluation for any data mechanism Y under consideration. In addition, it will depend on the quality q: so our privacy valuations be a function of the structure of our data mechanism, as well as the quality level. This will be denoted $m(Y, q)$. This framework is general enough to capture any quality-varying privacy-preserving mechanisms, and this generality is needed to be able to encompass the spectrum of possible privacy risks and information structures in IoT.

In Section 4.1, we outlined a general framework for quantifying the utility-privacy tradeoff in Internet of Things applications. Before covering an concrete example in the smart grid, we will discuss a new privacy metric, *inferential privacy*. Informally, inferential privacy is a guaranteed lower bound on the probability an adversary will correctly infer θ from the observations $(Y(q, t))_{t \in T}$.

Depending on the context, different privacy metrics may be more applicable than others. As argued in the philosophy of privacy [46], we believe that a plurality of privacy metrics and definitions is required to capture the essence of a concept as context-dependent and essentially contested as privacy [47].

Differential privacy gives a very powerful guarantee that is both agnostic to attacks and adversary's available side information; however, many practical applications require a particular structure of the uncertainty, such as additive independent Laplacian or Gaussian noise. For example, it is not clear how to consider how the level of differential privacy varies in a dynamical system when the sampling rate is adjusted.

By contrast, information theoretic metrics lend themselves nicely to the design of noise in a fashion that is oftentimes optimal with respect to some criterion. In [34], the authors are able to design an optimal noising scheme subject to a performance constraint in database estimation, and in [41], the authors consider compression schemes in the context of the smart grid, and provide a theoretical bounds on the information leakage subject to a distortion constraint.

Our work on privacy builds a hypothesis testing framework, which has been well studied in the information theory [36] and statistics [48] communities. Variational calculus methods for statistics were first introduced by Neyman and Pearson [49], and have been a fruitful way to find optimal estimators. In addition, a popular metric for critiquing the performance of an estimator is known as the minimax risk, which measures an estimator's expected loss against a worst-case distribution [50]; the minimax risk can act as a measure of the difficulty of a hypothesis testing problem. Alternatively, Fano was able to analyze the difficulty of the hypothesis testing problem by considering the entropy and mutual information between the parameter of interest and the observables [36, 51]; these results were extended to observations on the continuum in [52]. Each of these methods can provide a measure of the hypothesis testing problem's difficulty, which we use as a guarantee for privacy.

Throughout this section, we will be analyzing the privacy level for a fixed quality q. Naturally, one can vary q afterwards to see the effect of quality on privacy levels.

4.1.3 User and Data Mechanism Model

First, we introduce a model for how the private parameter θ influences the observed data $(Y(q, t))_{t \in T}$. We will allow $\mathcal{Y}$ to denote the possible values of $(Y(q, t))_{t \in T}$.

Definition 6. The private parameter θ follows a distribution P_θ. Similarly, $(Y(q, t))_{t \in T}$ given θ has a conditional distribution $P_{y|\theta}$.

We note that, formally, this assumption is quite succinct, but, in practice, determining these distributions are rarely trivial.

4.1.4 Adversary Model

Next, we introduce our adversary model.

Definition 7. Our adversary is able to observe the transmitted data $(Y(q, t))_{t \in T}$, and has knowledge of P_θ and $P_{y|\theta}$. In addition, this adversary has an arbitrary amount of computational power.

This adversary has access to the measured data signal, and also holds priors on the consumer's private information θ. He also knows how this private information affects the consumer's usage of IoT devices, $P_{y|\theta}$. Although this adversary has quite a bit of knowledge about the consumers, he does not hold arbitrary side information.

We note that it may not be realistic to suppose the adversary has access to P_θ and $P_{y|\theta}$. However, any adversary who tries to infer θ from y with less information will only do worse than our adversary model. Thus, this model provides a conservative estimate against all weaker adversary models.

4.1.5 Inferential Privacy Metric

Our privacy metric is the probability of error if an adversary tries to infer the private variable θ.

Definition 8. Under the usage model outlined in Assumption 6, a system is "α inferentially private" if, for any estimator $\hat{\theta} : \mathcal{Y} \to \Theta$, we have

$$\Pr(\hat{\theta}[(Y(q,t))_{t \in T}] \neq \theta) \geq \alpha \tag{24}$$

This estimator can be based on information in P_θ and $P_{y|\theta}$.

Here we note that this is in essence an ex ante privacy metric, that is, the privacy is spread across Θ according to P_θ. As often arises in many statistical estimation problems, an ex post privacy metric (i.e., a privacy metric that guarantees privacy for every type) is not a well-posed problem.

For example, suppose $\Theta = \{0, 1\}$, and consider the estimator $\hat{\theta} \equiv 0$. For any consumer of type $\theta = 0$, the adversary will correctly infer their type with this estimator. In other words, an adversary can always violate the privacy of one type of consumer by making the blanket assumption that everyone is a fixed type. In a sense, we gain privacy by noting that the adversary has to be successful across the different types Θ (weighted according to P_θ).

Regardless of the algorithm the adversary uses, we can bound the probability it will successfully breach a consumer's privacy. Furthermore, this formula allows us to vary the quality level q, such as how often data are collected and transmitted. We will examine this on a concrete example in Section 4.2. This guarantee is also simple for consumers to interpret, and can be used in the design of privacy contracts between the utility companies and consumers [53].

Remark 1. To instantiate the above general framework for utility-privacy tradeoff with our fundamental limits of NILM, we will use P_θ to denote our prior and $P_{Y|\theta}$ to denote our device models. Thus, if the fundamental limit from Proposition 2 is α, then we know our system is α inferentially private.

4.2 Example: Direct Load Control

In this section, we instantiate our utility-privacy framework in a concrete context. Specifically, we consider the privacy of DLC programs in the smart grid.

DLC has been a promising future direction for the smart grid for a variety of reasons. By controlling loads which can be modified without much impact on consumer satisfaction, we can allay many costs by shifting loads from peak demand and compensating for real-time load imbalances. In addition, as renewable energy penetration increases, the generation side of power is growing more uncertain and will require demand flexibility. In this section, we will consider the load imbalance signal as exogenous, and use a DLC scheme to try and compensate the imbalance.

In addition, such DLC policies are being deployed today. For example, Pacific Gas and Electric deployed the SmartAC program in Spring 2007 [54]. Another provider of demand response services has recruited over 1.25 million residential customers in DLC programs, and has deployed over 5 million DLC devices in the United States. In California, they have successfully curtailed over 25 MW of power consumption since 2007 [55]. As these programs are being deployed on a large scale, it is important to consider the privacy aspects of these programs [4].

In this chapter, we consider different sampling rates as a method of varying the quality of data q. Our motivations for this are twofold.

First, there are many cases where noise-free data are required, for practical, regulatory, performance, or economic reasons. For example, suppose random noise is added to your energy consumption signal before being transmitted to the utility company. A consequence of this mechanism is that the energy bill you receive will not be a deterministic function of your energy usage, but rather a random variable with a conditional dependence on your energy usage. Many consumers may be unhappy with this mechanism in which they may be billed for more energy than they used, and a lot of regulatory overhead would be necessary for a utility company to roll out such a mechanism, even in the face of statistical arguments that the effect of such a random mechanism is negligible in the long run.

Second, an analysis of the effect of sampling rates on operational performance is the first step in enacting the *data minimization* principle for dynamical systems. In the United States, the Obama Administration examined privacy issues in its June 2011 smart grid policy framework report [56]. The report recommends that State and Federal regulators should consider, as a starting point, methods to ensure that consumers' detailed energy usage data are protected in a manner consistent with federal Fair Information Practice (FIP) principles. One of the key principles is data minimization. This principle is consistent with the notion of privacy by design [57].

Similarly, the FIP principle of data minimization appears in smart grid privacy recommendations by the National Institute of Standards and Technology [58], the North American Energy Standards Board [59], the Department of Energy [60], the Texas Legislature and Public Utility Commission [61], and the California Public Utilities Commission (CPUC) [62].

The NISTIR 7628 [58] expresses the data minimization principle in the smart grid context as:

> Limit the collection of data to only that necessary for Smart Grid operations, including planning and management, improving energy use and efficiency, account management, and billing.

All these recommendations and policy proposals have been broad in coverage by necessity, as regulators do not want to burden electric utilities with specific limits on what they can collect. However, electric utilities who want to follow these privacy recommendations do not have a sound reasoning principle to help them decide how much data is too little or too much. Our goal in this section is to start discussing scientifically sound principles that can help determine how much data to collect in order to achieve a certain level of functionality of the grid, and how much privacy is granted to consumers under this data collection policy.

By analyzing the effect that sampling rate has on Smart Grid operations, we can begin to quantify the utility of data, a necessary first step to enacting data minimization. Intuitively, there should be a sampling rate where higher sampling frequencies have a negligible effect on the system's performance. For example, this could be due to the ability of the controller to leverage this high frequency data, or the time scales of the system itself. Conversely, there should intuitively be a sampling rate that is so low that the system's performance is comparable to the performance should the controller receive no measurements at all. Finding these regimes of operation is the goal of the first half of our framework.

As mentioned previously, there are several approaches to preserve the privacy of a consumer participating in an AMI, including adding noise to data, modifying how data are aggregated, and the duration of data retention [9–13, 35]. These quality-varying mechanisms are currently an active topic of research. We note that our work is complementary to these other privacy policies. Our analysis is meant to assist electric utilities in following privacy recommendations: we seek to determine how much data to collect and how often it should be collected. Once this is in place, encryption, anonymization, and aggregation techniques can be employed in tandem.

We evaluate the performance of a widely studied DLC scheme as a function of the sampling rate. As we will later show, increasing the sampling period is a means of improving the privacy of consumers. In particular, we focus on a DLC application using thermostatically controlled loads (TCLs) to manage load imbalances.

4.2.1 DLC Model

In this section, we consider one recently proposed DLC program for concreteness. We note that our contribution is a general framework for numerically analyzing the sensitivity of these DLC programs to different information collection policies. We consider this research to be complementary to other research in how parameters affect system performance [63, 64].

TCLs, which are often heating, ventilation, and air conditioning (HVAC) systems for buildings, are a promising avenue for the implementation of DLC policies [65, 66]. This is due to the fact that buildings have a thermal inertia and can, in essence, store energy. Moreover, power consumption can be deferred and shifted while resulting in an imperceptible change in temperature.

4.2.2 Thermostatically Controlled Load Model

There are several TCL and DLC models in the literature (e.g. [67–69]) and our analysis can easily be applied to any of these models. For concreteness, we consider the model presented in [69].

Let $\mathcal{I}$ denote the set of TCLs participating in a DLC program. We model the temperature evolution of each TCL $i \in \mathcal{I}$ as a discrete-time difference equation:

$$x_i(k+1) = a_i x_i(k) + (1 - a_i)[T_{a,i}(k) - m_i(k)T_{g,i}] + \epsilon_i(k) \tag{25}$$

In the earlier equation, $x_i(k)$ is the internal temperature of TCL i at time k, $T_{a,\,i}$ is the ambient temperature around TCL i, m_i is the control signal of TCL i, and ϵ_i is a noise process.[3] The term $a_i = \exp(-h_B/(R_i C_i))$, where h_B is the base sampling period,[4] R_i is the thermal resistance of TCL i, and C_i is the thermal capacitance of TCL i. The T_g term represents the temperature gain when a TCL is in the ON state, and $T_g = R_i P_{trans,\ i}$, where $P_{trans,\ i}$ is the energy transfer rate of TCL i. Let P_i denote the power consumed by TCL i when it is in the ON state.

We note that, in this framework, we model the variability of occupancy behavior in the $\epsilon_i(k)$ terms. When a large amount of historical data, as well as auxiliary features, are available, one can update the posterior distribution of $\epsilon_i(k)$ to reflect the available observations. However, for simplicity, in this chapter, we simply assume a fixed distribution.

The local control for TCL i is modeled by the variable m_i. We assume that the local controller performs an ON/OFF hysteresis control based on its setpoint and deadband. For a cooling TCL, this is defined as:

$$m_i(k+1) = \begin{cases} 0 & \text{if } x_i(k+1) < T_{set,i} - \delta_i/2 \\ 1 & \text{if } x_i(k+1) > T_{set,i} + \delta_i/2 \\ m_i(k) & \text{otherwise} \end{cases} \tag{26}$$

In these equations, $T_{set,\ i}$ and δ_i are the temperature setpoint and deadband of TCL i, respectively. If $m_i(k) = 1$, then we say that TCL i is in the ON state at time k, and similarly $m_i(k) = 0$ means that i is in the OFF state at k.

In the next few sections, we will assume that these local control signals can also be overridden by the direct load controller, replacing Eq. (26). We will introduce a privacy-aware sampling policy that only intermittently provides the controller access to observations $(x_i(k), m_i(k))$.

4.2.3 Direct Load Control Objective

We consider DLC policies that attempt to compensate for load imbalances and defer demands from peak times by switching TCLs between the ON state and the OFF state. The marginal cost of peak loads and unexpected load imbalances is responsible for a large portion of the preventable costs in the electricity grid; for a more detailed treatment of the benefits and impact of a DLC policy which can shave demand, we refer the reader to [70].

Formally, we consider the load imbalance as an exogenous variable. In particular, the centralized DLC controller is given some desired power trajectory P_{des}

---

[3] Our development focuses on air conditioning for notational simplicity, but similar statements can be made for heaters.

[4] Here, h_B denotes the time scale of the dynamics. Later on, we will introduce how often the direct load controller may receive fewer measurements to preserve privacy, and this subsampling period will be denoted h.

for the TCLs.[5] The goal of the controller is to minimize the error between the actual power consumed by the TCLs and the signal P_{des}, that is, it wishes to minimize $\sum_k \left| \sum_{i \in \mathcal{I}} P_i m_i(k) - P_{des}(k) \right|$.

4.2.4 Direct Load Control Capabilities

To achieve the DLC objective, we assume that the centralized DLC controller has the capability of telling TCLs to switch modes between ON and OFF when the temperature $x(k)$ is between $T_{set,\,i} - \delta_i/2$ and $T_{set,\,i} + \delta_i/2$. More explicitly, if the centralized DLC controller issues a command to a TCL to switch from OFF to ON, the TCL turns on its air conditioner earlier than it would have in the absence of a control command. This DLC command will override the local controller. We assume that the centralized DLC controller has no control authority when the temperature is outside of the deadband, with the local controller deterministically in the OFF state when $x(k) < T_{set,\,i} - \delta_i/2$ and in the ON state when $x(k) > T_{set,\,i} + \delta_i/2$.

Note that the control policy effectively tightens the deadband. In particular, this control policy maintains customer satisfaction in the sense that the effective deadband is never larger than the user-specified deadband.

Our model of a direct load controller is as follows. We assume the centralized DLC controller has access to the parameters $\beta = (a_i, T_{a,\,i}, T_{g,\,i}, T_{set,\,i}, \delta_i, P_i)$ for each TCL $i \in \mathcal{I}$. In other words, the controller knows the dynamics of each TCL. However, it is only able to observe the signals $(x_i(k), m_i(k))$ for certain values of k, determined by the privacy-aware sampling policy. One of the contributions of this paper is the extension of a DLC controller to situations where measurements are intermittent.

For the rest of this section, we will assume a privacy-preserving sampling policy that considers subsampling rates. In other words, our sampling policy is parameterized by a subsampling period $h \in \mathbb{N}$, and at time k, the centralized controller has access to the measurements $(x(k), m(k))_{k \in T_k}$, where the set $T_k = \{hl : l \in \mathbb{N}, hl \leq k\}$ denotes the time indices in which measurements are available.[6]

4.2.5 Direct Load Controller

In this section, we outline a DLC policy inspired by work in the recent literature [65, 69]. Our model of a direct load controller is as follows. First, the controller maintains an estimate of the thermal state of each TCL. Let $\hat{x}_i(k)$ and $\hat{m}_i(k)$ denote the estimates of $x_i(k)$ and $m_i(k)$, respectively.

[5]We consider this load imbalance signal exogenous. In future work, we hope to examine elements of generation, such as scheduling, and how it is influenced by these programs.

[6]For simplicity, we assume that either all the TCLs transmit their state information at time k or none of them do. More asynchronous transmissions can be handled with some additional notational baggage.

The estimator acts as follows:

$$\hat{x}_k(k) = \begin{cases} x_k(k) & \text{if } k \in T_k \\ a_i\hat{x}_k(k-1) + (1-a_i)[T_{a,i}(k-1) - \hat{m}_i(k-1)T_{g,i}] & \text{if } k \notin T_k \end{cases} \quad (27)$$

$$\hat{m}_i(k) = \begin{cases} m_i(k) & \text{if } k \in T_k \\ 0 & \text{if } k \notin T_k \text{ and } \hat{x}_i(k) < T_{\text{set},i} - \delta_i/2 \\ 1 & \text{if } k \notin T_k \text{ and } \hat{x}_i(k) > T_{\text{set},i} + \delta_i/2 \\ \hat{m}_i(k-1) & \text{otherwise} \end{cases} \quad (28)$$

At time k, the estimator uses the observation if it is available. If no measurement is available, it evolves the estimates according to the dynamics with known parameters β, under the assumption that $\epsilon_i(k) = 0$. Similarly, it supposes that a TCL does not switch states under the local controller, unless the estimate of the thermal state of the TCL leaves the deadband.

These estimates are used to issue control commands. Our controller takes a binning approach, as seen in recent research [65, 69]. Each TCL is assigned to a bin based on its thermal state relative to its deadband, and whether or not it is in the ON or OFF state.

Based on its estimate of how many TCLs are in each bin, the controller issues a command to each bin, stating what fraction of the TCLs in each bin should switch states. Here, for simplicity, we assume that every TCL consumes the same amount of power when on (i.e., $P_i = P$ for all $i \in \mathcal{I}$). For more details, we refer the reader to [71].

An example of this control algorithm is depicted in Fig. 1. In the top figure, we see how the TCLs are divided into bins, with $N_{\text{bin}} = 6$. The number in each bin denotes how many TCLs are actually in the bin, the number in parentheses denotes the estimated number of TCLs in the bin. In this example, we assume $P_i = 2.5$ kW for each TCL i. There are an estimated 495 TCLs on, so the estimated total power consumption of the TCLs is 1.2375 MW. Suppose, in an extreme case, we wish to decrease power consumption by 500 kW. Thus, we would have to turn off 200 TCLs. According to the estimate, if we tell every TCL in the (1, ON) bin (the top-left bin), 154 TCLs will turn off. Therefore, we must tell 46 TCLs in the bin (2, ON) to turn off as well, where there is estimated to be 170 TCLs. Thus, the control command issued to the bin (1, ON) is 1, to bin (2, ON) is $46/170 = 0.27$, and to all other bins is 0. In the bottom figure, the TCLs actually in each bin switch from the ON state to the OFF state according to a Bernoulli coin flip, with probability equal to the command issued, and the estimates are updated based on the expected number of TCL switches. The numbers inside the bin represent the actual number of TCLs in each bin after the switching is completed, and the estimated number of TCLs in each bin after the switching is completed.

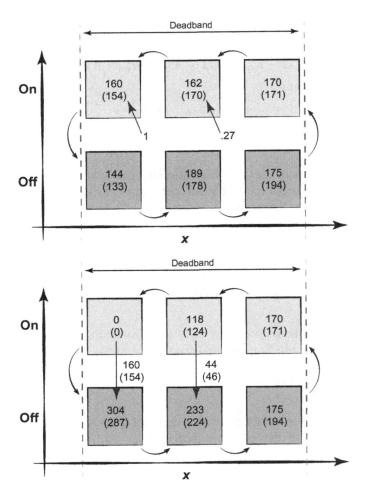

FIG. 1

An example execution of the DLC control law.

Closing the loop on this model development, we have the following model of the TCL with the control actuations by the centralized DLC controller. The closed loop dynamics are given by the following equation:

$$x_i(k+1) = a_i x_i(k) + (1 - a_i)[T_{a,i}(k) - \tilde{m}_i(k)T_{g,i}] + \epsilon_i(k) \tag{29}$$

Here, the parameters are the same as in Eq. (29). Note that the only difference in these dynamics and the open-loop dynamics without DLC is the modification of the $\tilde{m}_i(k)$ term. Furthermore, the mode of the TCL with DLC, $\tilde{m}_i(k)$ is given by:

$$\tilde{m}_i(k) = \begin{cases} 1 - m_i(k) & \text{with probability } c \\ m_i(k) & \text{with probability } 1 - c \end{cases} \qquad (30)$$

$\tilde{m}_i(k)$ will depend on the local control law and the centralized DLC law, with preference given to the centralized command. Here, $m_i(k)$ is the local control law as defined in Eq. (26).

4.2.6 DLC Model Simulations

For simulations, we assume each TCL consumes $P_i = 2.5$ kW when in the ON state, and we consider a DLC controller in control of 1000 TCLs. Parameters for each TCL i are drawn independently, from distributions based on recent studies of a 250 m^2 home [65, 69]. The time step h_B was chosen to be $h_B = 1$ min, and the number of bins $N_{bin} = 10$.

The ambient temperature $T_a = 32°C$ for all TCLs,[7] and the noise process $\epsilon_i(k)$ is independent across k and distributed according to an $N(0, 0.0005)$ distribution[8] for each k.

California Independent System Operator market signals are given in 5-min intervals [69, 72], so for simulations, the signal P_{des} is independently drawn from a $U(875$ kW, 1.35 MW$)$ distribution.[9] That is, $P_{des}(k)$ is uniformly drawn for $k \in \{0, 5, 10, ...\}$. For other values of k, we take the linear interpolation.

Simulations of the aggregate power consumption of all the TCLs are shown in Fig. 2 for the uncontrolled case, the case where $h = 1$ min, and the case where $h = 30$ min. Comparing the top plot with the middle and bottom plots, we can see that a DLC policy can reduce the load imbalance even when the controller does not always receive measurements. However, small unforeseen temperature deviations can cause the controller's performance to degrade if enough measurements are not provided, as seen by comparing the middle and bottom plots.

In addition, the thermal state of one TCL is shown in Fig. 3. We can see that the temperature inside the TCL remains inside the deadband, resulting in no loss of comfort to the consumer, in all three cases.

---

[7] For these simulations, we assumed that the ambient temperature is constant across 1 h, which can be reasonable for this short-time frame.

[8] This is the variance of the noise for one time step, so 0.0005 models the variance of temperature across $h_B = 1$ min.

[9] This framework can handle other distributions for the load imbalance signal, but a uniform distribution was chosen as a noninformative prior [48]. The parameters of the distribution were chosen as reasonable values for which energy consumption could be compensated. From simulations, we find that a larger interval is more difficult to track, as expected.

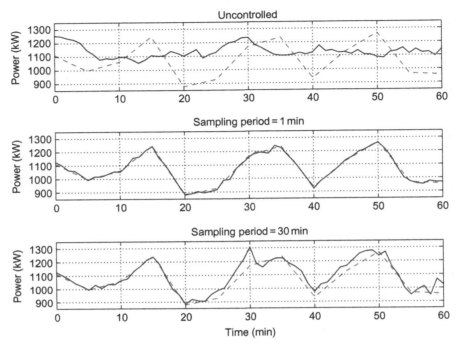

FIG. 2

A sample simulation of the aggregate power consumption of 1000 TCLs. The *solid line* represents the actual power consumption, and the *dotted line* represents the desired power consumption. The *top figure* shows the power consumption in the absence of any control commands, the *middle figure* shows the power consumption with a sampling period of $h = 1$ min, and the *bottom figure* shows the power consumption with a sampling period of $h = 30$ min.

In Fig. 4, we plot the error between the actual power consumption and the desired load imbalance compensation signal. First, we randomly drew a P_{des} signal and TCL parameters. Then, for this fixed P_{des} signal and TCL parameters, we ran 500 trials for each sampling period h, and we consider the empirical distribution of the difference between the actual power consumed by all the TCLs and the desired power signal: $\sum_{i \in \mathcal{I}} P_i m_i - P_{des}$. We used the ℓ_1 norm on the error signal, so, if we assume a fixed price for spot market electricity purchases/sales throughout the hour interval, this is directly proportional to the cost the utility company must pay.

4.2.7 DLC Privacy Analysis

As a counterpoint to the previous section, we take the theory in Section 3 and use them on real data to address several different problems. We used the emonTx wireless open-source energy monitoring node from OpenEnergyMonitor[10]

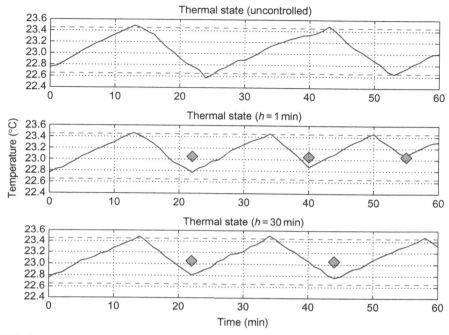

FIG. 3

The thermal state of one sample TCL. The *top graph* shows the thermal states of the TCL when there is no control. The *middle and bottom graphs* show the thermal states based on a controller that receives observations every $h = 1$ min and $h = 30$ min, respectively. The *dotted lines* indicate the deadband limits. The *diamonds* indicate when the DLC policy issued control commands to the TCL.

from several devices at 12 Hz. We used current transformer sensors and an alternating current (AC) to AC power adapter to measure the current and voltage, respectively, of the devices that we monitored. For each device we measured the root-mean-square (RMS) current, RMS voltage, apparent power, real power, power factor, and a UTC time stamp.

Data were recorded in a laboratory setting for a microwave, a toaster, a kettle, an LCD computer monitor, a projector, and an oscilloscope. As our sensors are highly accurate, we treat the measurements as noise free.

What is an upper bound for the probability of successfully distinguishing a toaster turning on and a kettle turning on, as a function of the sampling rate? We analyze how likely we are to distinguish the two devices as the sampling rate changes. This is shown in Fig. 5. We down-sampled the 12 Hz signal. In

[10]http://openenergymonitor.org/emon/emontx.

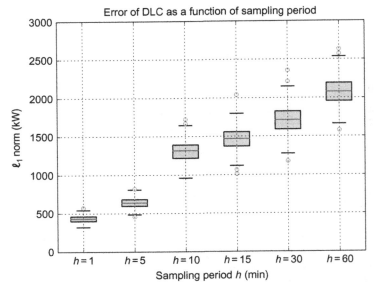

FIG. 4

A plot of how the error between the actual power consumed by the TCLs and the desired power consumption signal empirically varies with the sampling period h. The value we are plotting is $\| \sum_{i \in \mathcal{I}} P_i m_i - P_{des} \|_1$. The *whiskers* indicate all data points within 1.5 times the interquartile range. For reference, the error after 500 simulations of uncontrolled TCLs has an empirical mean of 5.39 MW with a standard error of 302 kW.

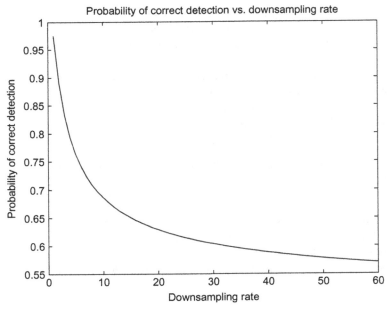

FIG. 5

The probability of successfully discriminating a toaster and a kettle as a function of the sampling rate. We fixed $\sigma^2 = 1$.

addition, if we down-sampled with rate K, we assumed it was equally likely that the signal would begin on any of the first K time steps.

It should be noted that a downsampling rate of K implies that we only receive 1/K as many measurements. Thus, if we sample for 1 s, our original problem would be a separation problem in $\mathbb{R}^{12,000}$, whereas the downsampled problem is a separation problem in $\mathbb{R}^{\lfloor 12,000/K \rfloor}$.

As expected, the probability of successful NILM decreases with the sampling rate. In addition, the performance degrades quite quickly, and we barely perform better than guessing when the downsampling rate is 60 (i.e., we sample every 5 s). This result allows us to determine a lower bound on the sampling rate necessary to achieve a certain effectiveness of NILM. It gives prescriptions on what hardware specifications and network capacity is needed in AMIs to achieve a certain goal.

In these examples, we considered a privacy guarantee on an adversary inferring the state of an individual device. In practice, we may frequently need to consider several different combinations of devices; for more details we refer the reader to [73, 74].

4.3 Closing Remarks on the Utility-Privacy Tradeoff

In this section, we have formalized a framework for modeling the utility-privacy tradeoff in energy systems. In addition, we instantiated this framework on a DLC example, with load imbalance correction as a control objective. As a privacy metric, we used the fundamental limits from Section 3 to bound the probability an adversary can correctly infer the in-home behaviors. We considered how a design parameter, the sampling rate, affects both the usefulness of the collected data and the privacy of energy consumers. With this tradeoff formalized, system designers can intelligently pick these design parameters based on the utility of data and the privacy of data.

5 CONCLUSIONS

The goal of this chapter is to provide insight into the privacy risks inherent in the collection of large amounts of energy data. First, we consider the fundamental limits in the inferences that can be made as a result of this big data: this provides a theoretical bound which no algorithm can ever break due to the statistical properties of the energy disaggregation problem. Second, we introduce a framework for comparing the control efficacy of collected data with the privacy issues of high-frequency, high-resolution data. This is done in generality at first, and then instantiated in a DLC example. We are able to simulate the change in load imbalance correction error that can be attained with higher-frequency samples. Similarly, we are able to analyze how the fundamental limits of NILM are affected: we can guarantee adversaries cannot correctly infer in-home behaviors with probability higher than α. This is an example of the influence big data has on operational performance as well as privacy of consumers, which is a form of analysis that will become more and more essential in the next generation of smart grid technologies.

References

[1] G.T. Gardner, P.C. Stern, The short list: the most effective actions U.S. households can take to curb climate change, Environment: Science and Policy for Sustainable Development, 2008.

[2] J.A. Laitner, K. Ehrhardt-Martinez, V. McKinney, Examining the scale of the behaviour energy efficiency continuum, European Council for an Energy Efficient Economy, 2009.

[3] K.C. Armel, A. Gupta, G. Shrimali, A. Albert, Is disaggregation the Holy Grail of energy efficiency? The case of electricity, Energy Policy 52 (2013) 213–234, https://doi.org/10.1016/j.enpol.2012.08.062.

[4] M.A. Lisovich, D.K. Mulligan, S.B. Wicker, Inferring personal information from demand-response systems, IEEE Secur. Privacy 8 (2010) 11–20, https://doi.org/10.1109/MSP.2010.40.

[5] R. Anderson, S. Fuloria, On the security economics of electricity metering, Ninth Workshop on the Economics of Information, 2010.

[6] G. Smith, Marijuana bust shines light on utilities, The Post and Courier (28 January), 2012.

[7] Government Accountability Office, Electricity Grid Modernization: Progress Being Made on Cybersecurity Guidelines, But Key Challenges Remain to Be Addressed, 2011.

[8] A.A. Cárdenas, S. Amin, G. Schwartz, R. Dong, S.S. Sastry, A game theory model for electricity theft detection and privacy-aware control in AMI systems, Proc. of the 50th Allerton Conf. on Communication, Control, and Computing, 2012, pp. 1830–1837. https://doi.org/10.1109/Allerton.2012.6483444.

[9] K. Kursawe, G. Danezis, M. Kohlweiss, Privacy-friendly aggregation for the smart-grid, Proc. of the 11th Int. Conf. on Privacy Enhancing Technologies, 2011, pp. 175–191. 978-3-642-22262-7.

[10] A. Rial, G. Danezis, Privacy-preserving smart metering, Proc. of the 10th Annu. ACM Workshop on Privacy in the Electronic Society, ACM, 2011, pp. 49–60. https://doi.org/10.1145/2046556.2046564. 978-1-4503-1002-4.

[11] G. Taban, V.D. Gligor, Privacy-preserving integrity-assured data aggregation in sensor networks, Int. Conf. on Computational Science and Engineering, vol. 3, 2009, pp. 168–175. https://doi.org/10.1109/CSE.2009.389.

[12] F. Li, B. Luo, P. Liu, Secure information aggregation for smart grids using homomorphic encryption, 1st IEEE Int. Conf. on Smart Grid Communications (SmartGridComm), 2010, pp. 327–332. https://doi.org/10.1109/SMARTGRID.2010.5622064.

[13] G. Acs, C. Castelluccia, I have a DREAM! (DiffeRentially privatE smArt Metering), Information Hiding, Lecture Notes in Computer Science, vol. 6958, Springer, Berlin, Heidelberg, 2011, pp. 118–132. https://doi.org/10.1007/978-3-642-24178-9_9. 978-3-642-24177-2.

[14] R. Dong, L. Ratliff, H. Ohlsson, S.S. Sastry, A dynamical systems approach to energy disaggregation, 2013 IEEE 52nd Annu. Conf. on Decision and Control (CDC), 2013, pp. 6335–6340. https://doi.org/10.1109/CDC.2013.6760891. ISSN 0743-1546.

[15] R. Dong, L.J. Ratliff, H. Ohlsson, S.S. Sastry, Energy disaggregation via adaptive filtering, 2013 51st Annu. Allerton Conf. on Communication, Control, and Computing (Allerton), 2013, pp. 173–180. https://doi.org/10.1109/Allerton.2013.6736521.

[16] A.J. Bell, T.J. Sejnowski, An information-maximization approach to blind separation and blind deconvolution, Neural Comput. 7 (6) (1995) 1129–1159.

[17] J. Cardoso, Infomax and maximum likelihood for blind source separation, IEEE Signal Process Lett. 4 (4) (1997) 112–114, https://doi.org/10.1109/97.566704.

[18] A. Belouchrani, K. Abed-Meraim, J.F. Cardoso, E. Moulines, A blind source separation technique using second-order statistics, IEEE Trans. Signal Process. 45 (2) (1997) 434–444, https://doi.org/10.1109/78.554307.

[19] S.B. Leeb, S.R. Shaw, J.L. Kirtley Jr., Transient event detection in spectral envelope estimates for nonintrusive load monitoring, IEEE Trans. Power Delivery 10 (3) (1995) 1200–1210, https://doi.org/10.1109/61.400897.

[20] S. Gupta, M.S. Reynolds, S.N. Patel, ElectriSense: single-point sensing using EMI for electrical event detection and classification in the home, Proc. of the 12th ACM Int. Conf. on Ubiquitous Computing, ACM, New York, NY, USA, 2010, pp. 139–148. https://doi.org/10.1145/1864349.1864375. 978-1-60558-843-8.

[21] J. Froehlich, E. Larson, S. Gupta, G. Cohn, M.S. Reynolds, S.N. Patel, Disaggregated end-use energy sensing for the Smart Grid, IEEE Pers. Commun. 10 (1) (2011) 28–39, https://doi.org/10.1109/MPRV.2010.74.

[22] J.Z. Kolter, M.J. Johnson, REDD: a public data set for energy disaggregation research, Proc. of the SustKDD Workshop on Data Mining Applications in Sustainability, 2011.

[23] J.Z. Kolter, T. Jaakkola, Approximate inference in additive factorial HMMs with application to energy disaggregation, Proc. of the Int. Conf. on Artificial Intelligence and Statistics, 2012, pp. 1472–1482.

[24] O. Parson, S. Ghosh, M. Weal, A. Rogers, Nonintrusive load monitoring using prior models of general appliance types, Proc. of the 26th AAAI Conf. on Artificial Intelligence, 2012, pp. 356–362.

[25] C. Dwork, Differential privacy, Proc. of the Int. Colloq. on Automata, Languages and Programming, Springer, 2006, pp. 1–12.

[26] K. Chaudhuri, D. Hsu, Sample complexity bounds for differentially private learning, COLT, 2011, pp. 155–186.

[27] J. Le Ny, G.J. Pappas, Differentially private filtering, IEEE Trans. Autom. Control 59 (2014) 341–354, https://doi.org/10.1109/TAC.2013.2283096.

[28] Z. Huang, S. Mitra, G. Dullerud, Differentially private iterative synchronous consensus, Proceedings of the 2012 ACM Workshop on Privacy in the Electronic Society, ACM, New York, NY, USA, 2012, pp. 81–90, https://doi.org/10.1145/2381966.2381978. 978-1-4503-1663-7.

[29] S.L. Warner, Randomized response: a survey technique for eliminating evasive answer bias, J. Am. Stat. Assoc. 60 (309) (1965) 63–69, https://doi.org/10.1080/01621459.1965.10480775.

[30] B.G. Greenberg, A.L.A. Abul-Ela, W.R. Simmons, D.G. Horvitz, The unrelated question randomized response model: theoretical framework, J. Am. Stat. Assoc. 64 (326) (1969) 520–539, https://doi.org/10.1080/01621459.1969.10500991.

[31] T. Dalenius, Towards a methodology for statistical disclosure control, Statistisk Tidskrift 15 (1977) 429–444.

[32] L. Sweeney, *k*-anonymity: a model for protecting privacy, Int. J. Uncertainty Fuzziness Knowledge Based Syst. 10 (5) (2002) 557–570.

[33] A. Narayanan, V. Shmatikov, Robust de-anonymization of large sparse datasets, Proceedings of the 2008 IEEE Symposium on Security and Privacy (SP '08), 2008, pp. 111–125.

[34] F. du Pin Calmon, N. Fawaz, Privacy against statistical inference, 2012 50th Annu. Allerton Conf. on Commun., Control, and Computing (Allerton), 2012, pp. 1401–1408. https://doi.org/10.1109/Allerton.2012.6483382.

[35] L. Sankar, S.R. Rajagopalan, H.V. Poor, Utility-privacy tradeoffs in databases: an information-theoretic approach, IEEE Trans. Inf. Forens. Secur. 8 (2013) 838–852, https://doi.org/10.1109/TIFS.2013.2253320.

[36] T.M. Cover, J.A. Thomas, Elements of Information Theory, Wiley-Interscience, Hoboken, NJ, 1991.

[37] S. Han, U. Topcu, G.J. Pappas, Differentially private distributed constrained optimization, IEEE Trans. Autom. Control 62 (1) (2014) 50–64.

[38] J. Hsu, Z. Huang, A. Roth, Z.S. Wu, Jointly private convex programming, Proceedings of the Twenty-Seventh Annual ACM-SIAM Symposium on Discrete Algorithms (SODA '16), Arlington, VA, January 10–12, 2016, pp. 580–599.

[39] Z. Huang, Y. Wang, S. Mitra, G.E. Dullerud, On the cost of differential privacy in distributed control systems, Proc. of the 3rd Int. Conf. on High Confidence Networked Systems, ACM, New York, NY, USA, 2014, pp. 105–114. 978-1-4503-2652-0.

[40] R. Dong, W. Krichene, A.M. Bayen, S.S. Sastry, Differential privacy of populations in routing games, 2015 54th IEEE Conference on Decision and Control (CDC), 2015, pp. 2798–2803, https://doi.org/10.1109/CDC.2015.7402640.

[41] S.R. Rajagopalan, L. Sankar, S. Mohajer, H.V. Poor, Smart meter privacy: a utility-privacy framework. 2011 IEEE International Conference on Smart Grid Communications (SmartGridComm), 2011, pp. 190–195, https://doi.org/10.1109/SmartGridComm.2011.6102315.

[42] R. Jia, R. Dong, S.S. Sastry, C. Spanos, Privacy-enhanced architecture for occupancy-based HVAC control (under review), 8th ACM/IEEE International Conference on Cyber-Physical Systems (ICCPS) 2016.

[43] W. Diffie, M.E. Hellman, Privacy and authentication: an introduction to cryptography. Proc. IEEE 67 (3) (1979) 397–427, https://doi.org/10.1109/PROC.1979.11256.

[44] C. Dwork, A. Roth, The algorithmic foundations of differential privacy, Foundations and Trends® in Theoretical Computer Science, vol. 9, No. 3–4, 2014, pp. 211–407, https://doi.org/10.1561/0400000042.

[45] J. Giraldo, A. Cárdenas, E. Mojica-Nava, N. Quijano, R. Dong, Delay and sampling independence of a consensus algorithm and its application to smart grid privacy. IEEE 53rd Annu. Conf. on Decision and Control, 2014, pp. 1389–1394, https://doi.org/10.1109/CDC.2014.7039596.

[46] H. Nissenbaum, Privacy as contextual integrity, Washington Law Rev. (2004).

[47] D.J. Solove, Conceptualizing privacy, Calif. Law Rev. 90 (2002) 1087.

[48] R.W. Keener, Theoretical Statistics: Topics for a Core Course, Springer, 2010.

[49] J. Neyman, E.S. Pearson, On the problem of the most efficient tests of statistical hypotheses, Philos. Trans. R. Soc. Lond. A 231 (1933) 289–337.

[50] L. Le Cam, Convergence of estimates under dimensionality restrictions, Ann. Statist. 1 (1) (1973) 38–53.

[51] B. Yu, Assouad, Fano, and Le Cam, Festschrift for Lucien Le CamSpringer, 1997, pp. 423–435.

[52] T. Han, S. Verdú, Generalizing the Fano inequality. IEEE Trans. Inf. Theory 40 (4) (1994) 1247–1251, https://doi.org/10.1109/18.335943.

[53] L.J. Ratliff, R. Dong, H. Ohlsson, A.A. Cárdenas, S.S. Sastry, Privacy and customer segmentation in the smart grid, IEEE 53nd Annual Conference on Decision and Control (CDC), 2014.

[54] M. Alexander, K. Agnew, M. Goldberg, New approaches to residential direct load control in California, ACEEE Summer Study on Energy Efficiency in Buildings2008.

[55] California Energy Commission, Docket No. 13-IEP-1F: increasing demand response capabilities in California., (2013).

[56] Obama Administration, A policy framework for the 21st century grid: enabling our secure energy future, (2011).

[57] A. Cavoukian, Privacy by design: strong privacy protection—now, and well into the future, A Report on the State of PbD to the 33rd International Conference of Data Protection and Privacy Commissioners, 2011, https://www.ipc.on.ca/wp-content/uploads/Resources/PbDReport.pdf.

[58] NISTR 7628 – Guidelines for Smart Grid Cyber Security: Vol. 2, Privacy and the Smart Grid, The Smart Grid Interoperability Panel Cyber Security Working Group, July 2010, https://www.smartgrid.gov/document/nistr_7628_guidelines_smart_grid_cyber_security_vol_2_privacy_and_smart_grid

[59] North American Energy Standards Board, NAESB Privacy Policy, (2015), https://www.naesb.org/privacy.asp.

[60] Department of Energy, Data access and privacy issues related to smart grid technologies., (2010).

[61] Public Utility Commission of Texas, Electric Substantive Rules [Chapter 25]., (2014).

[62] California Public Utilities Commission, Decision adopting rules to protect the privacy and security of the electricity usage data of the customers of Pacific Gas and Electric Company, Southern California Edison Company, and San Diego Gas & Electric Company., (2011).

[63] N. Lu, An evaluation of the HVAC load potential for providing load balancing service. IEEE Trans. Smart Grid 3 (3) (2012) 1263–1270, https://doi.org/10.1109/TSG.2012.2183649.

[64] N. Lu, Y. Zhang, Design considerations of a centralized load controller using thermostatically controlled appliances for continuous regulation reserves. IEEE Trans. Smart Grid 4 (2) (2013) 914–921, https://doi.org/10.1109/TSG.2012.2222944.

[65] D.S. Callaway, Tapping the energy storage potential in electric loads to deliver load following and regulation, with application to wind energy. Energy Convers. Manag. 50 (5) (2009) 1389–1400, https://doi.org/10.1016/j.enconman.2008.12.012.

[66] C. Perfumo, E. Kofman, J.H. Braslavsky, J.K. Ward, Load management: model-based control of aggregate power for populations of thermostatically controlled loads. Energy Convers. Manag. 55 (2012) 36–48, https://doi.org/10.1016/j.enconman.2011.10.019.

[67] N. Ruiz, I. Cobelo, J. Oyarzabal, A direct load control model for virtual power plant management. IEEE Trans. Power Syst. 24 (2) (2009) 959–966, https://doi.org/10.1109/TPWRS.2009.2016607.

[68] S. Moura, J. Bendtsen, V. Ruiz, Observer design for boundary coupled PDEs: application to thermostatically controlled loads in smart grids. IEEE 52nd Annu. Conf. on Decision and Control2013, , pp. 6286–6291, https://doi.org/10.1109/CDC.2013.6760883. ISSN 0743-1546.

[69] J.L. Mathieu, S. Koch, D.S. Callaway, State estimation and control of electric loads to manage real-time energy imbalance. IEEE Trans. Power Syst. 28 (1) (2013) 430–440, https://doi.org/10.1109/TPWRS.2012.2204074.

[70] D.S. Callaway, I.A. Hiskens, Achieving controllability of electric loads. Proc. IEEE 99 (1) (2011) 184–199, https://doi.org/10.1109/JPROC.2010.2081652.

[71] R. Dong, New Data Markets Deriving from the Internet of Things: A Societal Perspective on the Design of New Service Models (Ph.D. thesis), University of California, Berkeley, 2017. Technical Report No. UCB/EECS-2017-52.

[72] California Independent System Operators, Business practice manual for market operations., (2014).

[73] W. Kleiminger, F. Mattern, S. Santini, Predicting household occupancy for smart heating control: a comparative performance analysis of state-of-the-art approaches, Energy Build. 85 (2014) 493–505.

[74] W. Kleiminger, C. Beckel, S. Santini, Household occupancy monitoring using electricity meters, 2015 ACM International Joint Conference on Pervasive and Ubiquitous Computing (UbiComp 2015)2015.

Index

Note: Page numbers followed by *f* indicate figures, *t* indicate tables, and "*b*" indicate boxes.